Technology Transfer Systems in the United States and Germany

Lessons and Perspectives

H. NORMAN ABRAMSON, JOSÉ ENCARNAÇÃO, PROCTOR P. REID, AND ULRICH SCHMOCH, *EDITORS*

Binational Panel on Technology Transfer Systems in the United States and Germany

FRAUNHOFER INSTITUTE FOR SYSTEMS AND INNOVATION RESEARCH

NATIONAL ACADEMY OF ENGINEERING

NATIONAL ACADEMY PRESS
Washington, D.C. 1997

NATIONAL ACADEMY PRESS • 2101 Constitution Avenue, NW • Washington, DC 20418

NOTICE: The National Academy of Engineering was established in 1964, under the charter of the National Academy of Sciences, as a parallel organization of outstanding engineers. It is autonomous in its administration and in the selection of its members, sharing with the National Academy of Sciences the responsibility for advising the federal government. The National Academy of Engineering also sponsors engineering programs aimed at meeting national needs, encourages education and research, and recognizes the superior achievements of engineers. Dr. Wm. A. Wulf is president of the National Academy of Engineering.

The Fraunhofer Society was established in 1949 and obtained its present semipublic status of a federal research institution with the mission of applied research in 1973. Most of its nearly 50 institutes conduct research in various areas of technology. A considerable share of this work is contract research for industry. The Fraunhofer Institute for Systems and Innovation Research, which is responsible for this study, is active in the interdisciplinary field of technology, economy, and politics. It often works for the German government as an advisory body. Professor Dr.-Ing. H.-J. Warnecke is president of the Fraunhofer Society.

This report is the result of work conducted by an independent panel appointed by the Council of the National Academy of Engineering and the President of the Fraunhofer Society. The panel members responsible for the report were chosen for their expertise, with regard for appropriate balance.

Parts I and II of this publication have been reviewed by a group other than the authors according to procedures approved by a National Academy of Engineering report review process. Parts I and III have been reviewed by various external experts and internally by the Fraunhofer Institute for Systems and Innovation Research.

Funding for this effort was provided by the German-American Academic Council Foundation.

Library of Congress Catalog Card Number 97–76548

International Standard Book Number 0-309-05530-X

Copyright © 1997 by the National Academy of Sciences. All rights reserved.

No part of this book may be reproduced by any mechanical, photographic, or electronic procedure, or in the form of a phonographic recording, nor may it be stored in a retrieval system, transmitted, or otherwise copied for public or private use, without written permission from the publishers, except for the purpose of official use by the United States government.

Printed in the United States of America

BINATIONAL PANEL ON TECHNOLOGY TRANSFER SYSTEMS IN THE UNITED STATES AND GERMANY

German Panel

PROF. DR. JOSÉ ENCARNAÇÃO, *Chair*, Director, Fraunhofer-Institut für Graphische, Datenverarbeitung, Darmstadt

PROF. DR. OTTO H. SCHIELE, Former President, Arbeitsgemeinschaft industrieller, Forschungseinrichtungen (AiF), Köln

DR. GERHARD SELMAYR, Chancellor, Universität Karlsruhe

DR. SIGMAR KLOSE, Senior Vice President, Program Management New Systems Technologies, Boehringer Mannheim GmbH

DR. KNUT MERTEN, Former President and CEO, Siemens Corporate Research, Inc., Princeton, New Jersey; Director and Head of Department, Siemens AG, Central Department "Technology," München

PROF. DR. PETER C. LOCKEMANN, Director, Forschungszentrum Informatik, Karlsruhe

PROF. DR. BERND HÖFFLINGER, Director, Institut für Mikroelektronik (IMS), Stuttgart

PROF. DR. PETER H. HOFSCHNEIDER, Director, Max-Planck-Institut für Biochemie, Martinsried

DR.-ING. HERBERT GASSERT, Member of Science Council (Wissenschaftsrat), Köln; Former Executive President of BBC and Member of the Supervisory Board of ABB, Mannheim; President of German Federation of Technical-Scientific Associations (DVT)

U.S. Panel

DR. H. NORMAN ABRAMSON, *Chair*, Retired Executive Vice President, Southwest Research Institute

DR. ALEXANDER H. FLAX, Former Senior Fellow, National Academy of Engineering

DR. ROBERT C. FORNEY, Retired Executive Vice President, E.I. du Pont de Nemours & Company

DR. DAVID A. HODGES, Daniel M. Tellep Distinguished Professor of Engineering, University of California, Berkeley

DR. ARTHUR E. HUMPHREY, Professor of Chemical Engineering, The Pennsylvania State University

DR. WILLIAM F. MILLER, Herbert Hoover Professor of Public and Private Management and Professor of Computer Science Emeritus, Stanford University

DR. ALBERT NARATH, President, Energy and Environment Sector, Lockheed Martin Corporation

DR. WALTER L. ROBB, President, Vantage Management, Inc.

DR. WILLIAM J. SPENCER, Chairman and CEO, SEMATECH

Staff of the Fraunhofer Institute for Systems and Innovation Research, Karlsruhe (FhG-ISI)

DR. ULRICH SCHMOCH, Study Director, Senior Scientist
RAINER BIERHALS, Senior Scientist
VIOLA PETER, Scientist
RENATE KLEIN, Administrative Assistant

Staff of the National Academy of Engineering, Washington, D.C. (NAE)

DR. PROCTOR P. REID, Study Director, Associate Director, NAE Program Office
PENELOPE J. GIBBS, Administrative Assistant
SIMON GLYNN, Research Associate
GREG PEARSON, Editor

Preface

Increasingly, the value of science and engineering research to society is measured by how readily research results are translated into useful products and services. Fundamental to this process is technology transfer, which has been the subject of growing public discourse since the early 1980s and is now more than ever a focal point of policy interest. This renewed attention to technology transfer is occurring at a time when expanding international markets, global competition, and other pressures are forcing companies to rationalize or reengineer their operations, often in the face of increasingly constrained resources.

The following consensus study, prepared by a binational panel of German and American experts, documents the significance of effective technology transfer to industrial competitiveness in a global economy. The study's findings make clear that it is no longer appropriate to view technology transfer as a simple one-way transfer—from research performer to technology user—of processed knowledge and finished concepts. Rather, technology transfer should be understood as a mutual, multidirectional exchange—comprising many different forms and mechanisms—within and between nonindustrial research institutions and industry.

This comparison of the origins, framing conditions, instruments, and recent development of the German and American technology transfer systems reveals strengths and weaknesses in both countries. It also offers a starting point from which each nation can pursue new paths toward strengthening economic and technological performance, as well as cultivate more intensive, mutually advantageous international collaboration in technology transfer. The binational study panel, comprising experts from 18 scientific and technological institutions and enterprises in Germany and the United States, has articulated important, well-founded recommendations for improving and further developing the German and American technology transfer systems. I am confident that these recommendations will be given serious consideration by scientific, economic, and public policy communities on both sides of the Atlantic. Furthermore, the study speaks to several interesting possibilities for further dialogue between experts in both countries.

On behalf of the German-American Academic Council, I would like to thank the cochairmen, H. Norman Abramson and José Encarnação, and their colleagues on the binational panel for their considerable efforts on this project. I would also like to thank the staff of the U.S. National Academy of Engineering and the German Fraunhofer Institute for Systems Innovation Research who worked on the project, in particular the co-directors for the study, Proctor P. Reid and Ulrich Schmoch.

PROF. DR. HEINZ RIESENHUBER
President, German-American Academic Council Foundation

Acknowledgments

This study has three parts: an overview and comparative report (Part I), a U.S. country report (Part II), and a German country report (Part III). These three reports were prepared by staff of the U.S. National Academy of Engineering (NAE) and the German Fraunhofer Institute for Systems and Innovation Research (FhG-ISI) based on contributions of members of the binational panel, commissioned papers, and staff research. The findings and joint recommendations in Part I represent the consensus of the full binational panel. Part I also includes country-specific recommendations for which each of the respective national delegations to the binational panel is solely responsible. Each national delegation was responsible for developing, reviewing, and finalizing its country report. Each delegation's review of its national technology transfer system was greatly enriched by the insightful questions and comments of members of its counterpart delegation.

The editors would like to thank all members of the binational panel for their considerable efforts on this study. They would also like to thank FhG-ISI research staff members Rainer Bierhals and Viola Peter, and NAE researcher Simon Glynn for their valuable contributions to the two country reports. Renate Klein and Penelope Gibbs provided critical administrative and logistical support. Greg Pearson, the NAE editor, greatly improved the style and logical structure of the report and helped prepare it for publication.

Finally, the editors would like to express their appreciation to the German-American Academic Council for its generous financial support of the project.

Contents

PART I:
OVERVIEW AND COMPARISON   1

INTRODUCTION   2
   Understanding Technology Transfer, 2
   Factors Shaping National Technology Transfer Systems, 3
THE GERMAN AND U.S. R&D SYSTEMS   3
   Major Similarities, 4
   Major Differences, 5
COMPARISON OF MAJOR TECHNOLOGY TRANSFER INSTITUTIONS   9
   Functional Similarities: An Overview, 9
   Technology Transfer from Higher Education Institutions, 11
   Technology Transfer from Government Laboratories, 20
   Technology Transfer from Contract Research Institutes, 25
   Technology Transfer by Industrial R&D Consortia, 27
SELECTED TECHNOLOGY TRANSFER ISSUES IN A
   COMPARATIVE CONTEXT   29
   Role of Start-Up Companies in Technology Transfer, 29
   Technology Transfer to Small and Medium-Sized Enterprises
      in Technologically Mature Industries, 30
   Intellectual Property Rights and Technology Transfer to Industry, 32
   International R&D Collaboration and Technology Transfer, 34
CONCLUSIONS AND RECOMMENDATIONS   35
   General Conclusions and Challenges, 35
   Recommendations, 41
   Joint German-U.S. Recommendations for Fostering Transatlantic
      Collaboration in R&D and Technology Transfer, 48

Annex I:
Suggestions for Transatlantic/International Collaborative Projects   53

TRANSATLANTIC COOPERATIVE COMPUTER APPLICATIONS
   OVER GLOBAL NETWORKS   53
SUGGESTED DEVELOPMENT OF A TRANSATLANTIC NETWORK
   OF INFORMATION ANALYSIS CENTERS   54
COLLABORATION AMONG GERMAN AND U.S. INDUSTRIAL
   RESEARCH ASSOCIATIONS   55

PART II:
TECHNOLOGY TRANSFER IN THE UNITED STATES   61

INTRODUCTION   62
THE R&D ENTERPRISE   62
   R&D Funders and Performers, 63
   Distribution of Publicly Funded R&D, 70
   The Industrial R&D Enterprise, 79
   Technology Transfer to U.S. Industry in Context, 90
TECHNOLOGY TRANSFER FROM HIGHER EDUCATION
   TO INDUSTRY   91
   Distinguishing Characteristics of the Enterprise, 91
   History of University-Industry Relations, 96
   Technology Transfer by Research Universities and Colleges, 99
U.S. FEDERAL LABORATORIES AND TECHNOLOGY TRANSFER
   TO INDUSTRY   124
   Overview, 124
   Federal Laboratories by Major Mission Area, 126
   Federal Laboratories and Technology Transfer: History and Legislation, 133
   The Federal Laboratories and Technology Transfer: Mechanisms, 135
   Measuring the Performance of Federal Laboratory Technology Transfer, 144
   The Future of Federal Laboratory Technology Transfer, 147
   Conclusions, 149
TECHNOLOGY TRANSFER BY PRIVATELY HELD, NONACADEMIC
   ORGANIZATIONS   151
   Overview, 151
   Organizations That Create and Transfer Technology, 152
   Organizations That Transfer or Facilitate the Transfer of
      Technology Created by Others, 162
   Conclusion, 174

Annex II:
Case Studies in Technology Transfer   177

BIOTECHNOLOGY   177
   Simon Glynn and Arthur E. Humphrey

THE DEVELOPMENT AND TRANSFER OF MANUFACTURING AND PRODUCTION TECHNOLOGIES TO U.S. COMPANIES *Robert K. Carr*	193
MICROELECTRONICS *Simon Glynn and William J. Spencer*	213
SOFTWARE *Simon Glynn*	224
ELECTRIC POWER RESEARCH INSTITUTE: THE BOILER TUBE FAILURE REDUCTION PROGRAM *Jim Oggerino*	237

PART III:
TECHNOLOGY TRANSFER IN GERMANY — 241

EXECUTIVE SUMMARY	242
INTRODUCTION	245
THE GERMAN R&D ENTERPRISE	246
General Structures, 246	
Industrial R&D Structures, 250	
Impact of European Research, 263	
TECHNOLOGY TRANSFER FROM UNIVERSITIES	272
Universities, 272	
TECHNOLOGY TRANSFER FROM PUBLIC INTERMEDIATE R&D INSTITUTIONS	302
Max Planck Society, 302	
Helmholtz Centers, 312	
Blue List Institutes and Departmental Research Institutes, 319	
Fraunhofer Society, 320	
TECHNOLOGY TRANSFER BY INDUSTRIAL R&D CONSORTIA	332
Federation of Industrial Research Associations, 332	
TECHNOLOGY TRANSFER IN SELECTED AREAS	341
Technology Transfer in Information Technology, 341	
Technology Transfer in Microelectronics, 342	
Technology Transfer in Biotechnology, 343	
Technology Transfer in Production Technology, 345	
CONCLUSION: AN ASSESSMENT OF TECHNOLOGY TRANSFER IN GERMANY	346

Annex III:
Examples of Technology Transfer in Germany — 349

GTS-GRAL: TECHNOLOGY TRANSFER FROM UNIVERSITY TO A NEW TECHNOLOGY-BASED FIRM *G.E. Pfaff*	349
CO_2 DYEING PROCESS: INDUSTRIAL COOPERATIVE RESEARCH *Eckhard Schollmeyer*	351

PRODUCTION AUTOMATION: TRANSFER FROM A FRAUNHOFER
   INSTITUTE TO INDUSTRY   352
   M. Hägele
MEDIGENE: ESTABLISHMENT OF A START-UP COMPANY IN
   BIOTECHNOLOGY   353
   Peter Heinrich
TECHNOLOGY LICENSING BUREAU (TLB) OF THE HIGHER
   EDUCATION INSTITUTIONS IN BADEN-WÜRTTEMBERG   354
   Thomas Gering

APPENDIXES   361

Notes   363

References   381

Biographical Information for the Binational Panel   400

Index   409

Figures and Tables

FIGURES

1.1 German and U.S. industry R&D expenditures, percentage by industrial sector, 1973, 1983, 1993, 7
2.1 International total R&D expenditures, 1994, 63
2.2 Total and nondefense R&D spending as a percentage of GDP, by country, 64
2.3 National R&D expenditures, by performing sector and sources of funds, 65
2.4 National R&D expenditures, by performing sector, sector of funds, and character of work, 1995, 68
2.5 Federal obligations, by agency and type of activity, 1995, 74
2.6 Federal obligations for basic and applied research, by field, 75
2.7 U.S. industrial R&D spending, by sector, 1973, 1983, and 1993, 81
2.8 Number of new strategic technology alliances, by industry and region, 83
2.9 R&D spending by U.S. affiliates of foreign-owned firms as a percentage of all privately funded U.S. R&D, 1982–1994, 85
2.10 Distribution of U.S. scientific and technical articles, by sector, 1993, 100
2.11 University patents by broad fields, 108
2.12 All U.S. patents by broad fields, 109
2.13 UIRC foundings by decade, 1880–1989, for UIRCs existing in 1990, 111
2.14 Federal R&D funds by selected categories of performers, estimated values for fiscal year 1994, 126
2.15 Federal laboratory licensing activity, 1987–1994, 136
2.16 Active CRADAs at federal laboratories, 1987–1994, 137
2.17 New research joint venture announcements, 160
A-1 Allocation of R&D funds for different industries: product vs. process development, fiscal year 1994, 197

xiii

A-2   MEP appropriations, including 1995 recision and 1996 continuing resolution, 208
A-3   Equivalent availability loss due to boiler tube failure, 1985–1992, 240
3.1   Organization chart of the German R&D system, 247
3.2   Main R&D-performing institutions in Germany, expenditures in billion 1995 DM, 249
3.3   Specialization index of European Patent Office (EPO) patents of German origin in relation to the average distribution at the EPO for the period 1989 to 1991, 251
3.4   Specialization index of European Patent Office (EPO) patents of U.S. origin in relation to the average distribution at the EPO for the period 1989 to 1991, 253
3.5   Partners of SMEs in R&D and technology-related activities, by percent, 255
3.6   Participation structure in the Second Framework Program, by country, 1987–1991, 265
3.7   R&D expenditures of Germany (1992–1993) and the EU by sections of the Third Framework Program, 266
3.8   Number of participants in the Second Framework Program, by country, 1987–1991, 267
3.9   Volume of research conducted in areas of technology, as a percentage of total EUREKA financing, status as of 1995, 268
3.10 Number of EUREKA projects, including those with German participation, according to technology, status as of 1995, 269
3.11 Involvement of EUREKA participants by major organization type, status as of 1995, 269
3.12 Financing sources for JESSI, 1989–1996, 270
3.13 Program structures of JESSI, 271
3.14 Research funds of German universities in constant 1980 DM, 275
3.15 Distribution of research funds at universities, according to major areas, 1993, 277
3.16 External research funds of universities, according to major sources, 1980, 1985, 1990, 278
3.17 Relation of external, related infrastructure, and institutional base R&D funds of universities in selected areas in 1990 in current DM, 281
3.18 External funds from industry at the University of Karlsruhe, for selected areas, 1980 and 1990, in constant 1980 DM, 282
3.19 Patent applications to the German Patent Office by German university professors, 301
3.20 Specialization of German Patent Office patents of German university professors, in relation to the average distribution at the EPO for the period 1989 to 1992, 303
3.21 Max Planck institutes' expenditures in main supported areas, percent of total, 305

3.22   Budget structure of 30 consolidated Fraunhofer institutes in West Germany, 323
3.23   Industry-oriented activities of 30 consolidated Fraunhofer institutes in West Germany, 1994, 324
3.24   Budget structure of 30 consolidated Fraunhofer institutes in West Germany, by research area, in 1994, 325
3.25   Typical division of labor between Fraunhofer institutes and industry, 326
3.26   Share of FhG industrial contracts, according to research area, 327
3.27   Specialization of German Patent Office patents held by the FhG in relation to the average distribution at the EPO for the period 1989 to 1992, 331
3.28   Evaluation steps for publicly funded projects involving industrial cooperative research, 336
3.29   Public and industrial funds for cooperative research, 1986–1993, in constant 1980 DM, 337
3.30   Volume of public funds and industrial funds spent on cooperative research, 339

TABLES

1.1   German and U.S. R&D Expenditures, Percentage by Source of Funds and Performing Sector, 1994, 4
1.2   The Relative Scale of the German and U.S. Technology Transfer Systems in Context, 5
1.3   Distribution of Government R&D Budget Appropriations in the United States and Germany, by Socioeconomic Objective, 1994, 8
1.4   Functional Similarities Between Research Institutions in the United States and Germany, 11
1.5   Support for German and U.S. Academic R&D, Percentage Share by Contributing Sector, 1994, 13
1.6   Research Expenditures at Universities in the United States and Germany, Percentage by Disciplinary Field, 1993, 14
2.1   U.S. Expenditures, by Performing Sector and Source of Funds, 1995, 66
2.2   Support for U.S. Academic R&D, Percent Shares by Sector, 69
2.3   U.S. Defense-Related R&D, Various Comparisons, 71
2.4   Distribution of Government R&D Appropriations by Socioeconomic Objective in the United States, 1987 and 1994, 73
2.5   Federal and State Government Investment in Cooperative Technology Activities, by Type of Program, Fiscal Year 1994, 78
2.6   High-Tech Companies Formed in the United States, 1960–1994, 86
2.7   Top 20 Most-Emphasized U.S. Patent Classes for Inventors from the United States and Germany, 1993, 89
2.8   Industry-Sponsored Research as a Share of Total Academic Research Expenditures at the Top 20 Research Universities, Fiscal Year 1994, 93

2.9 R&D Expenditures at Universities and Colleges, by Science and Engineering Field, Fiscal Year 1994 (dollars in thousands), 94
2.10 R&D Expenditures at Universities and Colleges, Percent Share by Major Science and Engineering Field, Fiscal Year 1994, 96
2.11 UIRC Research by Discipline, 1990, 114
2.12 UIRC Research by Technology Area, 1990, 115
2.13 UIRC Research by Industry, 1990, 116
2.14 Distribution of UIRCs by Importance of Selected Goals, 117
2.15 Output per UIRC, 1990, 118
2.16 Active CRADAs by Federal Agencies and Laboratories, 1987–1994, 138
2.17 Distribution of 85 Large Independent R&D Institutes by Research Focus, 1994, 153
2.18 The Six Largest Independent, Nonprofit, Applied R&D Institutes in the United States, 154
2.19 Distribution of 35 Large Affiliated R&D Institutes by Research Focus, 199,4, 155
2.20 Primary Technical Areas of Joint Research Ventures (JRVs), 1985–1995, 161
A-1 Biotechnology Drugs in Development, 1989–1993, 178
A-2 Biotechnology Medicines or Vaccines Approved for Use by the Food and Drug Administration as of 1993, 179
A-3 Selected Nonmedical Uses of Biotechnology, 181
A-4 Use of New Technology in Manufacturing, Japan and the United States, 1988, 202
A-5 Revenue Trends and Forecasts, Customized Software and Services (dollars in billions), 1991–1997, 226
A-6 Global Spending for Prepackaged Software, 1991–1997 (dollars in millions), 227
A-7 Federal Funding for Computer Science and Engineering Research and All Science and Engineering Research, Fiscal Year 1991, 228
A-8 Agency Budgets by HPCC Program Components, Fiscal Year 1994, 230
3.1 Types of Knowledge Transfer from Academia to Industry, 256
3.2 Research Funds of German Universities (billions of DM), 276
3.3 Size and Response Rate of Survey Sent to German Universities, 290
3.4 Percent Share of University External Funds in Four Focal Areas, 1995, 291
3.5 Orientation of University R&D Activities, by Percent, 1995, 292
3.6 Channels of University Technology Transfer by Percent and Mean Score, 293
3.7 Benefits to University Researchers from Contacts with Industry, by Percent and Mean Score (percent total sample), 1995, 294
3.8 Barriers to Industry Contacts, by Percent and Mean Score, 1995, 294

FIGURES AND TABLES

3.9 Reasons for Industry Interest in University Research, by Percent and Mean Score, 1995, 295
3.10 Average Mean Scores in Major Question Groups, 295
3.11 Responses to the Survey of UIRCs, 1990, 296
3.12 Industrial Contributions to UIRCs, Percent Share by Area, 1990, 297
3.13 Orientation of R&D Activities at UIRCs, Percent Share, 1990, 298
3.14 Channels of U.S. UIRC and German University Technology Transfer, Mean Score in the Four Focal Areas, 299
3.15 Benefits of Industry Contacts at UIRCs, by Percent, and at German Universities, by Mean Score, 299
3.16 Average Number of Permanent Staff and Scientists at Max Planck Institutes, Main Sections, 1993, 305
3.17 Areas of Research at Max Planck Institutes, Percent by Expenditures and Scientists, 1994, 306
3.18 Budget Structure of the MPG, 1994, 308
3.19 Structure of Project Funds, 1993, 308
3.20 Spending, Percent Share of Total Budget, and Trend for Major Research Areas of the Helmholtz Centers, 1993, 314
3.21 Budgets and Staffing of Selected Helmholtz Centers That Emphasize Industrially Relevant Research, 1993, 316
3.22 Structure of the Food and Beverages Sector and Its Member Research Associations, 334
3.23 Importance of Cooperative Research in Different Industry Sectors in Germany, 1989, 338

PART
I
OVERVIEW AND COMPARISON

INTRODUCTION

Part I of this report presents an overview of the structure, operation, and performance of major sectors of the national technology transfer systems in Germany and the United States and identifies opportunities for the two national systems to learn from each other. It draws substantially on the two country reports prepared by the German and U.S. delegations to the binational panel on "Technology Transfer Systems in the United States and Germany: Lessons and Perspectives," with staff support from the National Academy of Engineering, Washington, D.C., and the Fraunhofer Institute for Systems and Innovation Research, Karlsruhe. The U.S. and German country reports, Parts II and III of the report, are freestanding documents that map the technology transfer landscape in each country in detail. This section of the report includes country-specific recommendations for which each of the two national delegations is solely responsible, as well as joint recommendations that represent the consensus of the full binational panel.

The focus of the panel's deliberations has been on systems and mechanisms involved in the transfer of technology (broadly defined) from organizations that perform research and development (R&D), but do not directly engage in the commercialization of technology, to organizations that use technology to produce commercial products and services. The principal organizations involved in this type of technology transfer are nonindustrial R&D performers: universities and affiliated institutions; government laboratories; and an array of public, private, and mixed (public and private) contract R&D institutes and consortia. The panel also looked at a diverse group of organizations (e.g., professional societies, industry associations, and technology brokers) that performs little, if any, R&D of its own, yet plays an important role in facilitating technology transfer between the nonindustrial R&D performers and private industry. Although private companies producing goods and service perform the vast majority of R&D and technology transfer in both countries, intrafirm and interfirm technology transfer by technology users lies beyond the scope of the panel study.

Understanding Technology Transfer

The panel defines technology transfer as the movement of technological and technology-related organizational know-how among partners (individuals, institutions, and enterprises) in order to enhance at least one partner's knowledge and expertise and strengthen each partner's competitive position. Technology transfer occurs throughout all stages of the innovation process,[1] from initial idea to final product. Like the innovation process proper, technology transfer is usually iterative, involving multiple transfer steps. Technology transfer can take place via informal interactions between individuals; formal consultancies, publications, workshops, personnel exchanges, and joint projects involving groups of experts from different organizations; and the more readily measured activities such as patenting, copyright licensing, and contract research. Technology transfer may be confined to specific regions, or it may span regions or nations within one continent or across several continents.

This definition of technology transfer encompasses direct and indirect forms. Direct technology transfer is linked to specific technologies or ideas and to more visible channels such as contract or cooperative research projects. Indirect technology transfer concerns the exchange of knowledge through such channels as informal meetings, publications, or workshops.[2] In early stages of the technology life cycle, indirect technology transfer predominates, so that it is often difficult to trace the origins of specific technologies or ideas. In the public debate, there is a strong tendency to look only at direct technology transfer. However, indirect and direct technology transfer are closely intertwined, and for the competitiveness of a country, it is important that both types of transfer be efficient. The following analysis is based on a broad interpretation of technology transfer and is not limited a priori to specific mechanisms.

Factors Shaping National Technology Transfer Systems

Technology transfer activities within a country are shaped by many different factors. Among the most important of these are the scale and technological intensity of the country's "home" market; the performance of domestic labor and capital markets; the volume and composition of public and private spending on R&D and technology transfer activities within the country; the extent of linkages to foreign sources of technology; the domestic intellectual-property regime; the endowment of human capital and R&D/technology transfer institutions; and a broad range of public policies and private practices and attitudes that shape a nation's collective outlook on innovation, change, and risk.

To examine comparatively the organization and performance of the two national systems, both the overview and the country reports focus on factors, policies, practices, and institutions most directly linked to R&D and technology transfer. It is important, however, not to overlook major international differences in the scale and nature of domestic markets, in the organization and performance of markets for labor and capital, and in other production factors that have profound consequences for domestic technology transfer systems. This report tries to address briefly at least the most significant of these issues, but a more detailed analysis of these structural economic factors is beyond the scope of the study.[3]

THE GERMAN AND U.S. R&D SYSTEMS

A comparison of the general size and structure of the German and American national R&D systems reveals a number of fundamental structural similarities and differences. These are the foundation for a more detailed examination of technology transfer activities and institutions within the two countries.

Major Similarities

Germany and the United States invested 2.3 and 2.5 percent, respectively, of their gross domestic product in R&D activity in 1994. Public and private shares of total R&D funding were similar in the two countries in 1994 (roughly 40 percent public and 60 percent private in each) as were the shares of total R&D performed by industry and by higher education and affiliated institutions (66 to 71 percent by industry and 15 to 19 percent by higher education and affiliated institutions) (Table 1.1). Both countries have roughly the same broad institutional categories of R&D and technology transfer performers: universities, government laboratories, public and private affiliated and independent "intermediary" R&D institutions; and a range of organizations that do not perform R&D but do facilitate technology transfer. Moreover, both countries possess highly diversified public- and private-sector R&D portfolios that span the full spectrum of science and engineering disciplines and a wide range of technologically evolving industries.

TABLE 1.1 German and U.S. R&D Expenditures, Percentage by Source of Funds and Performing Sector, 1994

Sector	Germany		United States	
	R&D Fund Sources	R&D Performers	R&D Fund Sources	R&D Performers
Industry	61.5	66.9	58.9	70.8[a]
Government	38.0[b]	15.2[c]	37.0[d]	10.2
Higher Education	—	17.5	2.3[e]	15.5[f]
Private nonprofit	0.5	0.4	1.8	3.5[g]

[a]Includes industry-administered federally funded research and development centers (FFRDCs).

[b]State and federal government funds, as well as funds of the German Research Association and other quasipublic organizations.

[c]Includes Helmholtz Centers, the Max Planck Society, the Fraunhofer Society, Blue List institutes, departmental institutes, state institutes, and similar publicly chartered institutions.

[d]Includes $61 billion of federal funds and $1.6 billion of state and local funds specifically targeted for R&D.

[e]Includes general-purpose state or local government appropriations, general-purpose grants from industry, foundations and other outside sources, tuitions and fees, endowment income, and unrestricted gifts.

[f]Includes university-administered FFRDCs.

[g]Includes FFRDCs administered by other nonprofit institutions.

SOURCES: Bundesministerium für Bildung, Wissenschaft, Forschung und Technologie (1996) and National Science Board (1996).

Major Differences

The U.S. R&D and technology transfer system is roughly four times the absolute size of its German counterpart, whether measured by the volume of R&D spending, the number of R&D-performing institutions, the size of the R&D workforce, or the volume of high-technology production, patenting, and research publications (Table 1.2). This difference in scale reflects the relative size of the two nations' economies and populations.[4]

This size difference does not mean that the two countries are not comparable. Behind the United States and Japan, Germany is the third-largest country in the world in terms of the absolute size of its R&D budget. With that investment, Germany is able to exploit all relevant areas of science and technology, unlike many smaller countries. Nevertheless, the larger scale of the U.S. research enterprise has certain advantages. Compared with its German counterpart, the large U.S. population of R&D performers means more opportunities for synergy[5] and specialization among R&D institutions as well as more intense competition for research funds. The U.S. domestic market has many large-scale, technology-intensive segments that are much more homogeneous than those in the German market in terms of regulation and consumer demand. Thus, the U.S. market offers more opportunities for new high-tech products. The Common Market of the European Union has an absolute volume comparable to the U.S. market, but the actual integration of the different European national markets is still quite limited in comparison with the U.S. market.

In the United States, operational responsibility for R&D and technology transfer is more widely distributed among a larger and more diverse population of institutions than it is in Germany. There also appears to be greater diversity and autonomy among U.S. technology transfer agents within each of the major technology transfer sectors than is true in Germany. This diversity is manifested in terms of size (research budgets, staff), ownership and management types (private,

TABLE 1.2 The Relative Scale of the German and U.S. Technology Transfer Systems in Context

	Germany	United States
R&D Employment (1993)	229,800	962,700
R&D Spending (1994[a])	$36.8 billion	$168.5 billion
High-technology manufacturing production (1992[b])	$175.2 billion	$640.2 billion
Domestic utility patent applications by nationals (1994)	36,800	107,233
Scientific and technical articles, all fields (1993)	27,902	140,588

[a]Calculated with purchasing power parity exchange rates.
[b]Measured in constant 1980 dollars.

SOURCES: Deutsches Patentamt (1995), National Science Board (1993, 1996), Organization for Economic Cooperation and Development (1996b,c), U.S. Patent and Trademark Office (1995).

public, state, federal, for profit, not for profit, etc.), research and technology transfer portfolios, and productivity. In other words, the German system is more uniform across industrial sectors, scientific fields, and regions than its U.S. counterpart. It is also relatively more uniform than the American system in terms of the patterns of federal, state, and private shared funding practices across these sectors, fields, and regions.

INDUSTRIAL R&D PORTFOLIOS

There are important differences in the industrial R&D portfolios of the two countries. As the data in Figure 1.1 indicate, for the past 20 years, German industrial R&D has remained concentrated in traditional manufacturing industries in which German firms have long excelled, namely the automotive, electrical and nonelectrical machinery, electronic and communication equipment, and industrial-chemicals sectors. Over the same period, the distribution of U.S. industrial R&D activity among sectors has changed significantly (Figure 1.1). U.S. industrial R&D has long been more heavily concentrated in high-tech (R&D-intensive) industries than that of its German counterpart. Nevertheless, the U.S. industrial R&D enterprise has seen a rapid increase in the share of total industrial R&D accounted for by several major nonmanufacturing industries[6] as well as a dramatic decline in the share accounted for by the electrical machinery and aerospace sectors, particularly since the mid-1980s.

These differences in industrial R&D activity are reflected in the industrial output, exports, and patent portfolios of the two nations. According to patent and trade statistics, U.S. industry excels in the fields of information technology, chemistry and chemical engineering, biomedical engineering, pharmaceuticals, and biotechnology. German industry specializes in several types of mechanical engineering, as well as in civil engineering and some types of chemistry and chemical engineering (Gehrke and Grupp, 1994).

ALLOCATION OF PUBLIC R&D FUNDS

There are several important differences between the two countries in terms of how they allocate public R&D monies. In the United States, more than half of all public R&D spending is committed to national defense, and an additional 11 percent of the total supports civilian space exploration. By contrast, defense and

---

FIGURE 1.1 *(opposite)* German and U.S. industry R&D expenditures, percentage by industrial sector, 1973, 1983, 1993. NOTE: The category "electrical machinery and apparatus" includes electrical motors, transformers, distribution, accumulators, and lighting. "Electronic and communication equipment" is comprised of electronic components, television and radio equipment, telecommunications, and audio-visual apparatus. SOURCE: Organization for Economic Cooperation and Development (1996a).

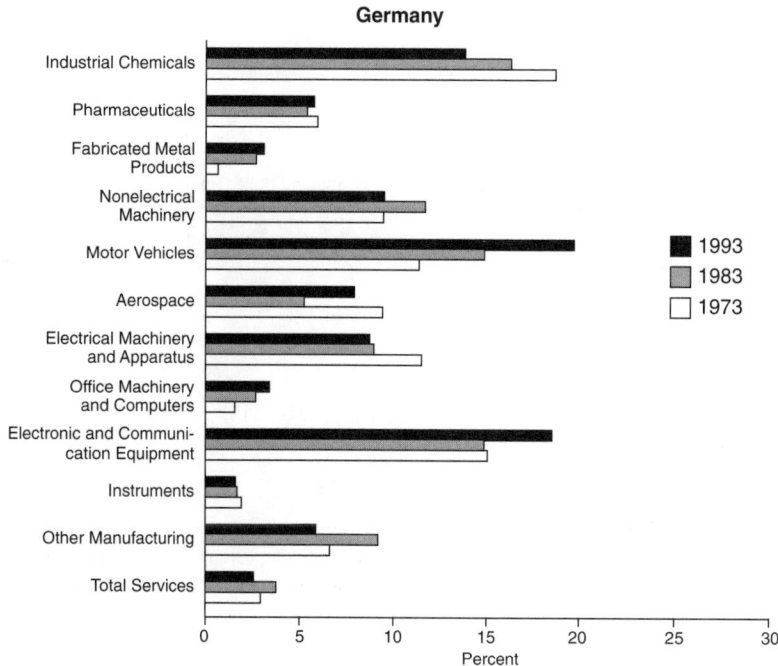

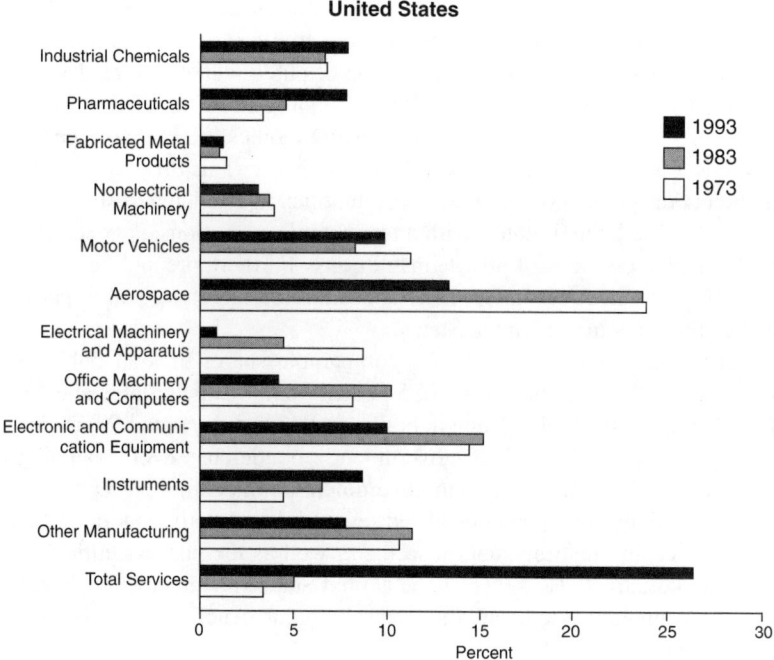

TABLE 1.3 Distribution of Government R&D Budget Appropriations in the United States and Germany, by Socioeconomic Objective, 1994

	Percent of Public R&D Funds			
	Total Funds		Civilian Funds	
Objective	United States	Germany	United States	Germany
Agriculture	2.5	2.6	5.6	3.0
Industrial development	0.6	15–20	1.3	17–23
Energy	4.2	3.8	9.4	4.4
Infrastructure	2.9	1.5	6.5	1.7
Environmental protection	0.8	4.2	1.8	4.9
Health	16.5	13	36.9	15
Civilian space	10.9	5.6	24.4	6.5
Defense	55.3	8.4	—	—
Advancement of research	4.0	13.8	8.9	16.0
General university funds	—	22–27	—	25–31
Not elsewhere classified	2.3	5.2	5.1	6.0

SOURCES: Bundesministerium für Bildung, Wissenschaft, Forschung und Technologie (1996), National Science Board (1996), calculations by the Fraunhofer Institute for Systems and Innovation Research.

space exploration claim only 8 percent and 6 percent, respectively, of total German public R&D expenditures (Table 1.3). In the United States, the areas of computer science and information technology, microelectronics, and aeronautics, in particular, have benefited from the high volume of public defense-related R&D.[7]

A special focus of German public R&D spending is "industrial development," which receives 15 to 20 percent of public R&D monies. Less than 1 percent of U.S. public R&D support goes toward such activity. This difference reflects a more direct engagement of German research policy in civilian industrial technology, which in the United States, with a few notable exceptions, is generally considered to be the province of private institutions. Furthermore, public funding of industrially relevant R&D in Germany appears to be more targeted to particular industries than it is in the United States.

Health-related R&D accounts for comparable shares of total public R&D expenditures in the two countries: 16.5 percent in the United States and 13 percent in Germany. It should be noted, however, that health-related R&D claims a much larger fraction of the U.S. government's nondefense R&D spending (37 percent) than it does of the German government's nondefense R&D budget (15 percent). Publicly funded health-related research has greatly benefited the biomedical device and instrumentation sector as well as the pharmaceutical industries in both countries. However, in the United States, the therapeutics and diagnostics biotechnology sector also has been a major beneficiary of government R&D investments.

In Germany, the category of "general university funds" has been introduced in Table 1.3 because of statistical problems with associating university base funds with specific socioeconomic objectives.[8] At the same time, the large share of this category is suggestive of the preeminent position universities occupy among all German research institutions with respect to their draw on the public R&D purse. In this context, it is noteworthy that about 43 percent of all German public R&D funds goes to universities, whereas in the United States, universities and colleges receive only 20 percent of public R&D funds. The relatively smaller claim of the U.S. academic research enterprise on U.S. public R&D funds is largely explained by the fact that roughly half of U.S. government R&D spending is directed to the development, testing, and evaluation of weapons and other systems for military use, work not done predominantly within academia.

The relatively large share of German public R&D funding allocated to the advancement of research, or basic research, is partly a consequence of differences in the classification of the German and U.S. data. Because most U.S. public funds are channeled through federal mission agencies, a large portion of U.S. government-funded basic research is statistically subsumed under specific socioeconomic objectives. Nevertheless, the separate classification of basic research in the German public R&D portfolio and the large claim of German academic research on total public R&D funds testify to the comparatively heavy emphasis German research policy places on basic research.

All in all, German public R&D funds are more evenly distributed among major socioeconomic objectives than are U.S. public monies. Moreover, there is more equal involvement of the German federal government and the state governments in funding and shaping the nation's public R&D portfolio than is the case in the United States. In the United States, a small amount of federal R&D funding is passed through state governments. Many U.S. states allocate a significant fraction of their budgets to doctorate-granting state universities for support of research. Such funds provide faculty salaries and support research facilities associated with faculty-directed research activity. Direct support by U.S. states of applied-research projects has, up to now, been concentrated in selected fields, including agriculture, transportation, labor relations, and public health. However, over the past decade, some states have moved to broaden their support to industrial fields (Coburn, 1995). By contrast, in Germany, the central government and the federal states are often jointly engaged in funding and administering public research institutions.

COMPARISON OF MAJOR TECHNOLOGY TRANSFER INSTITUTIONS

Functional Similarities: An Overview

Despite the many significant structural differences between the German and American R&D systems, many individual elements of the two national systems are functionally comparable and appear to face similar challenges and opportunities.

German and American research universities have the primary functions of education and research, with a focus on basic research. They perform most of same technology transfer functions and wrestle with many of the same issues, such as the tension between new and traditional university missions, conflict-of-interest concerns, differences between academic and industrial cultures, and so on.

The large German Helmholtz Centers and large federal laboratories in America perform basic and applied research in areas of public interest. They face similar demands to diversify their research portfolios and downsize in response to a contraction of their original national missions. The German Blue List (*Blaue Liste*) institutes, departmental research institutes, and independent state institutes seem to be comparable to the smaller U.S. federal and state-level laboratories.

The Max Planck institutes are characterized by their public base funding and their near-exclusive basic research orientation. Although they have no real institutional counterpart in the United States, their functional equivalents can be found in several publicly funded, U.S. university-affiliated basic research institutes (including federally funded research and development centers), basic research activities of certain U.S. federal laboratories (e.g., National Institutes of Health), and private institutes, such as the Howard Hughes Medical Institutes. Again, there is a uniformity of German institutions across the spectrum of research fields that is not found in the United States.

The highly networked, semipublic German Fraunhofer institutes conduct primarily applied research and development and pursue the mission of technology transfer to industry. There is no single institutional counterpart (public or semipublic) to the Fraunhofer Institutes in the United States. Instead, many of the contract R&D and technology transfer functions of the Fraunhofer institutes are performed in the United States by a large, diverse, and dispersed population of public and privately held for-profit and nonprofit organizations. Most prominent among the latter are the large independent engineering research institutes. Fraunhofer institutes and U.S. independent engineering research institutes also face several comparable challenges, for example meeting the technology transfer needs of small companies and competing and cooperating effectively with other nonindustrial technology transfer institutions in their respective national R&D systems (e.g., national laboratories, university-affiliated institutes).

In the two countries, industrial enterprises conduct the largest share of R&D, primarily applied research and development. Cooperative industrial research, including that conducted by R&D consortia, is well established in both Germany and the United States. The technology transfer challenges faced by industrial R&D consortia in the two countries are similar in many respects.

Table 1.4 gives an overview of these functional counterparts in German and U.S. research institutions. The following sections provide a more detailed comparative analysis of technology transfer institutions and mechanisms in the two countries.

TABLE 1.4 Functional Similarities Between Research Institutions in the United States and Germany

Primary Functions	United States	Germany
Education, basic research	Universities	Universities
Basic research	University-affiliated institutes Select federal labs and federally funded R&D centers Some independent research institutes	Max Planck institutes
Public mission, public interest	Large national laboratories Smaller federal laboratories State-level institutes	Helmholtz centers Blue List institutes Departmental institutes State-level institutes
Applied research, technology transfer	Independent engineering research institutes UIRCs	Fraunhofer institutes An-Institutes BMBF's cooperative programs
Applied research, development	Industrial consortia Industrial R&D collaborations	AiF cooperative research Industrial R&D collaborations

NOTE: UIRCs = university-industry research centers; An-Institutes = Institute an der Universität (literally, institutes at the university); BMBF = Bundesministerium für Bildung, Wissenschaft, Forschung und Technologie (Ministry for Education, Science, Research and Technology); AiF = Arbeitsgemeinschaft industrieller Forschungsvereinigungen (Federation of Industrial Research Associations).

Technology Transfer from Higher Education Institutions

The primary missions of German and U.S. universities are education and research directed at the advancement of knowledge. However, engineering schools and engineering departments in both countries have long performed a considerable amount of applied as well as basic engineering research. The principal contribution of universities to the technical needs of industry is human capital, that is, well-educated, learning-skilled science and engineering graduates. Thus, movement of science and engineering graduates to other sectors of a nation's innovation system must be considered the most important technology transfer channel of universities. Although German and U.S. universities are involved increasingly in the generation and licensing of intellectual property, the primary research output of German and American academic research remains nonproprietary new knowledge that is disseminated widely through publications and conferences. Because of the nonproprietary, or "public goods" nature, of much of its output, academic research is funded primarily by the public sector in both countries.

DIFFERENCES OF SCALE AND STRUCTURE

There are major differences in the scale of the German and American academic research enterprises. The volume of U.S. academic R&D is roughly four times that performed by German institutions of higher education. Likewise, the output of U.S. academic researchers, measured in terms of the publication of their work in the world's science and engineering literature is about four times that of their German counterparts. (See Table 1.2, above.) Again, however, these differences reflect the different sizes of the two countries' economies and populations.

The U.S. academic research enterprise also appears to be much more heterogeneous and decentralized in its administration and management than its German counterpart. The "nonsystem" of U.S. research universities and colleges is a highly autonomous population of public and private institutions, each established and developed in response to some unique combination of local, regional (state), and national needs and opportunities. These institutions vary considerably in the size of their research budgets, the general orientation of their research (some are more basic, others are more applied), the reputation (quality and productivity) of their research activities, the scope and intensity of their technology transfer activities, and their administration and accounting practices. By comparison, German academic research institutions are fewer in number, larger, and more homogeneous in size, administration, and management as well as in the overall breadth of their research portfolios. With few exceptions, all German universities have a public status.

FUNDING OF ACADEMIC RESEARCH

Government funds the vast majority of academic research in both Germany and the United States (Table 1.5). However, in Germany, about 74 percent of all academic research is supported by general-purpose, or base-institutional, funds provided by state (*Länder*) governments, which are responsible for education under the nation's constitution and fund the public universities within their jurisdiction. By contrast, agencies of the U.S. federal government contribute the largest share of U.S. academic research funding (60 percent); the share of support provided by U.S. state and local governments is significantly smaller.[9] U.S. private universities, which represent an important, highly productive part of the nation's academic research enterprise, depend much more heavily on federal R&D funds than do public universities, which receive both targeted research funding and general-purpose appropriations from state governments. In 1993, federal agencies funded roughly 56 percent of all research at public universities and 74 percent of research performed at U.S. private universities (National Science Board, 1996). Private companies provide comparable shares of total research funding at German and U.S. universities. U.S. academic institutions rely more heavily on nonprofit organizations for research support than do their German counterparts.

TABLE 1.5   Support for German and U.S. Academic R&D, Percentage Share by Contributing Sector, 1994

Sector	Germany	United States
Federal government	17.2	60.1
State and local governments[a]	73.5	7.6
Industry	8.3[b]	6.9
Academic institutions[c]	—	17.9
All other sources[d]	1.0	7.4

[a]German states (Länder) provide general-purpose funds to their universities for education and research. About DM 10 billion of these general-purpose funds were used for academic research in 1994. Only 10 percent of all research support provided for German universities by the Länder was project related. By contrast, the percent share of total U.S. academic R&D funded by U.S. state and local governments reflects only funds targeted specifically for academic R&D activities and does not include general-purpose appropriations used for separately budgeted research or to cover unreimbursed indirect costs. See note c, below. (National Science Board, 1996)

[b]Includes industry-financed foundations. Without foundations, the share of industry support is 7 percent.

[c]Since German universities determine how much of the general-purpose funding they receive from state governments will be spent on research and in which fields, most German state and local government R&D funding could be classified alternatively as academic institutional funds. The major sources of U.S. academic institutional R&D funds are general-purpose state or local government appropriations; general-purpose grants from industry, foundations, or other outside sources; tuition and fees; endowment income; and unrestricted gifts. (National Science Board, 1996)

[d]Includes grants from nonprofit organizations and international organizations, restricted gifts by private individuals, and other sources not elsewhere classified.

SOURCES: Bundesministerium für Bildung, Wissenschaft, Forschung und Technologie (1996), National Science Board (1996).

Since 1970, funding for U.S. academic research from the industrial sector has increased faster than funding from other sources. U.S. academic institutions also increased their share of total research funding during this period, while the shares of funding from federal and state government declined. In Germany, like in the United States, industrial support of academic research has increased faster since 1970 than has support from other sources. Recent statistical data show that in Germany, industrial funding of academic research has increased substantially since 1989, to roughly an 8-percent share of total academic R&D support. Funding provided by the German federal government for university research is primarily project oriented and has increased considerably since 1970.

As of the early 1990s, the distribution of academic research expenditures among major research fields in Germany and the United States reveals several

TABLE 1.6 Research Expenditures at Universities in the United States and Germany, Percentage by Disciplinary Field, 1993

Discipline	United States	Germany
Engineering	15.8	19.0
Life sciences	54.4	37.0
Other natural sciences[a]	21.7	23.0
Social sciences, humanities[b]	6.3	19.0
Other sciences	1.9	2.0

[a]The other natural sciences encompass the physical, mathematical, computer, and environmental sciences. Unlike the U.S. data, German data on academic research in the other natural sciences cannot be disaggregated by discipline.

[b]For purposes of comparison, research spending in psychology is included in this category.

SOURCES: Bundesministerium für Bildung, Wissenschaft, Forschung und Technologie (1996); National Science Board (1996); Wissenschaftsrat (1993a); calculations by the Fraunhofer Institute for Systems and Innovation Research.

significant differences (Table 1.6). These data have to be interpreted with caution, however, as the matching of U.S. and German disciplinary fields is problematic. In any case, the life sciences (i.e., the agricultural, biological, and medical sciences) accounted for 54 percent of U.S. academic research expenditures and only 37 percent of German university research spending. The social sciences and humanities claimed a significantly larger share of total German expenditures than was the case in the United States, and engineering and "other natural sciences" (i.e., the chemical, physical, mathematical, computer, and environmental sciences) accounted for slightly larger shares of total academic research spending in Germany than they did in the United States.

Public Funding of Academic Research

In Germany, a majority of academic R&D is financed with general, or base institutional, funds provided by the states (Länder). Contract and grant funds for German academic R&D come primarily from the German Research Association (*Deutsche Forschungsgemeinschaft* [DFG]), the Ministry for Education, Science, Research, and Technology (*Bundesministerium für Bildung, Wissenschaft, Forschung und Technologie* [BMBF]), and, increasingly, the European Commission. Contracts and grants generally cover only direct costs of personnel and additional equipment. As a result, the overhead costs related to research supported by these funds must de facto be covered by institutional base funds provided by the states. Therefore, in terms of personnel and time, academic research depends heavily on external sources: roughly half of all German academic research relies on contracts and grants if the related overhead funds covered by the states are included.[10]

The share of public base funds in U.S. universities is quite low. Instead, the vast majority of U.S. academic research in science and engineering is sponsored directly by nonacademic institutions, primarily federal government agencies, via grants or contracts that include money for overhead costs. The main funding sources are the National Institutes of Health, the National Science Foundation (NSF, comparable to the German DFG), the Department of Defense, the National Aeronautics and Space Administration, the Department of Energy, and the Department of Agriculture. The National Institutes of Health is the largest federal contributor to university research, accounting for almost half of these funds.

Most research performed by higher education institutions in Germany and the United States is either basic or long-term applied research. Nevertheless, because federal agencies sponsor the vast majority of academic research in the United States, even basic research in many academic fields is directed toward the applied needs of federal agencies.

Both German and U.S. academic researchers must compete for research funding on a project-by-project basis via peer-reviewed proposals. However, U.S. researchers depend on this competitive process for a significantly larger share of their total research support than do their German counterparts.[11] This competition for research grants requires a great deal of paperwork and grant management (i.e., non-research-related) effort by the principal investigator, who may serve as both a grant applicant and "volunteer" reviewer of the grant proposals of other researchers. This system both encourages intensive competition among researchers and fosters rapid dissemination of research ideas within the research community.

In contrast to the situation in the United States, until recently German academic researchers could rely on a relatively high and stable level of base funds to support the majority of their research activities. Hence, they were able to devote most of their nonteaching time to self-determined, long-term research and spend minimal effort seeking research support. However, in recent years, the pressure on German academic researchers to tap external sources of funding has increased due to stagnation in the growth of public base funding. Contracts and grants currently fund about 40 percent of all direct research costs (excluding overhead) in the engineering sciences. This share is likely to increase over time. As already noted, German universities have to cofinance the overhead costs related to research contracts and grants with institutional base funds. This means that 70-to-80 percent of all costs associated with engineering research activity at German universities depends on contracts and grants, if the related overhead costs covered by institutional base funds are included.

Industry Funding of Academic Research

Both German and U.S. universities receive research funding from private industry. In 1994, industry support accounted for about 8 percent of total German academic R&D expenditures and 7 percent of U.S. academic R&D expenditures. However, the nature of this funding differs in the two countries. The preferred mechanism of German industrial support for academic research is a research contract with clearly specified deliverables. In the United States, most industrial funding of academic R&D takes the form of grants, more open-ended arrangements without specifically defined research deliverables but with intensive involvement of the sponsors in decision making about the research orientation. Considering these differences and the perspectives of its university-based members, the panel judges university-industry research interaction in Germany to be more heavily oriented toward short-term, incremental problem solving (or less engaged in basic or long-term applied research) than university-industry linkages in the United States.

Although no aggregate data exist regarding the distribution of industrial funding of academic research by industry in the two countries, patterns of university patenting in Germany and the United States reveal significant differences in the industrial orientation of German and American university research. U.S. university patents are concentrated in the areas of biotechnology, medical technology, pharmaceuticals, and agriculture and food process technology, whereas most German university-related patents are in various fields of mechanical engineering and chemistry (Henderson et al., 1994; Schmoch et al., 1996). In other words, the composition of industrially relevant research at U.S. and German research universities reflects the relative specialization of U.S. and German industries as revealed in industrial patent and trade statistics.

TYPES OF UNIVERSITY-INDUSTRY TECHNOLOGY TRANSFER

German and American universities have moved more aggressively in the last 2 decades to develop closer ties with industry through establishment of patent licensing and technology transfer offices, affiliated institutes and research centers, high-tech incubators, and research parks. U.S. government policies have contributed to this trend in both countries. In consequence, German and American universities today are engaged extensively in technology transfer to private industry and have developed a wide range of mechanisms to execute or facilitate that transfer.

Informal Contacts, Consulting, and Personnel Exchange

In both systems, informal contacts between university researchers and industry researchers and managers via meetings, telephone conversations, and so forth are critical to successful technology transfer. Such contacts promote the discussion and exchange of research results and lay the groundwork for more formal types of cooperation such as grants or contracts.[12]

Similarly, consulting by faculty members is an important channel of technology transfer in both countries. U.S. and German science and engineering faculty are allowed to spend roughly 20 percent of their time on outside activities, including consulting with industry. Faculty in both countries see consultancies with industry as important personal learning opportunities. Consultancies also enable faculty to earn extra income, cultivate industrial funding sources of research, and create opportunities for graduate student theses. In the United States, regular consultancy by professors with multiyear contracts seems to be an effective means for establishing long-term relationships with industrial partners. In Germany, faculty consulting generally has more of a short-term orientation; that is, it is usually directed at solving discrete technical problems of a firm. This observation applies especially to professors at German polytechnical schools.

Arguably, there are greater incentives for U.S. university-based researchers to consult with and seek research support from private industry than there are for German university researchers. U.S. faculty are paid on a 9-month basis and are expected to make up the 3-month salary gap as well as fund most of their research with grants. German faculty are civil servants paid on a 12-month basis, and their research is supported in large part by base institutional research funds. Nevertheless, the opportunity for additional personal income serves as a strong incentive to German faculty members to engage in secondary consulting activity.

Another effective instrument of university-industry technology transfer is the exchange of research staff. Such exchanges are done differently in Germany than in the United States. In the United States, leading research universities often engage in temporary exchange of research personnel with private industry in the context of collaborative research projects. In Germany, however, such exchanges are rare, in large measure because of German civil-service and public employment regulations. Additional disincentives to this type of transfer activity include the relatively high job security of public employees compared with those in industry and the fact that it is impossible to transfer supplementary public pension entitlements (*Versorgung des Bundes und der Länder* [VBL]) to private employment. At the same time, German technical universities have a long-standing tradition of appointing as professors high-level researchers from industry. Once appointed, these faculty members maintain close ties to their industries of origin through consultancies and contract research. This practice leads to more practice-oriented education and close relations between universities and industry.

Finally, at both German universities and leading U.S. research universities, industrial research personnel often play valuable roles as technical advisors to masters- and doctoral-level students and as members of advisory groups for whole departments. However, in the United States, university faculty almost always assume primary supervisory responsibility for their students. At German institutions, industrial personnel may be more directly involved in supervising student research.

Cooperative Research

At German and American universities, cooperative research with private companies is an increasingly important means of technology transfer to industry.

Public-sector funding has been a major catalyst of university-industry cooperation in both countries.

The institutional framework that has structured most university-industry research collaboration in the United States in recent years is the university-industry research center (UIRC) (Cohen et al., 1994). UIRCs are an organizationally diverse set of institutions that facilitate industry access to university research results, engage industry in the definition of a research portfolio, and otherwise promote technology transfer to participating firms in exchange for sustained general or targeted funding (primarily grants) from companies. In most instances, this support is at least matched by funds from public sponsors. Among the many state and federal government entities that support these collaborative research centers, the NSF has assumed the leading role. Indeed, the U.S. panel considers the expanding networks of NSF-sponsored industry-university cooperative research centers, engineering research centers, and science and technology centers to be very cost-effective mechanisms for forging university-industry research partnerships.[13]

UIRCs account for roughly 50 percent of all industrial funding of U.S. academic research and rely on public funds for nearly half of their research budgets. The centers provide a framework for ongoing collaborative research and technology transfer relationships between universities and industry. UIRCs vary considerably with respect to their research orientation (i.e., basic or applied) as well as their disciplinary or technological focus. However, most industrial support of UIRCs appears to be directed at more basic and long-term applied research. The more autonomous university-affiliated research institutes are concentrated in the health and life science fields, where the distinction between basic and applied research is more blurred.

While many leading U.S. research universities have developed effective policies, practices, and institutional frameworks (such as UIRCs) for engaging private companies in mutually beneficial cooperative research, there is ample evidence that a great many more U.S. research universities are still struggling to put effective policies and practices in place (Government-University-Industry Research Roundtable, 1996).

In Germany, the dominant form of collaborative research is cooperation of regular university institutes with industrial enterprises on projects funded by the BMBF. The number of such cooperative projects funded by BMBF has increased considerably during the past decade, paralleling the proliferation of UIRCs in the United States. Recently, researchers involved in the special research areas (*Sonderforschungsbereiche*) funded by the DFG in university engineering departments have been encouraged to collaborate with industrial partners. In both the BMBF- and DFG-sponsored research, collaboration is directed generally at more long-term, precompetitive activities.

In Germany, a specific institutional response to the growing demand for increased technology transfer from academia to industry are the An-Institutes

(*Institute an der Universität*; literally, institutes at the university), whose budgets are equivalent to about 15 percent of the total external R&D funding of the universities.[14] An-Institutes are legally defined as independent entities in order to achieve more administrative flexibility than regular university institutes. As a result, they can adapt more easily to the needs of industry. Both An-Institutes and UIRCs rely on industry funding for roughly one-third of their total research support. However, in contrast to UIRCs, most industrial support of An-Institute research takes the form of contracts, not grants.[15]

In summary, UIRCs and other U.S. university-affiliated research institutes and the German An-Institutes represent similar institutional responses to the opportunities for increased interaction with industry and the constraints, or problems, associated with pursuing such activities within the traditional framework of academic departments. Industry-sponsored research at the German An-Institutes seems to be more oriented toward short-term applied research and problem solving and is more contract-driven than is true for industry support of American UIRCs.[16]

Patent Licensing

During the last 2 decades, German universities with technical faculties and U.S. research universities have begun to establish many new technology transfer units. Since passage of the 1980 Patent and Trademark Amendments (P.L. 96-517), more commonly know as the Bayh-Dole Act,[17] most American universities with substantial research activities have established special offices that support the patenting of inventions and the active marketing of these patents. There is great diversity among U.S. research universities with respect to their approach to patenting and technology licensing. Some universities, public institutions in particular, lay claim to all research output generated in their labs; others are more flexible in negotiating the disposition of intellectual property resulting from research on their campuses. Likewise, some institutions look to their technology licensing offices to generate revenue, and others see these units as instruments for building long-term relationships with private companies as research patrons or partners. Only a small number of institutions can claim success concerning either objective. Many research universities are still searching for effective ways to manage and grow their R&D and technology transfer activities with industry.

At present, most German universities are equipped with technology transfer offices. However, their primary function is to build relationships between small and medium-sized enterprises and faculty members, not to license patents. Only a few of these offices are actively engaged in licensing activities. The current lack of a broad patenting and licensing function at German universities is due to various factors. Among the most important are that under German law, the right to exploit inventions resulting from university-based research supported by institutional base funds rests exclusively with the individual professor or inventor involved, not with the inventor's host institution;[18] most universities have neither funds nor infrastructure to support patenting and licensing activities; inventions resulting from federally funded academic research generally can only be licensed on a nonexclusive basis to interested industrial partners; and a portion of any licensing income earned from inventions developed with federal government funds must go to the funding agency.

Start-Up Companies

With the option of establishing or working for a high-tech start-up company, U.S. academic researchers have an additional important vehicle through which they can transfer as well as have a direct hand in commercializing the results of their own research or technologies originating elsewhere—a vehicle largely unavailable to their German counterparts.[19] This means of technology transfer has proved to be very effective in highly science-based, technically dynamic industries such as software and biotechnology. This difference between the German and American transfer systems is very important and is discussed in more detail in the section "Selected Technology Transfer Issues in a Comparative Context," later in this overview.

Technology Transfer from Government Laboratories

OVERVIEW OF MAJOR PLAYERS

Government laboratories perform 8 percent of all German R&D and 15 percent of all American R&D.[20] To date, their measurable contribution to technology transfer to private industry has been small relative to the size of their R&D budgets. However, government laboratories are seeking to play a more important role in technology transfer in both countries.

The U.S. federal government maintains about 720 laboratories. However, fewer than 100 of these laboratories have sufficient resources and capabilities to engage in significant technology transfer to the civilian economy. The major laboratory sponsors and administrators are the Department of Defense, the Department of Energy, the Department of Agriculture, the National Aeronautics and Space Administration, the Department of Health and Human Services/National Institutes of Health, and the National Institute of Standards and Technology. The budget for federally funded R&D facilities represents about one-third of the total federal R&D expenditure. More than half of the budget of U.S. federal R&D facilities is spent for defense purposes.

The 16 German Helmholtz Centers (formerly called *Großforschungseinrichtungen* [GFEs]) are comparable in size and organization, though not necessarily in research portfolios, to the large federal laboratories in the United States. The BMBF is responsible for these large facilities. They receive 90 percent of their base funds from the federal government and the remaining 10 percent from the states. The total spent by Helmholtz Centers on R&D is equivalent to about 20 percent of all public support for nonuniversity research and development. The German counterparts to the smaller and medium-sized U.S. federal and state laboratories are the departmental research institutes (*Ressortforschungseinrichtungen*), which serve the missions of specific federal ministries, and the institutes of the Blue List (*Blaue Liste*), the latter being independent bodies financed equally by the federal government and the states.[21] R&D spending by these institutions represents another 11 percent of public research and development spending outside universities. The institutes of the Max Planck Society (*Max-Planck-Gesellschaft* [MPG]) also are financed equally by the federal government and the states; in this regard, they are comparable to the Blue List institutes. The MPG budget is equivalent to 7 percent of total public R&D spending outside universities.

TECHNOLOGY TRANSFER ACTIVITIES

In recent years, many government-financed national research facilities in the United States and Germany have experienced increasing pressure to engage in technology transfer to private companies. In particular, the large mission-oriented national laboratories in both countries (U.S. national laboratories and German Helmholtz Centers) are in the stage of considerable reorientation and restructuring. In the United States, declining funding for defense and civilian nuclear research and demands from Congress during the 1980s to harness the federal laboratories more effectively in the service of industrial competitiveness and environmental technologies encouraged national laboratories to diversify into new research areas, including more commercially relevant fields, and to become more involved in technology transfer to private industry. Likewise, in Germany, the Helmholtz Centers have been encouraged to diversify their research portfolios and expand their interactions with private companies in response to declining demand and funding for research in fields related to civilian nuclear energy, the former primary mission of the largest Helmholtz Centers.

U.S. Federal Laboratories

For U.S. federal laboratories, the Cooperative Research and Development Agreement (CRADA) is the most heavily used mechanism for engaging in cooperative R&D with industrial partners. The CRADA, created by acts of Congress in 1986 and 1989, has a number of important advantages over other types of cooperative R&D agreements that were used prior to its creation. Foremost among these is the authority it gives participating laboratories to protect from disclosure any intellectual property relevant to the agreement. CRADAs constitute the only mechanism by which the federal government can define in advance the disposition of intellectual property rights in government-industry collaborations not involving a government contract. In addition, CRADAs authorize laboratories to contribute staff and equipment to a CRADA project with a private-sector partner. Importantly, participating firms can contribute staff, equipment, and funds for CRADA-related activities, but laboratories cannot transfer CRADA funds to a private-sector partner. An interesting aspect of CRADAs is that they can be initiated by industry and do not necessarily have to be defined by a federal laboratory. Department of Energy laboratories have accounted for a majority of CRADAs negotiated by federal agencies since 1987.[22]

In addition to cooperative research, U.S. federal laboratories employ other instruments of technology transfer, in particular licensing of intellectual property rights. These licensing activities are based on the Stevenson-Wydler Technology Innovation Act (P.L. 96-480) and Bayh-Dole Act of 1980 and subsequent legislation that allow federal laboratories to grant exclusive licenses and to use license revenues for their own purposes.[23] A primary thrust of recent U.S. technology transfer laws has been toward providing these laboratories and individual laboratory researchers with more incentives and entrepreneurial-like decision-making powers for technology transfer. For example, federal inventors now get at least a 15-percent share of the royalties, and the responsible department also receives some licensing income. Thus, the internal incentives for patenting are high. In addition to the acts mentioned above, many other regulations for supporting technology transfer have been introduced (e.g., the requirement that each federal laboratory has to set aside 0.5 percent of its budget for technology transfer activities). As a general tendency, the laboratories and the individual researchers have got more incentives and entrepreneur-like decision-making competencies for technology transfer. Currently, the National Institutes of Health account for the lion's share of all licensing revenues earned by federal agencies for technologies developed within their laboratories.

German Helmholtz Centers, Blue List Institutes, and Departmental Institutes

The main instrument of technology transfer for German Helmholtz Centers is formal cooperation with industrial partners on projects of common interest. Each partner pays for the work it performs; that is, the Helmholtz Centers generally receive no funding from their industrial collaborators. However, public R&D budget constraints are placing increased pressure on Helmholtz Centers to attract additional contract and grant funding.

In recent years, several Helmholtz Centers have instituted patent policies, and some Helmholtz Centers have established their own patent and licensing offices, which actively market their technologies to private firms. This development was initiated in the early 1980s by the introduction of new regulations concerning licensing of Helmholtz-generated intellectual property. Prior to that time, license revenues did not increase the Helmholtz Centers' budgets, because public base funds were reduced by the same amount. At present, two-thirds of license income can be used for technology transfer projects (e.g., for the industry-oriented development of technical concepts). One-third of license income still has to be transferred to the government; however, the Helmholtz Centers are presently seeking a ruling that will allow them to use all license income for technology transfer. Present regulations generally limit the exclusiveness of licenses to 5 years, a period so short that in most cases it acts as an impediment for industrial cooperation with Helmholtz Centers. In practice, therefore, most exclusive licenses are extended.

In the case of Blue List institutes and independent state institutes, patent and license regimes are comparable to those of the Helmholtz Centers, but only few institutes have instituted a more active patent policy.

The departmental research institutes have no common and consistent policy for technology transfer to industry. Only a few of these institutions have attempted to foster technology transfer. In many respects, their current legal situation with respect to intellectual property rights is comparable to that of federal laboratories in the United States prior to the Stevenson-Wydler and Bayh-Dole Acts of 1980. Specifically, these institutions may not grant exclusive licenses to industrial firms and must transfer any license revenue back to the government. Thus, neither these institutions nor private companies have much incentive to engage each other in technology transfer through patent licenses. However, some departmental institutes have close relations with industry and are performing effective technology transfer in an informal way.[24]

Max Planck Society

The German MPG primarily conducts long-term basic research and, to a lesser extent, applied research in various areas to achieve and maintain scientific excellence, advance knowledge, and serve German societal goals. In this way, it has a research orientation comparable to that of universities, but the research teams and facilities are generally larger and it has no higher-education obligation. (However, many Max Planck scientists hold university professorships.) The largest part of Max Planck institute budgets comes from public base institutional funds. Thus, the institutes are able to set their research agendas independently according to researchers' interests (within the general framework of a discipline).[25]

Although the primary mission of the MPG is maintaining German excellence in all fields of basic research, the requirement for technology transfer has recently begun to play an increasing role. The society has a special patent and licensing office that actively looks for appropriate industrial partners to exploit the society's research results. In addition, many Max Planck research projects in strategic technological areas, such as biotechnology, material sciences, and organic chemistry, are conducted in cooperation with industry. The success of these industry contacts, however, depends largely on the initiative of the individual Max Planck scientists.

In the United States, most of the functions performed by Max Planck institutes are distributed among research universities, select federal laboratories, and a diverse population of privately held university-affiliated and independent research institutes.

FUTURE CHALLENGES

The diversification strategies of German and American government laboratories, particularly their efforts to engage in more collaborative R&D with private industry, have spawned intense policy debate in both countries. In the United States, there is general recognition that, in principle, many labs are a valuable element of the nation's R&D enterprise. However, no consensus exists regarding how these capabilities might be matched to the needs of private industry.[26] This matching will be highly dependent on defining the mission of the laboratories, something that has yet to be done. For this reason, U.S. policies regarding the technology transfer activities of federal laboratories are likely to remain in a state of great flux for the foreseeable future. In Germany, the importance of all types of national R&D institutions in areas of public interest is widely recognized. However, as the severe restructuring process related to the decline of nuclear energy has shown, continuous reflection about the content, extent, and orientation of public missions is necessary. Furthermore, there is disagreement about the appropriate level of collaboration with private industry and the division of labor with other institutions.

A question that must be answered is whether it makes more sense for the large U.S. defense-oriented federal laboratories and the German Helmholtz Centers to be downsized to fit their reduced traditional public missions or to be diversified or reoriented instead. Clearly, these laboratories are equipped with large numbers of highly trained R&D personnel and, in many cases, are unique facilities housing valuable equipment. Some of these labs are at the forefront in areas of basic and applied research that are relevant to both public missions and private industry and have successfully engaged in technology transfer to private companies. Many, however, have traditionally performed R&D that has little direct relevance to most civilian industries, have had limited experience dealing with private companies as clients,[27] and are likely to remain more bureaucratically encumbered than other major R&D performers by virtue of their continuing public-mission focus and management structures. At a time of increasingly constrained public R&D budgets and shifting national R&D needs, maintaining these laboratories at or near their present size denies resources to other public R&D performers that may be better equipped to take on the new R&D priorities. Given the large size of these facilities in Germany and the United States and their resulting economic and employment importance to their host states or regions, however, the political impediments to their downsizing are likely to remain formidable.

Given the present trends on the part of government laboratories in both Germany and the United States toward research diversification and increased interaction with private companies, it is not surprising that the other major technology transfer sectors in both countries are concerned about the impact this reorientation of mission will have on their long-term ability to compete fairly for public and private research funding. Perhaps more than any other issue facing the two countries' technology transfer systems, the fate of German and U.S. government laboratories has focused the attention of policymakers on the need to reconsider the traditional division of labor among the major elements of their national R&D systems. Ultimately, both nations will have to assess the relative strengths and weaknesses of these competing sectors of their technology transfer systems and seek to define the most productive role for each sector, while attending carefully to the potential for greater collaboration among them.

Technology Transfer from Contract Research Institutes

Germany and the United States have a variety of research institutes that perform contract research for both industrial and government clients. However, in Germany, these institutes play a considerably larger role in serving the R&D needs of private industry than do their counterparts in the United States.

THE FRAUNHOFER MODEL

In Germany, contract research is conducted mainly by the 46 institutes of the semipublic Fraunhofer Society, which receives about 1 percent of the total national R&D budget. Fraunhofer institutes receive between 20 and 30 percent of their budgets in the form of base institutional funds from the federal government; the exact amount depends on their success in generating sufficient contract work for public and private clients. Thus, the research orientation of the Fraunhofer institutes is heavily demand driven. Another characteristic feature of Fraunhofer institutes is their close relationship to universities, institutionalized through the joint appointment of Fraunhofer directors as university professors.[28] Thus, the Fraunhofer society is a significant bridging institution between academic and industrial research.

Other typical channels of technology transfer from Fraunhofer institutes are on-the-job training of graduate students and an active patent policy. In recent years, Fraunhofer institutes have assumed a more active role in the establishment of spin-off companies, a highly effective yet still relatively underutilized instrument of technology transfer in Germany. Presently, the Fraunhofer Society is seeking to develop new instruments for technology transfer, especially through the establishment of for-profit "innovation centers," each associated with a nonprofit institute. The mission of the innovation centers is to develop the research results of the institutes further to industrial products, and to introduce them into the marketplace.

The competence of the institutes is largely sustained and advanced by research projects for public clients that are medium or long-term in orientation and by public institutional base funds used for self-determined research in new strategic areas. The success of the Fraunhofer model depends on the roughly equal contribution of institutional base funds, contracts for public clients, and contracts for industrial clients to the institutes' research budgets.

The organization of the Fraunhofer institutes into one society allows for strategic cooperation among different institutes working in the same technological cluster and joint investment in costly facilities (e.g., demonstration centers). In the special case of microelectronics, the six related institutes cooperate closely with industry through the microelectronics alliance, which coordinates their research activities with the needs of potential applicants.

In the Fraunhofer institutes, Germany has a dense infrastructure of publicly funded contract R&D institutions that are geared toward serving the R&D needs of both traditional manufacturing and new high-tech industries. These institutes are geographically distributed, networked, and perform a lot of general and industry-targeted production and manufacturing R&D (e.g., industrial engineering, mechanical engineering, materials engineering) as well as R&D in highly dynamic technology areas. Decisions regarding the reallocation of roles and missions is generally the responsibility of the individual institutes, which have to continually adapt their research portfolios to the needs of the market. This adaptation is also coordinated between different Fraunhofer institutes.

The An-Institutes, previously discussed in the context of universities, also perform contract research for firms and engage in activities similar to those of Fraunhofer institutes. However, An-Institutes are not organized in a network.

INDEPENDENT AND AFFILIATED RESEARCH INSTITUTES IN THE UNITED STATES

In the United States, there is no system of industry-oriented contract R&D institutions (publicly or privately funded) that is truly comparable to the German Fraunhofer institutes. Instead, most of the combined contract R&D and technology transfer functions performed by the Fraunhofer institutes in Germany are carried out in the United States by a large, diverse, and highly dispersed population of nonprofit and for-profit R&D organizations, including a plethora of privately held affiliated and independent nonprofit institutions, several large private R&D and management consulting firms, and the research units of some U.S. industrial consortia.

The vast majority of U.S. privately held affiliated and independent research institutions receive most of their funding from federal mission agencies or private foundations and conduct primarily basic research. More than half are concentrated in the health and medical fields, and these organizations collectively account for a significant share of all health and medical R&D performed in the United States. While the R&D activities of many institutions constitute a critical link in U.S. drug testing and evaluation and directly benefit health- and medicine-related industries in many other ways, the research agendas of these institutes are not driven or shaped to any significant extent by the day-to-day R&D needs of these industries.

Even within the relatively small population of private independent and affiliated engineering R&D organizations, many of which were established originally to serve the needs of regional industries, there are today relatively few whose R&D activities are substantially geared to the applied R&D needs of private industry. Five of the seven largest independent engineering R&D institutions perform the vast majority of their R&D to address the needs of federal agency missions, not the needs of private companies. Unlike the Fraunhofer institutes, these independent engineering institutions are sustained exclusively by contract research, are not networked, and are only marginally linked to U.S. universities. Moreover, U.S. independent engineering research organizations appear to be less targeted or specialized in terms of areas of technical expertise than are the Fraunhofer institutes.[29]

Technology Transfer by Industrial R&D Consortia

Cooperative industrial research, whereby independent industrial enterprises join together to conduct research projects of common interest, is an important vehicle of technology transfer in Germany, the United States, and other parts of the world. Although no hard data are available on the volume of cooperative R&D in Germany and the United States, such activity is estimated to represent in excess of 4 percent of the total industrial R&D in both countries.[30] Formal industrial R&D consortia, though responsible for only a subset of all cooperative R&D performed by German and American companies, are nonetheless substantial ways of technology transfer in both countries. R&D consortia have a longer history and a more established role in Germany than they do in the United States.

Consortia appear to be organized in different ways in the two countries. In Germany, there are about 100 industrial research associations, representing about 50,000 enterprises, joined under the umbrella organization of the Federation of Industrial Research Associations (*Arbeitsgemeinschaft industrieller Forschungsvereinigungen* [AiF]). A characteristic feature of the AiF is its bottom-up approach to selecting research projects. A group of companies, generally small and medium-sized enterprises, define a project of common interest. Each project is suggested by a different group of companies. The projects are selected by the associations and carried out by the most appropriate research establishments. About half of the projects are executed within the associations' own institutions, the rest are contracted out to universities and other public or private organizations. Two-thirds of the projects are financed by the associations and one-third by the Federal Ministry of Economics. For projects sponsored by the Federal Ministry of Economics, AiF assumes responsibility for evaluating the project and administering the funds. The results of the projects are published and made available to all members of an association. However, the enterprises involved directly in the definition and execution of AiF projects tend to profit the most from this type of technology transfer. The cooperative projects of the industrial research associations represent about 1 percent of total German industrial R&D spending.

In the United States, there has been unprecedented growth in industrial cooperative research since the early 1980s. As of December 1995, 575 joint research ventures had been registered with the U.S. Department of Justice, and evidence suggests that these account for but a fraction of the research alliances U.S. companies have entered into since the early 1980s (Hagedoorn, 1995; Vonortas, 1996).[31] Compared with the level of company participation in Germany, however, U.S. firms are less involved in consortia-related activity.

U.S. consortia are likely to include independent R&D organizations, universities, and federal laboratories in addition to private firms. During the early-to-mid 1990s, a growing number of consortia across a wide spectrum of industries were organized around the technological capabilities of U.S. federal laboratories. This growth has leveled off in recent years, however, as federal laboratories have refocused on their core missions.[32] Many U.S. industrial R&D consortia receive at least some public funding. A few, such as Semiconductor Manufacturing Technology Research Corporation (SEMATECH) and the National Center for Manufacturing Sciences, receive core funding from the federal agencies that helped establish them in the 1980s. The vast majority of U.S. industrial consortia, however, are strictly private-sector undertakings that derive most of their research support from member companies. These consortia (e.g., Electric Power Research Institute, Gas Research Institute) compete for federal research funding on a project-by-project basis but receive little or no core funding from the federal government. The mechanisms or channels by which U.S. industrial R&D consortia solicit R&D funding from the federal government are more ad hoc, decentralized, and diverse than those used by AiF member companies.

Because of the many different organizational types of U.S. consortia, it is difficult to generalize about the way U.S. consortia define and execute R&D projects or transfer technology. U.S. consortia appear to differ from the AiF in that they involve the cooperation of a relatively stable group of firms that define a series of common research projects, which are then carried out or outsourced by a separate consortium-managed research institution.

Another aspect of U.S. consortia that distinguishes them from AiF consortia is that the research entity established by a U.S. consortium can itself suggest new research projects to the consortium membership. The advantage of this type of organization is that it facilitates the building of mutual confidence among consortium members, a decisive prerequisite for successful technology transfer. However, consortium research organizations can also develop an agenda and a dynamism of their own that lead them to generate projects of less interest to their constituent member firms.

Finally, in part because of their diversity and highly autonomous natures, there has been remarkably little sharing of organizational and operational practices among the rapidly growing population of U.S. R&D consortia.[33] By contrast, the more comprehensive, institutionalized character of the German AiF appears to facilitate organizational learning among participating industries and consortia.

SELECTED TECHNOLOGY TRANSFER ISSUES IN A COMPARATIVE CONTEXT

Role of Start-Up Companies in Technology Transfer

Start-up companies play a critical role in the transfer and commercialization of fast-moving, science-based technologies in the United States via movement, or "spin-out," of researchers and technology from universities, large established companies, and government laboratories. There is no counterpart in Germany to the prominent role that start-up companies perform in the commercialization of new technology in the United States.

Many factors have enabled high-tech start-up companies to perform their unique roles in the U.S. innovation system.[34] The following are among the most important.

- The existence of sophisticated financial markets, particularly access to a large volume of venture capital and highly developed public equity markets.
- The large scale and technological intensity of relatively homogeneous segments of the U.S. domestic market.
- The large size, high mobility, accessibility, and entrepreneurial orientation of the U.S. technical workforce.
- The sheer scale and accessibility of U.S. publicly funded nonproprietary research, particularly university-based research.
- The scale of federal government procurement combined with explicit preferences or set-asides for small and medium-sized vendors and suppliers.
- A history of regulatory and public policy commitments conducive to high-tech start-up companies, including the competition-oriented or technology-diffusion-oriented enforcement of intellectual property rights and antitrust law (competition policy), as well as the relatively risk-friendly system of company law, particularly bankruptcy law.
- A highly individualistic, entrepreneurial culture nurtured in industry and many U.S. research universities by private practices, public policies, and various institutional mechanisms such as technology business incubators and venture capital firms that encourage risk taking.

Many, if not most, of these supporting factors are either muted or nonexistent within the German innovation system. German venture capital markets and public equity markets are underdeveloped.[35] Entrepreneurial activity and career mobility of much of the German technical workforce are circumscribed by civil service regulations and institutional practices (e.g., reward structures, compensation schedules, conflict of interest restrictions) that govern university and other public-sector scientists and engineers. Unlike their counterparts in U.S. universities or federal laboratories, most university- or public-laboratory-based German researchers/entrepreneurs lack access to the institutional resources needed to pursue and defend patents.[36] German public policies in the areas of company law (including bankruptcy law), taxation, capital markets, and so forth are decisive disincentives to the establishment of high-tech start-up companies and entrepreneurial risk-taking behavior in general.

The U.S. experience shows that many start-up companies fail and only a very few are extremely successful. Therefore, venture capital firms have to invest in a sufficiently large number of start-up companies to produce a "winner" and must count on the few highly profitable outliers to compensate them for losses incurred throughout the rest of their portfolio (Scherer, 1996). This type of high-risk–high-potential yield strategy is rare in the German business culture.

Technology Transfer to Small and Medium-Sized Enterprises in Technologically Mature[37] Industries

The R&D and technology transfer needs of German small and medium-sized enterprises (SMEs) in more technologically stable manufacturing industries are supported by a dense, comprehensive, and highly institutionalized network of industry-oriented R&D institutes and non-R&D-performing technical organizations. These institutions support the technology transfer, technology commercialization, and industrial modernization requirements of many SMEs. By contrast, the U.S. R&D and technology transfer infrastructure serving SMEs in these industries is relatively piecemeal, fragmented, and weak.

German SMEs in technologically mature industries are served by highly networked, publicly funded R&D institutions and industry-organized R&D consortia that are heavily oriented toward the incremental product and process R&D needs of a national industrial base dominated by technologically mature industries. While many of these publicly funded R&D institutions serve the needs of technologically dynamic industries, the institutes of the Fraunhofer Society, state laboratories, and institutes based at or affiliated with universities also perform near-term, industry-specific, applied contract research for large companies and SMEs in traditional areas of German industrial strength such as mechanical and electrical engineering. Moreover, through participation in robust industrial associations, which have a significant influence on public R&D policy at the state, federal, and European Commission levels, German SMEs are considerably involved in the shape and resource allocation of their national R&D enterprise.

A large population of industry-led organizations, including the Chambers of Industry and Commerce, industrial associations, Technical-Scientific Associations (*Technisch-wissenschaftliche Vereine und Gesellschaften*), the Organization for the Rationalization of German Industry (*Rationalisierungskuratorium der Deutschen Wirtschaft*), and the Steinbeis Foundation, provide SMEs with a wide range of industry-tailored technology-related services. However, various studies show that SMEs still make insufficient use of these rich opportunities for support (Beise et al., 1995).

At least some of the German institutions that support SMEs have counterparts in U.S. professional and technical societies. Services provided by the American societies include technical and business consulting, technology brokering, workforce training, and apprenticeships, as well as testing and evaluation facilities and the establishment of new-business incubators. However, compared with their German counterparts, U.S. SMEs in technologically mature manufacturing industries operate on the periphery of the nation's R&D enterprise. The R&D portfolios of U.S. research universities, federal laboratories, and most nonprofit research institutes do not overlap very much with the process and product R&D needs of U.S. SMEs (or of large U.S. firms, for that matter) in these industries. Factors that have helped disconnect SMEs in many industries from the nation's research enterprise include the high-tech, public-mission orientation of federal R&D funding; the fragmented structure and low levels of industrial self-organization of many technologically mature U.S. industries; and changes in the industrial composition of the U.S. economy (i.e., the increasing shares of total U.S. industrial output accounted for by service and high-tech manufacturing industries).

Similarly, the technology transfer infrastructure supporting U.S. SMEs in more stable industries appears to be much less well developed than its German counterpart. Indeed, the poor performance of U.S. companies relative to firms (SMEs in particular) based in other advanced industrialized countries in adopting advanced manufacturing technology and production techniques has been widely documented (National Academy of Engineering, 1993; National Research Council, 1993). U.S. SMEs in most manufacturing industries have traditionally relied on large industrial customers, vendors of hardware and software, and to a lesser extent on private consultants as primary sources of new technology, technical assistance, and advice. For the most part, U.S. industrial and trade associations and chambers of commerce have provided very little in the way of technical-extension and industrial-modernization services to their memberships.

In recent years, several industry-led initiatives, some with limited public funding, have begun to address innovation and technology diffusion challenges, particularly those related to manufacturing, that face SMEs as well as larger firms in a number of technologically mature U.S. industries. For example, in response to new "lean" retailing strategies enabled by advances in information technology, segments of the U.S. textile and apparel industry have orchestrated (through increased industry self-organization and support from federal agencies and university-based researchers) a revitalization of their entire design, supply, and marketing chain through effective application of modern information technology (Abernathy et al., 1995).[38] Similarly, many of the manufacturing challenges facing the U.S. automotive industry in the late 1970s and early 1980s have been addressed effectively through a combination of firm-specific and industrywide initiatives, often in partnership with federal agencies or academic researchers.[39] Other examples of successful or promising industry-led efforts to meet the manufacturing and other technology diffusion needs of SMEs include the National Center for Manufacturing Sciences and SEMATECH's work with semiconductor equipment and material manufacturers.[40]

Of the many industry-led initiatives in this area to date, the committee considers the technology road mapping exercise of the Semiconductor Industry Association begun in the early 1990s to be a particularly promising instrument for advancing both the development and diffusion of new technology in industries where technological advance is more evolutionary than revolutionary (Rea et al., 1996). By inventorying the industry's sources of technology and forecasting technological needs throughout the industry's value-added chain, the semiconductor industry technology road map has been successful, in the view of U.S. panel members, at focusing the attention and resources of the industry and the federal government on a shared conception of technological challenges and opportunities.

In addition to these industry-specific initiatives, state and federal governments have attempted to strengthen the existing but relatively weak network of private and public service providers with more comprehensive industrial-modernization and technical-extension programs.[41] To date, however, the level of public resources dedicated to these programs and their current reach measured in terms of the number of companies they serve remain quite modest (National Academy of Engineering, 1993; Shapira, 1997).

Intellectual Property Rights and Technology Transfer to Industry

A wide range of government laws and policies shape the dynamic of technology transfer in Germany and the United States. These include, among others, R&D and technology transfer policies proper, bankruptcy law, competition policy, intellectual property law, different regulatory environments, labor law, and laws structuring capital markets. The U.S. and German country reports consider how these policies and laws interact in different ways in different sectors. Of the many public policies that affect technology transfer, those concerning intellectual property rights have a particularly important impact.

The U.S. and German governments have taken steps since the early 1980s to remove legal and administrative impediments to private-sector commercialization of technology developed with public funds. However, to date, the U.S. government's actions in this regard have been more comprehensive and, arguably, more effective than those of its German counterpart. In the judgment of U.S. panel members, the Bayh-Dole Act of 1980, the Technology Transfer Act of

1986, and subsequent U.S. legislation affecting the disposition of intellectual property developed with public funds in research universities, federal laboratories, and other R&D institutions have removed impediments to and provided an important stimulus for technology transfer and R&D collaboration between U.S. public R&D performers and U.S. companies.

In Germany, the initiatives for removing legal and administrative barriers to the transfer of publicly funded R&D results from universities and government laboratories to private industry have been less aggressive and less consistent. On the one hand, German university professors are allowed to exploit their inventions privately, if the inventions are the result of research financed by base funds. In particular, they can sell their patents or give exclusive licenses to industrial firms. On the other hand, they can grant only nonexclusive licenses if the research was funded by the federal government, especially the BMBF.[42] Furthermore, licensing income earned on inventions based on research funded by the federal government must be partly transferred to the original funding agency. The contradictory requirements also apply to Helmholtz Centers, Blue List institutes, and departmental research institutes.[43] These restrictive policies regarding the transfer of intellectual property rights are obviously not consistent with the explicit focus of many public R&D programs on industrial technology and technology transfer.

An important advantage of the U.S. system is the existence of a grace period for patent applications, a particular advantage for researchers, who often publish first and decide to patent later (Becher et al., 1996; Straus, 1997). As the European patent system has no grace period, even U.S. researchers cannot use their national grace period if they intend to file their patents abroad.[44] German panel members believe that the absence of a grace period in Germany is a significant barrier to technology transfer from scientific institutions to industry in fields where proprietary rights are considered critical to the subsequent development and commercialization of innovations by private firms.

Since the early 1980s, the U.S. government has taken a number of steps both domestically and in international forums to strengthen the legal claims of patent and copyright holders and develop more effective (sui generis) legal protection for new types of intellectual property in areas such as software, biotechnology, and microelectronics. These efforts have been paralleled in Europe by efforts to strengthen intellectual property regimes at the national and European Commission levels. Yet, in some technology fields, most notably software and biotechnology, significant differences remain in the extent of protection for intellectual property rights in the United States and Europe.

There are other general differences between the German/European and U.S. patent systems, such as the U.S. first-to-invent versus the European first-to-file approach, differences in the interpretation of patent claims, and differences in disclosure requirements. Although both the European and American systems provide effective incentives for innovation and technology transfer, their differences create obstacles to transatlantic technology transfer. Because of these differences, inventors who wish to seek intellectual property protection on both sides of the Atlantic are required to accommodate the often conflicting/competing legal requirements and standards of the two patent systems.

A decisive shortcoming of the European patent system is the fragmented responsibility of national and international authorities. For patent applications at the European Patent Office, only the application and granting procedures are recognized transnationally; the granted patents are valid only in the designated countries. As a result, there are no central European courts with jurisdiction over cases of patent challenges or infringement; national courts are responsible. Furthermore, the European Patent Organization is not an organ of the European Union; other non–European Union countries are members. Therefore, the European Commission has an advisory, rather than an executive, function and has a limited influence on the development of European patent protection. Legal changes are within the competence of the member countries of the European Patent Organization; however, in many cases, forging consensus among member countries in support of such changes has proved extremely difficult. This fragmentation of authority leads to administrative barriers, legal uncertainty, and enormous costs associated with patent protection, for both European and non-European patent applicants (Straus, 1997).

International R&D Collaboration and Technology Transfer

Comparative analysis of the technology transfer systems of the United States and Germany has underscored the potential for mutually beneficial transnational collaboration in various areas of R&D and technology transfer. The cooperation between the two countries cannot be viewed in isolation from the general process of internationalization, which was not the focus of the present study. Therefore, this topic will be addressed only briefly.

The internationalization of R&D and technology transfer is a trend with considerable momentum. There are many past and current examples of successful international technology transfer and R&D collaboration. The recent history of the internationalization of industry through foreign direct investment, trade, and the proliferation of transnational technical alliances is rife with examples of technology transfer and collaborative R&D involving firms based in different countries. Growth of international collaboration among university-based researchers in science and engineering is well documented by the explosion in the number of jointly authored research papers. Government-to-government collaboration is well established in certain areas of basic research such as the human genome project, fusion, and global climate change. Development of international standards and conformity assessment regimes by collaborating public and private standards bodies from different countries also has a long history. There has even been limited international collaboration involving governments and industry in precompetitive research such as the intelligent manufacturing systems initiative.

The international production and diffusion of knowledge by public and private research institutions are important and growing. In some areas (e.g., worldwide computer networks), the frontier between national and international activity is completely blurred. National R&D and technology transfer policies have to take this situation into account.

There are many benefits to international collaboration in R&D and technology transfer. These include synergies as well as economies of scale and scope in R&D; risk and cost sharing; accelerated diffusion of new technology; more open markets with an accompanying stimulus to international trade, investment, and economic development; and the ability to tackle cross-border public R&D challenges (e.g., global environmental challenges).

This comparative study suggests that there are promising opportunities for mutually beneficial collaboration between public and private German and U.S. R&D organizations in areas of precompetitive applied research and technology. At the same time, there are significant impediments to international collaboration in these areas, particularly when public and private partners are directly or indirectly involved. Indeed, the negotiation of successful collaborative projects between public- and private-sector entities in areas of industrially relevant research is complex enough within the confines of a single national innovation system. Issues that must be resolved relate to such things as the appropriate division of labor between public and private institutions, who pays for which part of the related R&D activities, and who will own the results of the research and under what conditions. However, when such collaboration is pursued at the international level, many other obstacles come into play. These include lack of awareness of opportunities due to inadequate exchange of information, differences in intellectual property regimes and practices, and the existence of different funding, accounting, and administrative requirements and practices.

In this context, new public policy challenges include capturing the economic benefits of public R&D spending, ensuring equitable access to other countries' public R&D investments, and ensuring adequate quid pro quo from foreign firms and foreign countries that gain access to a country's public R&D (National Academy of Engineering, 1996b). Given these factors, it is not surprising that funding for transatlantic or international projects oriented toward applied research or technology transfer is scarce. The absence of more explicit international policies or agreements is a significant impediment to transatlantic technology transfer.

CONCLUSIONS AND RECOMMENDATIONS

General Conclusions and Challenges

This comparative assessment of major sectors of the German and U.S. technology transfer systems reaffirms that many deeply interrelated factors shape the organization, conduct, and performance of technology transfer in an advanced industrialized country. Among the most important of these factors are the size and technical intensity of the nation's domestic market, the composition and dynamics of its industrial base, the organization and performance of its capital and labor markets, its societal goals and priorities as expressed in its allocation of public monies, its public policies and established private-sector practices, and the culture of its people.

The comparative study has helped the binational panel better appreciate similarities and differences of function, form, and context of the R&D and technology transfer activities in the two countries. Further, it has enhanced the panel's understanding of the relative strengths and weaknesses of the two national systems. In so doing, it has focused the panel's attention on a limited number of country-specific as well as shared opportunities for enhancing the performance of the two national systems.

OBSERVATIONS REGARDING THE NATURE OF TECHNOLOGY TRANSFER

Effective technology transfer is greatly facilitated by the close interaction of individuals involved in the development and/or application of the technology transferred. In spite of the global communications revolution and the proliferation of multinational companies, technology transfer remains first and foremost a "contact sport," involving close, sustained interaction of individual scientists and engineers from different organizations.[45]

Mobility of technical personnel among institutions is an important facilitator of technology transfer. Indeed, one of the most important means of technology transfer in rapidly evolving technical fields is the movement of people (scientists and engineers) among organizations—whether through temporary exchange or permanent transfer.

Technology transfer is highly industry and technology specific. The preferred mechanisms of technology transfer vary depending on the characteristics of the technology being transferred, the industry involved, and the rate of technological change affecting the industry at the time. For example, patent licensing is a critical instrument of technology transfer in sectors where the time to commercialization is long (e.g., biotechnology). However, patent licensing is relatively less important in microelectronics, where current technology life cycles are short. Research publications, conferences, and the movement of research personnel from academic and related institutions to industrial research organizations are primary modes of technology transfer in highly dynamic, science-based industries such as biotechnology and software. However, they are relatively less important avenues of technology transfer in technologically more slowly advancing or mature sectors such as the automotive or electrical machinery industries. In these areas, industrial associations and contract research are effective instruments for the diffusion of knowledge.

MAJOR SIMILARITIES BETWEEN THE GERMAN AND U.S. SYSTEMS

There are many similarities between the German and U.S. technology transfer systems. The R&D portfolios of both countries are highly diversified and span the full spectrum of science and engineering fields. Technology transfer systems in both countries support a wide range of technologically nascent and well-established industries. The overall structure of the two national systems, as well as the functional roles of many of the major institutional players in each system, are roughly comparable.

Comparable institutions in the two countries face many similar challenges. German and American research universities wrestle with reconciling new and traditional university missions, conflicts of interest, and tensions between academic and industrial research cultures. To cope with these problems, universities in both countries have established special transfer-oriented research entities in addition to supporting traditional faculty research. In the United States, these include a diverse population of university-industry research centers. In Germany, the An-Institutes were created as legally independent bodies that could operate outside of the heavily regulated environment of universities.

In both countries, the industry-oriented research institutes must try to meet the transfer needs of small companies, adapt to new demands of their clients, build competencies in emerging areas, and compete and cooperate effectively with other nonindustrial technology transfer institutions within their respective national R&D systems.

The large German Helmholtz Centers and large U.S. national laboratories (Department of Defense and Department of Energy) are both under pressure to diversify their research portfolios and downsize in response to a contraction of their original national missions. These concerns were addressed at great length (and inconclusively) in several major recent studies[46] but were considered by the binational panel to be beyond the scope of the current study. Nevertheless, the panel believes it is only logical that laboratories, which in the pursuit of their public missions conduct R&D of potential relevance to industry, have policies and procedures in place that facilitate technology transfer and collaborative R&D. In both countries, the large public laboratories have made positive strides toward improving the organizational and policy framework for collaboration with industry. However, the German and U.S. delegations believe that much more needs to be done in this area.

These common challenges suggest the potential for greater mutual learning among functionally comparable institutions in the two countries.

MAJOR DIFFERENCES BETWEEN THE GERMAN AND U.S. SYSTEMS

At the same time, comparison of the two nations' technology transfer systems underscores major differences in their scale, structure, organization, operation, and performance.

The scale and openness of the U.S. economy and innovation system overall, as well as the large size and technological intensity of relatively homogeneous segments of the U.S. domestic market, create more opportunities than exist in Germany for technology-driven firms. Despite significant progress toward integration of national economies within the European Union during the past decade, European markets for many technology-intensive products and services still remain much more fragmented by different national policies and practices than do U.S. markets. Per capita consumption of high-tech products and services is significantly higher in the United States than it is in Germany or the European Union as a whole.

The German technology transfer enterprise is relatively stable, structured, and homogeneous within sectors compared to its U.S. counterpart. In general, the German technology transfer system relies heavily on organizations and institutions that are national in scope, highly differentiated by industry and technology area, and, in many instances, well interconnected or coordinated. These characteristics are reinforced by the relatively stable composition of the German industrial production and R&D base over time. The integrated structure and stability of the German system (compared with the U.S. system) have yielded enhanced communication and cross-institutional learning among organizations, as well as rapid incremental innovation and technology diffusion in several technologically mature industries. To summarize: The German innovation system is organized to excel in the application of new technologies that increase the performance of existing industries.

The U.S. technology transfer enterprise is more highly diversified, less coordinated, more flexible, and more rapidly evolving than its German counterpart. Technology transfer in the United States relies heavily on the movement of individual researchers between organizations and within and between different sectors of the nation's technology enterprise. Operational responsibility for technology transfer in the United States resides within a very large, highly distributed, rapidly evolving, and weakly coordinated set of organizations. The structure and dynamics of the U.S. system afford it greater flexibility than its German counterpart and thus make it better able to react to changing conditions. This flexibility has provided advantages to the United States, particularly in fast-moving fields. The major shifts in the sectoral composition of industrial production and industrial R&D in the United States during the past 20 years attest to the system's dynamism. To summarize: The U.S. innovation system is structured to excel in opening up new technological frontiers and launching new industries.

In general, the science and engineering R&D workforce is more mobile in the United States than it is in Germany, particularly when movement between public and private institutions is considered.[47] This reflects broad differences in the structure, culture, and regulation of labor markets in the two countries. In the United States, public policies and private strategies pose fewer barriers and offer more incentives and opportunities for the movement of research scientists and engineers between academic, government, and commercial organizations than they do in Germany. In Germany, the inflexible regulation of civil service and public employment is substantially responsible for the low mobility of the science and engineering workforce. To a certain extent, the German system compensates for the limited intersector mobility of scientists and engineers through more extensive interinstitutional linkages, such as the dual appointment of outstanding researchers as research institute directors and university professors.

High-technology start-up companies play a critical role in the transfer and commercialization of technology developed in universities, government laboratories, private-research institutes, and large private companies in the United States but not in Germany. Many factors have contributed to the special role of start-up companies in the United States. Among the most important of these have been a highly mobile technical workforce; the availability of private venture capital; the existence of sophisticated public equity markets; and the large scale and high-technology intensity of relatively homogeneous segments of the U.S. domestic market. There is no counterpart in Germany to the prominent role that start-up companies perform in the commercialization of new technology in the United States. This is due to a range of social, legal, structural, and cultural factors in Germany, many of which have very little to do with the performance of German R&D and technology transfer institutions proper. These factors include weak venture capital and limited public equity markets; unsupportive labor, company, and bankruptcy laws; and impediments related to the tax code and other regulations.

Small and medium-sized enterprises (SMEs) in technologically mature industries have greater access to new technology in Germany than they do in the United States. The R&D and technology transfer infrastructure for technologically evolving industries in Germany is relatively comprehensive in scope and well coordinated. It is sustained by a high degree of industrial self-organization and a significant level of public-sector support. By contrast, the U.S. R&D and technology transfer infrastructure for technologically mature industries is much more uneven, fragmented, and diversified in its organization, conduct, and performance.

German industrial R&D consortia are generally structured within the framework of broad industry associations and organizations. The industrial R&D consortia in the United States are organizationally more differentiated, less widespread, and less comprehensive in scope than their German counterparts. They are ad hoc in character and, until recently, have been more limited to capital-intensive industries than have consortia in Germany.

The German contract research institutes are more tightly networked and uniform with respect to size, organization, and administration across technology areas than their U.S. counterparts. Moreover, individual contract research institutes in Germany appear to be more targeted to particular industries than such institutes are in the United States. The semipublic German Fraunhofer Society is an effective institution of technology transfer that has no direct organizational counterpart in the United States. Only a few of the private independent engineering research and technology transfer institutions in the United States are oriented toward industrial clients to a comparable extent. The Fraunhofer institutes are more targeted to specific areas of technical expertise, more networked, and more closely linked to universities than the U.S. independent engineering institutes.

Overall, the federal and state governments in the United States are much less involved financially and organizationally in industrially relevant, civilian applied R&D and technology transfer and diffusion than is the German public sector. Public attitudes toward the role of government in civilian R&D and technology transfer differ greatly in Germany and the United States. Government support of R&D in the United States is heavily concentrated in the defense and health areas (and associated industrial sectors). Outside of the health area, the U.S. federal government has invested few resources in direct support of the diffusion and use of industrially relevant civilian technology. The German public R&D portfolio is distributed over a larger number of public missions and industrial sectors than is the U.S. one. Civilian industrial development is a major objective of German public R&D spending. Moreover, German government support of effective diffusion and use of technology by industry is extensive.

The technology transfer activities of publicly funded individual R&D-performing institutions in Germany appear to be more targeted to particular industries than is true in the United States. Even the relatively small volume of state and federally funded R&D and technical-extension activities in the United States directed explicitly at the objective of industrial development tends to be oriented more toward regional economies and multiple broad technical fields than toward the needs of specific industries. Individual U.S. federal laboratories subordinate support for any particular industry to their broader public mission and tend to serve a broader population of industries through Cooperative Research and Development Agreements, licensing, and other technology transfer activities than their German counterparts. In Germany, the semipublic Fraunhofer institutes have the explicit mission of technology transfer; many departments of the large Helmholtz Centers work in areas that are strategically relevant for industry, and many R&D programs of the Federal Ministry for Education, Science, Research, and Technology have an explicit focus on technology transfer through university-industry collaboration.

CHALLENGES TO THE GERMAN AND U.S. SYSTEMS

Drawing on their combined assessment of the relative strengths and weaknesses of the German and U.S. technology transfer systems, the two delegations to the binational panel each identified priority challenges for their respective national systems.

The German delegation sees one primary challenge:

- to change aspects of the nation's cultural and legal framework that currently discourage innovation, commercialization of new technology, risk taking, R&D workforce mobility, and entrepreneurial activity generally within Germany.

The U.S. delegation sees two primary challenges:

- to preserve and further leverage the core strengths of the U.S. technology transfer system, namely, its flexibility, adaptability, and diversity; and
- to better meet the technology transfer needs of SMEs and the "infrastructural" and "pathbreaking" R&D needs of technologically mature industries.[48]

The recommendations of each delegation, presented in the next section, address specific aspects of these general challenges.

Recommendations

This comparative study has shown that the national institutional framework for technology transfer is very rich and diverse in both Germany and the United States. More important, the study has increased the binational panel's understanding of the relative strengths and weaknesses of the two nations' technology transfer systems and highlighted important opportunities for both German and U.S. stakeholders to further diversify and strengthen their respective systems. Accordingly, the German and U.S. delegations to the panel have each arrived at a short list of recommended actions (addressed to public and private policymakers within their respective countries) designed to exploit the opportunities and address a number of the major challenges facing their respective national technology transfer systems. In addition, the full binational panel has developed a limited number of joint recommendations aimed at expanding mutually beneficial international collaboration in basic and applied R&D and technology transfer among Germany, the United States, and other nations.

RECOMMENDATIONS OF THE GERMAN DELEGATION

The German delegation to the panel notes that efficient technology transfer is very important for the competitiveness of an economy and suggests a variety of related measures for Germany. At the same time, the delegation acknowledges the fundamental importance of basic, noncommercial research and the need to guarantee its future as prerequisites of successful and innovative technology transfer.

General Recommendations

1. The delegation recommends that German research policymakers foster cooperation between German and U.S. R&D institutions in the area of applied, transfer-oriented research.

The delegation is aware of various funding sources for supporting German-U.S. cooperation in the area of basic research. However, obtaining support for binational applied R&D and transfer-oriented projects is problematic. To solve this problem, the German-American Academic Council Foundation should support these types of projects more intensively and establish a permanent committee to look for appropriate funding sources. (Specific suggestions for joint German-U.S. projects are made in the "Joint German-U.S. Recommendations" section, below.)

2. The delegation recommends that, beyond improving institutional mechanisms of technology transfer, changes be made in German cultural and legal frameworks in order to encourage individual professional mobility, risk taking, entrepreneurial activities, and acceptance of technology.

Changes should be considered in such areas as the social security system, enterprise law, and tax regulations. In addition, the delegation recommends that the government and industrial associations organize public campaigns to improve public acceptance of new technology, encourage more risk taking, and develop an improved entrepreneurial spirit.

Specific Recommendations

3. The delegation believes that new technology-based firms are an important instrument for fostering the transfer and commercialization of technologies that find their origins in universities, research institutes, and established companies. Therefore, the delegation suggests that incentives to foster such spin-offs be improved through changes in the German legal, tax, and financial frameworks.

The delegation recommends the creation of special tax deductions for and public reinsurance of venture capital, equivalent to the existing system for foreign investment. In addition, the private liability of enterprise founders should be abolished, and the opportunities for small firms to go public should be improved. Furthermore, the delegation recommends introducing special tax deductions or even tax exemptions for new technology-based firms, especially in the first few years after their establishment.

4. The delegation believes that technology transfer is enhanced significantly by the temporary transfer of personnel from research institutes and universities to industrial laboratories and vice versa. To foster this professional mobility, the delegation recommends that German regulations regarding civil service and public employment be made more flexible.

Existing regulations affecting the ability of civil service and public employees to take temporary leave to work in private institutions are very rigid. The related administrative procedures are complex and a clear disincentive for this very efficient instrument of technology transfer. Improved regulations in this area should include the possibility for civil servants to participate actively in private enterprises.

> 5. *The delegation believes that the U.S. model of industrial grants to universities fosters creativity in research and technology transfer. Therefore, the delegation recommends that special tax incentives be introduced to support this form of open university-industry cooperation in addition to the presently dominant system of contract research.*

The delegation believes that industry grants in support of academic research have certain advantages over contract-supported R&D. For example, although deliverables may not be specified ahead of time, grant-supported research typically involves the sponsors in determining the orientation of the research. Special tax incentives should be introduced to foster research grants. These should include some mix of tax deductions (deductions from taxable income) and tax credits (direct deductions from taxes owed) for sponsored research, similar to the tax treatment of donations to political parties. In addition, the upper tax limits for donations to scientific institutions should be raised.

A very interesting American model is the Industry-University Cooperative Research Centers program, which is cofunded by the National Science Foundation and a group of industrial enterprises through grants. German decision makers should investigate whether a similar model can be introduced at German universities.

> 6. *The delegation recommends that contract research at universities and other research institutions be stimulated by special tax deductions.*

The recent decision of the Federal Financial Court (*Bundesfinanzhof*) that no tax deductions can be applied to contract research at public research institutions is very detrimental to technology transfer. The related regulations (i.e., *Abgabenordnung*) should be amended as soon as possible in order to avoid problematic long-term effects.[49] The delegation does not agree with some policymakers who want to impose sales taxes on the contract research of universities.

> 7. *The delegation sees the need to enhance the entrepreneurial spirit in public and semipublic research institutions. To achieve this goal, the delegation recommends increasing the flexibility of the public wage system.*

The delegation sees the need for more incentives for research staff to work on contract research, as efficient technology transfer requires entrepreneurial behavior of the involved researchers, which has to be supported by appropriate wage structures. The entrepreneurial spirit of young researchers should also be supported by the establishment of more chairs and professorships for innovation management. Up to now, only a few university business departments have established these positions.

> 8. *The delegation supports German initiatives to increase the commitment of all public or semipublic R&D institutions to conduct contract research. The delegation, however, recommends that equitable conditions for all these institutions be ensured to prevent unintentional distortions of the contract research market.*

Present political initiatives encourage various types of institutions to engage more actively in contract and cooperative research. The delegation believes that the competition provided by these institutions will enhance the quality of technology transfer. However, these institutions have different frameworks regarding base funding, overhead costs, research infrastructure, facilities, and so forth. A balanced framework will be necessary to prevent unintentional distortions of the research market—in particular through cross-subsidies of contract research at institutions with high levels of base funds.

> 9. *The delegation suggests that legal and financial requirements be changed to support a more active role for universities and public research institutions in the area of intellectual property rights and licenses.*

Due to lack of an appropriate infrastructure and financial incentives, most German universities do not protect adequately the intellectual property developed by their researchers, nor do they promote effectively the commercial exploitation of research results emanating from their laboratories. The delegation believes that universities and polytechnical schools need to get special funds and personnel to actively apply for patents and market licenses. In addition, the government should consider whether university professors should share in license revenues, according to the law for employed inventors (*Arbeitnehmererfindungsgesetz,* ArbEG), instead of being subject to the present personal exploitation privilege.

Like American universities, German universities should be able to obtain full title to inventions developed with public support, be completely relieved of return flows of license fees to publicly sponsored institutions, and should be allowed to license patents exclusively to private enterprises.

The delegation also recommends a more active patent policy for departmental research institutes and Blue List institutes, based on measures similar to those suggested for universities. In Helmholtz Centers, the present patent policy can be improved by allowing centers to completely retain license revenues and giving them more flexibility to use those revenues.

For general support of patents at research institutions, the grace period for patents should be reintroduced, comparable to the present patent law in the United States.

RECOMMENDATIONS OF THE U.S. DELEGATION

This comparative study has deepened the U.S. delegation's understanding of the fundamental strengths and weaknesses of the U.S. technology transfer system. Specifically, the delegation recognizes that the greatest strengths of the U.S. system are its large size, diversity, and decentralized and distributed organization; the high mobility of resources critical to R&D, technological innovation, and technology transfer (i.e., science and engineering personnel, capital, and other factors of production); and the system's extraordinary flexibility and adaptability to changing circumstances. At the same time, comparison with Germany has heightened the U.S. delegation's awareness of the relative weakness of the U.S. system in meeting the R&D and technology transfer needs of technologically mature industries—small and medium-sized enterprises in particular.

In an effort to reinforce and leverage the fundamental strengths of the U.S. enterprise and address areas of weakness, the U.S. delegation makes six specific recommendations.

1. The U.S. delegation recommends that the federal government do more to facilitate privately led initiatives aimed at strengthening the R&D base and technology transfer infrastructure of both technologically emerging and technologically mature industries. In particular, the government should seek to further encourage collaborative industrial R&D, including industrial R&D consortia, in technology areas that are precompetitive, noncompetitive, or related to standards.

The U.S. R&D and technology transfer infrastructure for American industry, particularly for technologically mature industries, is highly uneven, fragmented, and diversified in its organization, conduct, and performance. Recently, through self-organization often augmented by limited financial and technical support from government and universities, several major technologically mature U.S. industries have revitalized their R&D-base and technology transfer infrastructure. However, there are many other U.S. industries in which such a revitalization is needed.

Steps toward this objective could include the development of technology road maps by federal agencies, working in collaboration with industry, that would encourage private industry to engage in similar exercises[50] (see Recommendation 2, below); working with industry associations to identify regulatory changes that might promote the collective definition and diffusion of best practices in regulated industries; and amendments to the R&D tax credit that would foster collaborative R&D in technology areas, including those defined above.

2. The U.S. delegation encourages more U.S. industries to prepare dynamic technology road maps similar to the one prepared by the U.S. semiconductor industry.

In 1994, the U.S. Semiconductor Industry Association published *The National Technology Roadmap for Semiconductors*, which inventoried the industry's various sources of technology (e.g., raw material suppliers, equipment vendors, universities, federal labs), forecast technological needs and developments for all elements of the industry's value-added chain, and prioritized the industry's basic, precompetitive, noncompetitive, and standards-related research needs. Both the development of the road map and the ongoing follow-up to the initial exercise have been deemed successful and constructive by all parties involved (see Rea et al., 1996). The U.S. delegation believes that for many industries (particularly industries based on technologies that are more established and advancing incrementally) technology road map exercises such as this can be very useful for focusing the industry's attention on changing technological opportunities and challenges as well as for encouraging public- and private-sector investments to exploit or address them.[51]

> *3. The U.S. delegation recommends that industrial leadership organizations such as the Council on Competitiveness, the Business Roundtable, and the National Association of Manufacturers do more to assess and disseminate lessons regarding the value and effective organization and operation of industrial R&D consortia in the United States.*

During the past decade, industrial R&D consortia have become increasingly important vehicles of industrial technology transfer in the United States. Despite their proliferation across a growing number of industries, there has been remarkably little knowledge transfer concerning successful organizational and operational practices among the diverse, highly distributed, and relatively autonomous population of consortia. While the task of cross-industry or cross-institutional learning is complicated by the sheer diversity of the organizational types of such consortia, the documented experiences of a few highly publicized consortia suggest that there are many generalizable lessons to be learned. In the absence of a national "clearinghouse" for cross-industry learning in this area, like the German AiF, the U.S. delegation believes that U.S. organizations with significant industry participation such as the Council on Competitiveness, the Business Roundtable, and the National Association of Manufacturers are particularly well equipped to document, assess, and disseminate lessons learned to U.S. industry.

> *4. The U.S. delegation encourages all U.S. research universities to define policies to guide technology transfer and research collaboration with industry that are transparent and predictable. Such policies must be attuned to the diverse needs and circumstances facing different industrial sectors as well as different academic institutions (e.g., medical schools versus engineering schools) and the schools and departments within them. To this end, the delegation recommends that the National Academies' Government-University-Industry Research Roundtable address the issue of establishing a*

national effort to collect and disseminate information regarding the diversity of university good practices in this area.

A number of U.S. research universities or constituent graduate schools appear to have developed effective policies and practices for managing their R&D and technology transfer interactions with industrial partners. However, considerable evidence suggests that many more of the nation's research universities are still struggling to develop policies and practices that will engage industry more extensively in mutually beneficial research and technology transfer. For this reason, the delegation believes that academic research institutions need to work together to identify, develop, and disseminate the variety of good policies and practices that support effective university-industry collaboration.

The U.S. delegation considers the current diversity and decentralized nature of the U.S. university research enterprise to be sources of strength. Accordingly, the delegation does not believe that a one-size-fits-all approach to best practices in university-industry technology transfer or collaborative R&D makes sense. At the same time, the delegation recognizes that the very diversity and decentralized character of the university research enterprise make cross-institutional learning difficult to achieve. The delegation believes that the charter of the Government-University-Industry Research Roundtable and its previous work, particularly in the area of university-industry partnerships, well equip it to take the initiative in advancing the learning process in this area.[52]

5. The U.S. delegation recommends that the National Science Foundation expand the existing network of collaborative university-industry research centers (e.g., Industry-University Cooperative Research Centers, Engineering Research Centers, Science and Technology Centers) and continue to experiment with other institutional and organizational forms.

The U.S. delegation believes that university-based multidisciplinary centers sponsored by the NSF have on the whole proved to be cost-effective mechanisms for promoting university-industry research collaboration. For a relatively small federal investment, they can provide high educational value, contribute to the research needs of a broad spectrum of industries (both technologically dynamic and technologically mature), and provide valuable technical-extension services to regional industry. The federal government should be relatively unconcerned about some industries being served by more than one center, provided there is sufficient industry support for additional centers and there is a clear demand for more graduates in the field supported. Moreover, NSF should experiment and encourage others to experiment with new institutional and organizational modes that promote university-industry research collaboration. In this regard, the structure, conduct, and performance of the German independent, university-linked An-institutes and Fraunhofer institutes would seem worthy of further study and analysis.

6. The U.S. delegation recommends that all federal agencies—no matter the extent to which their mission relates directly to private industry—develop more effective mechanisms for ensuring appropriate industrial input into the formulation, execution, and evaluation of federal laboratory R&D and technology transfer activities.

The U.S. delegation recognizes that virtually all federal agencies with R&D portfolios are coming to rely more and more heavily on private-sector R&D capabilities and outputs to advance their public missions. Irrespective of whether a given federal laboratory considers the development and diffusion of industrially relevant technology a central, ancillary, or negligible part of its overall public mission, it is increasingly critical that the laboratory develop the capability for drawing effectively on private-sector R&D and R&D management and evaluation capabilities, as well as contribute to those capabilities.

Joint German-U.S. Recommendations for Fostering Transatlantic Collaboration in R&D and Technology Transfer

The binational panel believes significant opportunities exist for building on the learning process initiated by this study. The panel recommends that steps be taken in both countries to facilitate and enhance the exploitation of these opportunities. Specifically, the panel encourages mutually beneficial collaboration between German and U.S. public- and private-sector institutions in areas of basic and applied R&D and technology transfer.

The panel acknowledges that various impediments to transatlantic collaboration in the area of technology transfer remain. These include simple lack of awareness of opportunities; obstacles posed by different funding, accounting, legal, and administrative practices in different countries; differences in intellectual property regimes and practices; and structural as well as public-policy-induced barriers. The panel believes that many of these impediments are surmountable and that transnational R&D arrangements are, in the long run, a necessary component of free trade in a global economy.

The following three recommendations are aimed at fostering mutually beneficial transatlantic (European Union–United States) as well as global cooperation in R&D and technology transfer.

1. The panel recommends that the German and U.S. governments catalyze transatlantic efforts to develop a set of mutually agreed-upon principles that will help the United States and member states of the European Union to recognize and implement arrangements for mutually beneficial transatlantic collaboration involving public and private institutions in R&D and technology transfer.

The panel recognizes that many of the seemingly intractable impediments to enhancing collaboration between Germany and the United States involve manifold differences in national law and public policy in such areas as intellectual property rights; competition and antitrust policy; corporate law; health, safety, and environmental regulations; international trade and investment policy; and R&D, technology, and industrial policies. Many of these policy differences and the associated obstacles they pose to international collaboration are often poorly understood. Moreover, the international policy challenges raised by these national differences are weighted differently by different countries, and viewed as tightly interconnected.

For these reasons, the panel believes that these policy-related impediments will and should be addressed at the bilateral or multilateral level as interconnected policy challenges.[53] The ongoing negotiations among member states of the Organization for Economic Cooperation and Development to develop a Multilateral Agreement on Investment could be an appropriate location for the development of such principles. The panel believes that successful negotiations within these forums would serve as a useful starting point for more comprehensive global negotiation of these policy challenges within the World Trade Organization, the World Intellectual Property Organization, international standards-setting organizations, and other relevant international organizations.

2. The panel strongly urges public- and private-sector institutions that perform and fund R&D in Germany and the United States to adhere mutually to the generally accepted principles of national treatment (nondiscrimination on the basis of nationality) and transparency (full disclosure of terms) in their interactions with organizations based in other countries.

The panel believes that mutual observance of these principles by German and U.S. organizations will help overcome many of the impediments to mutually beneficial transatlantic collaboration in R&D and technology transfer identified above and will encourage R&D-funding and R&D-performing institutions based in other countries to follow their lead.

3. The panel recommends that research and technology transfer organizations in Germany and the United States be encouraged to experiment with a number of different types of transatlantic collaborative activity. The panel believes that, in the near term, experimentation with specific participant-driven collaborative projects (i.e., learning by doing) is the most promising approach to understanding and overcoming obstacles to enhanced binational, transatlantic, and global collaboration, particularly in the areas of applied research and technology transfer. In this context, the panel believes that the German-American Academic Council Foundation could play a constructive role in both fostering and evaluating participant-driven experiments with transatlantic collaboration, as well as in identifying and disseminating best practices in this area.

To this end, the panel has identified several types of binational, transatlantic, and international collaboration, both prospective and ongoing, that it considers particularly promising in the German-U.S. context. In some cases, the panel recommends specific actions to facilitate experimentation with a particular type of collaboration. In others, the panel merely identifies an opportunity or a proof of existence to illustrate the potential of a particular mode of collaboration.

Establishment of transatlantic networks of R&D institutes for the shared use of facilities and infrastructure, especially information infrastructure. As an example of what form such collaboration might take, two panel members developed a brief prospectus for a joint project to facilitate involvement of small and medium-sized companies in the development of computer-supported cooperative work (CSCW) applications to exploit emerging high-speed global telecommunications networks. (See Annex I.) Here again, the panel views limited transatlantic initiatives such as this as a useful starting point for more extensive global initiatives.

Transatlantic collaboration of university-based institutions in the areas of applied, transfer-oriented research. In particular, the panel sees promising opportunities for mutual learning through collaboration between the U.S. UIRCs and the German An-Institutes. Both sets of institutions represent efforts to overcome some of the administrative and cultural barriers to closer university-industry research collaboration. Admittedly, there are significant organizational and funding differences between these two groups of institutions (not to mention profound differences among the highly heterogeneous population of U.S. centers). An-Institutes are legally independent of the universities and rely primarily on research contracts with industry and government agencies. The diverse assembly of U.S. UIRCs, by contrast, tends to be more closely tied administratively with host universities and rely almost exclusively on research grants to support its work. Because of these differences, the panel views transatlantic collaboration between UIRCs and An-Institutes as a particularly promising opportunity for cross-institutional learning.

Development of a transatlantic network of technical information centers. The panel believes that Germany, the United States, and other countries would benefit significantly from the development of a transatlantic network of technical information centers that could develop common technical standards and interfaces, enable interconnection, foster collaboration in database building, and so forth. Binational and multinational initiatives linking centers based in the United States, Germany, and the rest of the European Union could serve as a stimulus (demonstration projects) to enhanced interconnection of technical information centers based in other countries worldwide. (See, for example, the panel's suggestion in Annex I for developing linkages between the U.S. Department of Defense, Information Analysis Centers, and corresponding German technical information centers.)

International collaboration in R&D and technology transfer related to health, safety, environmental, and other technical standards. International collaboration in these areas currently involves a multiplicity of public- and private-sector organizations worldwide, including national and international standards bodies, professional societies, government agencies, industrial associations, and international organizations. An example of this type of collaboration is the industry-sponsored International 300 mm Initiative, which is developing standards for 300-mm wafers for microcircuits.

Transatlantic collaboration of industrial R&D consortia and industrial research associations on research projects of common interest. The panel believes there are opportunities for mutually beneficial binational and international collaboration of industrial R&D consortia and industrial research associations in areas of generic application-oriented research, generally of a precompetitive character. A binational effort to enhance information exchange regarding cooperative industrial R&D activities in Germany and the United States could serve as a stimulus to further transatlantic as well as global collaboration among national consortia and research associations on generic precompetitive research projects of common interest. A member of the German delegation has prepared a list of specific projects suggestions in this area. (See Annex I.)

Transatlantic collaborations of small and medium-sized enterprises and R&D institutions. Considering the relative strengths and weaknesses of the German and American technology transfer systems, the panel believes that transatlantic collaboration involving SMEs in one country with R&D institutions based in the other could be mutually beneficial. Large companies are already engaged in transatlantic R&D collaboration of this type. However, SMEs are generally less well informed about the capabilities of foreign research and technology transfer institutions, let alone specific opportunities for collaboration with them. A binational effort to disseminate information to German and American SMEs regarding the research and technology transfer capabilities of R&D institutions in the two countries could facilitate mutually beneficial collaboration as well as encourage other nations to engage in such information exchanges.

Concluding Observations

The results of this joint study confirm the many important roles that the public sector plays in the R&D and technology transfer systems of advanced industrialized nations. *Of the many technology-transfer-related issues facing the German and U.S. governments, the panel believes that special attention should be given to the issue of public funding for R&D that is of an infrastructural, or long-term nature.* In the panel's view, this type of R&D activity is critical to a nation's economic growth and industrial competitiveness but tends to be underfunded by private-sector organizations because its outputs—new knowledge, know-how, skills, and generic technology—are broadly diffused without direct compensation to the R&D performer.

Furthermore, given the pace and magnitude of efforts currently under way in both Germany and the United States to adapt their respective technology transfer systems to rapidly changing political, economic, and technological circumstances at home and abroad, *the panel encourages the German-American Academic Council to consider sponsoring a reexamination of the two countries' technology transfer systems 5 years hence and expand the comparison to include one or two other countries.*

The present study has documented the role that effective technology transfer institutions, mechanisms, policies, and practices play in the competitiveness of industrial nations. The comparison of the German and U.S. technology transfer systems has helped the panel appreciate and understand the strengths and weaknesses of the two nations' systems and has suggested new approaches to specific technology transfer challenges faced in each country. Equally important, the binational study has underscored the potential for continued mutual learning between U.S. and German technology transfer institutions and has identified several promising opportunities for mutually beneficial transatlantic and international collaboration in applied research and technology transfer.

ANNEX I

Suggestions for Transatlantic/International Collaborative Projects

TRANSATLANTIC COOPERATIVE COMPUTER APPLICATIONS OVER GLOBAL NETWORKS

The ability of an international team to rapidly design, prototype, and manufacture a product is a key requirement for globally active firms. The new pressures on business include:

- *Reduced product life cycles.* Time-to-market is becoming an ever more significant factor contributing to the ability to achieve market share, profitability, and even survival.
- *Increased cost pressures.* The need to control costs, with the corresponding desire to improve productivity, continues unabated with renewed emphasis on the productivity of knowledge workers.
- *Increased demand for quality and customer service.* As competition builds, the increase in customers' expectations for responsiveness and personalized support is beginning to change the culture and operation of many industries.
- *Changing markets.* The only constant for business is that things will change. The need and ability to respond rapidly to changing market forces continues to push firms to adopt and implement technology.
- *New business models.* Constant change is now pushing into the very core of many corporations with corresponding new business models emerging for the way in which organizations and people work together. These include telecommuting, virtual corporations, collaborative product development, and integrated supply-chain management.

There is a need to develop computer-based methods and applications to address the challenges of a global marketplace. One major challenge is to demonstrate the feasibility of conducting business over transatlantic Asynchronous Transfer Mode (ATM) and Integrated Service Digital Network (ISDN) networks and in particular to show the real-world benefits of conducting the entire international product development in such an environment. A second challenge is to enable SMEs to become world-class companies by providing them the capabilities to effectively synchronize, manage, and develop their resources throughout the world.

The goal of such a project should be to show to small and medium-sized enterprises (SMEs) in the United States and Germany that private ATM- and ISDN-based intranets will provide a reliable and secure means of transporting data, voice, and video services among distributed firms. This will include integration of existing technologies to support the concurrent engineering method as it is applied to the product development process. These technologies include telemedicine, enterprise resource planning, CAD modeling, computer-supported cooperative work, user interface design, process management and documentation, virtual reality, and ATM and ISDN wide-area networking.

A set of realistic, factory-oriented product development scenarios and quantitative metrics for evaluating success should be developed, and the scenarios and metrics using the technology base developed for concurrent engineering should be implemented. Furthermore, distributed and collaborative virtual prototyping practices for future enterprise models should be demonstrated.

An important activity within such a project might be the development of a distributed decision-support system for medical diagnosis and training. This should address the following problems:

- Utilization of distributed competence centers for different diagnosis tasks on demand;
- Utilization of intelligent data handling and diagnosis;
- Integration of different data sources and diagnosis methods; and
- Remote training support.

Several institutions have already shown interest in such a project (e.g., in Germany, the *Fraunhofer Institut für Graphische Datenverarbeitung,* Darmstadt, and the *Gesellschaft zur Förderung Angewandter Informatik e.V.,* Berlin, and in the United States, the International Computer Science Institute [ICSI] in Berkeley, California, and the Center for Research in Computer Graphics, in Providence, Rhode Island).

There are many companies in both Germany and the United States interested in getting involved in and prototyping and evaluating applications developed to run on such a platform.

SUGGESTED DEVELOPMENT OF A TRANSATLANTIC NETWORK OF INFORMATION ANALYSIS CENTERS

The U.S. Department of Defense (DOD) has a long history of active support for specialized information analysis. The DOD Information Analysis Centers

(IACs) are unique organizations with the mission of collecting and disseminating information to practicing scientists and engineers in both the government and the private sector. As of October 1991, there were 23 IACs in operation.

The IACs' information collection activities systematically identify, catalog, and collect published information in specified subject areas. Many of the subjects would not be suitable for a joint U.S.-German effort; however, many of the IACs are active in areas which have broad technical interests in both countries. To explore further the opportunities and obstacles related to joint U.S.-German activity in this area, a study is proposed that would:

- examine the mission and authority of the existing IACs to determine if it is within current authority to expand the client base to foreign scientists and engineers.
- survey existing German research institutions to identify similar information-analysis organizations in Germany;
- design a collaborative effort for information analysis on a multinational basis;
- build collaborative relationships between U.S. Information Analysis Centers and their German counterparts; and
- evaluate the results of the collaborative effort and identify communications methods that contributed to success.

It is anticipated that a multinational information analysis center would achieve greater effectiveness with a larger client base and realize economies of scale. The long-term objectives of the German-American binational panel will be advanced as a result of the experiment and collaborative effort.

COLLABORATION AMONG GERMAN AND U.S. INDUSTRIAL RESEARCH ASSOCIATIONS

The following suggestions of possible collaborative projects were collected by a German delegation member from member associations of the German AiF. The project ideas differ as to their level of concreteness (i.e., concerning the detailed definition of the project methodology and aims, the potential for support by German and American sponsors, etc.). In any case, they show a broad interest by German and American institutions in collaborating in the area of pre-competitive applied research.

Project title: Integrated Supply Chain Management Program
R&D partners: Gesellschaft für Verkehrsbetriebswirtschaft und Logistik, Nürnberg
Lehrstuhl für Logistik der Universität Erlangen-Nürnberg
Center for Transportation Studies, MIT, Cambridge
Sponsor(s): Amoco, AT&T, CVS, Monsanto, Proctor & Gamble, Roadway Logistics Services, XEROX, Siemens Volkswagen, Quelle

Project title: Traffic information system
R&D partners: Forschungsinstitut für Logistiksysteme in Ballungsräumen, Herne
California State Department of Transportation
Sponsor(s): California State Department of Transportation

Project title: Customer information system
R&D partners: Forschungsinstitut für Logistiksysteme in Ballungsräumen, Herne
California State Department of Transportation
Sponsor(s): California State Department of Transportation

Project title: Value management/value engineering as an instrument of innovation management in small and medium-sized enterprises
R&D partners: Lehrstuhl Industriebetriebslehre, Universität Dortmund
Society of the American Value Engineers (SAVE), Franklin, Pennsylvania
Sponsor(s): AiF, SAVE

Project title: Standard production planning and control system for globally operating production networks
R&D partners: Forschungsinstitut für Rationalisierung, RWTH Aachen
BDO Seidmann Ltd., Atlanta
N. N., U.S. research institute; N. N., production company
Sponsor(s): Possibly AiF for the German part

Project title: Tools for the design of flexible organization structures
R&D partners: Forschungsinstitut für Rationalisierung, RWTH Aachen
Laboratory for Manufacturing and Productivity, MIT, Cambridge
Sponsor(s): Possibly AiF for the German part

Project title: Analysis of cooperation in supply chains in the area of production technology on the basis of an American-German comparison
R&D partners: Forschungsinstitut für Rationalisierung, RWTH Aachen
Sloan School of Management, MIT, Cambridge
Sponsor(s): Possibly AiF for the German part

Project title: Friction reduction at the forming edge by compressed air lubrication
R&D partners: Institut für Umformtechnik und Umformmaschinen, Universität Hannover
Alcoa Technical Center, Pennsylvania
Sponsor(s): Possibly AiF for the German side

ANNEX I

Project title: Pressure control of blank-holder in relation to drawing force
R&D partners: Institut für Umformtechnik und Umformmaschinen, Universität Hannover
Alcoa Technical Center, Pennsylvania
Sponsor(s): Possibly AiF for the German side

Project title: Pulsating pressure of blank-holder
R&D partners: Institut für Umformtechnik und Umformmaschinen, Universität Hannover
Ohio State University, Columbus
Sponsor(s): Possibly AiF for the German side

Project title: Determination of elastic recovery in deep drawing
R&D partners: Institut für Umformtechnik und Umformmaschinen, Universität Hannover
Ohio State University, Columbus
Sponsor(s): Possibly AiF for the German side

Project title: Smoke development in arc spraying processes
R&D partners: Forschungsvereinigung Schweißen und Schneiden, Düsseldorf
Lehrstuhl für Werkstofftechnologie, Universität Dortmund
High Temperature Laboratory, University of Minnesota
Thermal Spray Lab, University of New York, Stony Brook
Sponsor(s): Possibly AiF for the German side

Project title: Joining of plastic and light metal parts
R&D partners: Forschungsvereinigung Schweißen und Schneiden, Düsseldorf
Fraunhofer-Institut für Angewandte Materialforschung, Bremen
Center for Composite Materials, University of Delaware, Newark
Sponsor(s): Possibly AiF for the German side

Project title: Orbital welding of tubes with tungsten-inert gases
R&D partners: Forschungsvereinigung Schweißen und Schneiden, Düsseldorf
Schweißtechnische Lehr- und Versuchsanstalt, München
Astro Arc Polysoude, Sun Valley, California
Sponsor(s): Possibly AiF for the German side

Project title: Bolt welding with lift ignition for aluminum
R&D partners: Forschungsvereinigung Schweißen und Schneiden, Düsseldorf
Schweißtechnische Lehr- und Versuchsanstalt, München
TRW NELSON Stud Welding Division, Elyria, Ohio
Sponsor(s): Possibly AiF for the German side

Project title: Use of neural networks for controlling welding processes
R&D partners: Institut für Schweißtechnische Fertigungsverfahren, RWTH Aachen
N. N. (Negotiations with various American partners)
Sponsor(s): Possibly AiF for the German side

Project title: Structure and characteristics of glass melts
R&D partners: Hüttentechnische Vereinigung der Deutschen Glasindustrie, Frankfurt/M.
Center of Glass Research, Alfred University, Alfred, New York
N.Y. State College of Ceramics, Alfred University
Sponsor(s): Presently: Deutsche Forschungsgemeinschaft
National Science Foundation
Problem: Orientation too basic because of requirements of funders.
Interest in AiF support for more practice-oriented research

Project title: Revolution control of ring spinning machines
R&D partners: Institut für Textil- und Verfahrenstechnik, Denkendorf
Milliken & Company
Sponsor(s): Milliken & Company

Project title: Development of composite textiles and garment ensembles for foul-weather protection with optimized comfort
R&D partners: Bekleidungsphysiologisches Institut, Hohenstein
Adidas, Herzogenaurach
W. L. Gore
Malden Mills
Pearl Izumi
In Sport
Sponsor(s): Adidas, W. L. Gore, Malden Mills, Pearl Izumi, In Sport

Project title: Production of emulsions by membranes
R&D partners: Institut für Lebensmittelverfahrenstechnik, Universität Karlsruhe
Department of Food Science, University of Massachusetts, Amherst
Sponsor(s): Possibly AiF for the German part
Forschungskreis der Ernährungsindustrie, Bonn

Project title: Improvement of reaction flavors with respect to the original food flavors
R&D partners: Deutsche Forschungsanstalt für Lebensmittelchemie, Garching
The State University of New Jersey, Rutgers
Sponsor(s): Forschungskreis der Ernährungsindustrie, Bonn
Flavor Extract and Manufacturing Association

Project title: Effects of fatty trans-acids in food on the composition of fatty acids in human blood plasma and fatty tissue
R&D partners: Institut für Biochemie und Lebensmittelchemie, Universität Hamburg
U.S. Food and Drug Administration
Center for Food Safety & Applied Nutrition, Washington, D.C.
Sponsor(s): Forschungskreis der Ernährungsindustrie, Bonn
U.S. Food and Drug Administration

Project title: Effects of phase transitions in food on deterioration of quality
R&D partners: Institut für Lebensmitteltechnologie, Technische Universität Berlin
Center for Advanced Food Technology, Rutgers University, New Brunswick, New Jersey
Sponsor(s): Forschungskreis der Ernährungsindustrie, Bonn

Project title: Application of high electric field pulses in food processing and preservation
R&D partners: Institut für Lebensmitteltechnologie, Technische Universität Berlin
Department of Food Science, Ohio State University, Columbus
Sponsor(s): Forschungskreis der Ernährungsindustrie, Bonn
The Ohio Research Foundation

Project title: Authenticity control of fruit juice flavors
R&D partners: Institut für Lebensmittelchemie, Universität Frankfurt
Citrus Research and Education Center, University of Florida, Lake Alfred, Florida
Sponsor(s): Forschungskreis der Ernährungsindustrie, Bonn

Project title: Inactivation kinetics and molecular modeling of conformational transitions in enzymes at high pressure
R&D partners: Institut für Lebensmitteltechnologie, Technische Universität Berlin
Center for Advanced Food Technology, Rutgers University, New Brunswick, New Jersey
Sponsor(s): Forschungskreis der Ernährungsindustrie, Bonn

Project title: Investigation of the nutritional balance of cereal-based products made from unconventional raw material
R&D partners: Institut für Lebensmitteltechnologie, Technische Universität Berlin
Department of Grain Science, Kansas State University, Manhattan

Sponsor(s): Forschungskreis der Ernährungsindustrie, Bonn
American Association of Cereal Chemistry, St. Paul, Minnesota

Project title: Investigation of the role of resistant starch in cereal-based products from a nutritional and a technological point of view
R&D partners: Institut für Lebensmitteltechnologie, Technische Universität Berlin
Department of Grain Science, Kansas State University, Manhattan
Sponsor(s): Forschungskreis der Ernährungsindustrie, Bonn
American Association of Cereal Chemistry, St. Paul, Minnesota

Project title: Investigation of the shelf-life of cereal-based products enriched with high unsaturated fatty acids
R&D partners: Institut für Lebensmitteltechnologie, Technische Universität Berlin
Department of Grain Science, Kansas State University, Manhattan
Sponsor(s): Forschungskreis der Ernährungsindustrie, Bonn
American Association of Cereal Chemistry, St. Paul, Minnesota

Project title: Investigation of the functional and nutritional properties of β-glucans in cereal-based products
R&D partners: Institut für Lebensmitteltechnologie, Technische Universität Berlin
Department of Grain Science, Kansas State University, Manhattan
Sponsor(s): Forschungskreis der Ernährungsindustrie, Bonn
American Association of Cereal Chemistry, St. Paul, Minnesota

PART II
TECHNOLOGY TRANSFER IN THE UNITED STATES

INTRODUCTION

The collective capacity of the United States to deploy technology and technical know-how constitutes the nation's technology transfer enterprise. The enterprise involves all of the individuals, public- and private-sector institutions, and other resources (financial and physical capital) involved in the movement of technology within and among organizations operating in the United States. As in other market economies, most of the resources and operational intelligence of this enterprise resides in private companies and is organized and driven by the logic of markets. In 1995, industry performed over 70 percent of all U.S. R&D and employed more than 90 percent of all U.S. scientists and engineers. Similarly, the volume of technology transfer that takes place within and between private firms dwarfs that which takes place between industry and all other sectors of the R&D enterprise combined.[1] Indeed, the annual patent royalty income of just one large U.S. high-tech company such as IBM is greater than that of all nonindustrial sectors together. Nevertheless, the structure, goals, and performance of the U.S. technology enterprise are profoundly shaped by the contributions of a spectrum of nonindustrial R&D performers that are not themselves directly engaged in the commercialization of technology.

The specific focus of this report is on the institutions and mechanisms involved in the transfer of technology from nonindustrial R&D performers to private firms, which then use this technology to create new products and services. These institutions include nonindustrial R&D performers: universities and affiliated institutions, federal laboratories, and an array of public, private, and mixed (public/private) contract R&D institutes and consortia. Also implicated are a diverse group of organizations that perform little, if any, R&D of their own, yet play an important role facilitating technology transfer between the nonindustrial R&D performers and private industry.

THE R&D ENTERPRISE

A major distinguishing feature of the U.S. R&D enterprise is its colossal size. In 1994, the United States spent roughly $169 billion, or 2.5 percent of its gross domestic product (GDP), on research and development. Calculated in constant 1987 dollars, this sum equaled the combined R&D expenditures of Japan, Germany, France, and the United Kingdom (Figure 2.1). As of 1993, there were roughly 963,000 scientists and engineers engaged in R&D work in roughly 41,000 U.S.-based companies, 720 federal laboratories, 875 colleges and universities, and upwards of 2,300 other nonprofit R&D-performing organizations (e.g., research institutes, hospitals, consortia, etc.) (National Science Board, 1996; National Science Foundation, 1996C).

As a percentage of GDP, R&D spending in the United States compared favorably with that of most of its major trading partners in 1994 (Figure 2.2). How-

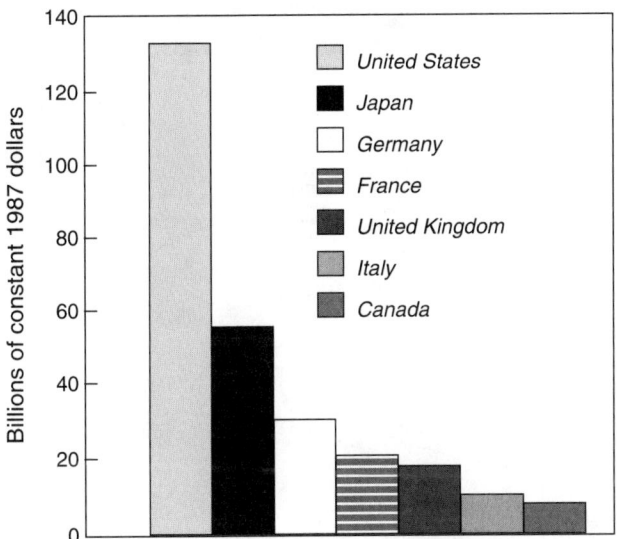

FIGURE 2.1 International total R&D expenditures, 1994. SOURCE: National Science Foundation (1996b).

ever, relative U.S. investments in R&D, estimated at 2.4 percent of GDP in 1995, have been declining since 1991, as have those of Germany and Japan. Moreover, international comparisons of the civilian R&D intensity of national economies (nondefense R&D as a percentage of GDP) reveal a persistent gap between the United States and other major industrialized countries.

R&D Funders and Performers

For statistical purposes, the U.S. R&D enterprise is divided into four major sectors: (1) government (federal, state, and local), (2) private industry, (3) nonprofit colleges and universities, and (4) other private nonprofit R&D funders or performers.

GOVERNMENT

Prior to 1980, the federal government was the leading source of R&D funds, accounting for as much as 66 percent of the nation's R&D spending in the early 1960s. During the past decade, however, with the end of the Cold War and declining defense budgets, the federal government's share has declined rapidly, amounting to only 35.5 percent of the total, about $61 billion, in 1995 (Figure 2.3). Less than one-third of federal R&D funds ($16.7 billion) were used to

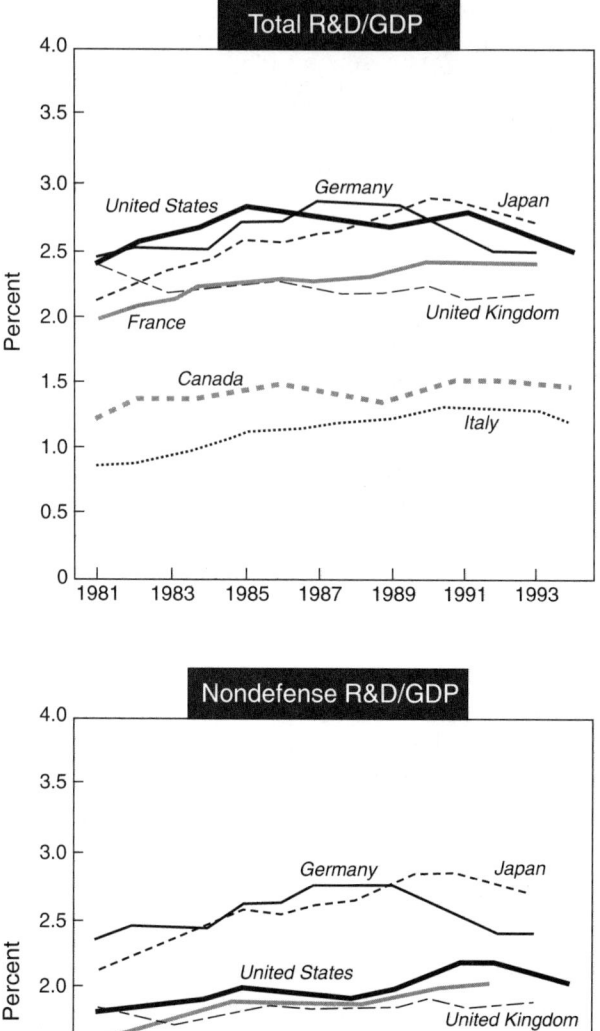

FIGURE 2.2 Total and nondefense R&D spending as a percentage of GDP, by country.
SOURCE: National Science Board (1996).

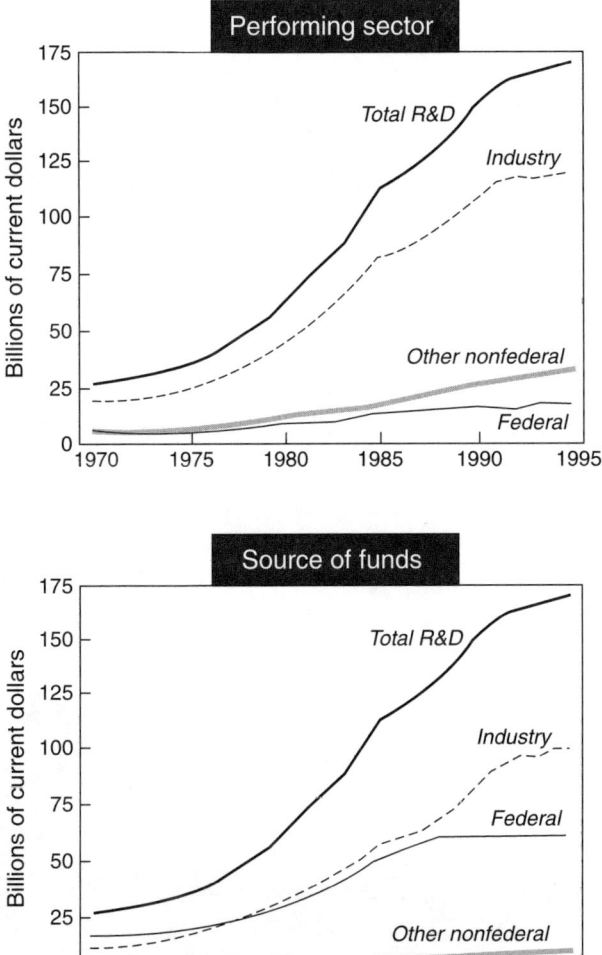

FIGURE 2.3 National R&D expenditures, by performing sector and sources of funds. SOURCE: National Science Board (1996).

support intramural R&D (i.e., R&D performed by the roughly 750 federal agency-operated research laboratories in 1995). The remaining two-thirds of the federal R&D budget supported R&D performed by private industry ($20.3 billion), universities and colleges ($13 billion), a collection of industry- and university-administered Federally Funded Research and Development Centers (FFRDCs) ($8 billion),[2] and other nonprofit institutions ($2.7 billion) (Table 2.1).

The federal government funded roughly 58 percent of all U.S. basic research, 36 percent of all applied research, and 29 percent of all development work in

TABLE 2.1 U.S. Expenditures, by Performing Sector and Source of Funds, 1995

Performing sector	Total	Industry	Federal Government	Universities and colleges[a]	Other nonprofit institutions	Percent distribution, performers
			Millions of dollars			
Total	171,000	101,650	60,700	5,500	3,150	100.0
Industry	119,600	99,300	20,300	—	—	69.9
Industry-administered FFRDCs[b]	1,800	—	1,800	—	—	1.1
Federal government	16,700	—	16,700	—	—	9.8
Universities and colleges	21,600	1,500	13,000	5,500	1,600	12.6
University-administered FFRDCs[b]	5,300	—	5,300	—	—	3.1
Other nonprofit institutions	5,100	850	2,700	—	1,550	3.0
Nonprofit-administered FFRDCs[b]	900	—	900	—	—	0.5
Percent distribution, sources	100.0	59.4	35.5	3.2	1.8	

FFRDC = federally funded research and development center; — = unknown, but assumed to be negligible.

NOTE: Data are estimated.

[a]Includes an estimated $1.6 billion in state and local government funds provided to university and college performers.
[b]FFRDCs conduct R&D almost exclusively for use by the federal government. Expenditures for FFRDCs, therefore, are included in federal R&D support, although some nonfederal R&D support may be included.

SOURCE: National Science Board (1996).

1995 (Figure 2.4).[3] That year, federal government laboratories performed 9.1 percent of all U.S. basic research ($2.7 billion), 12.3 percent of applied research ($4.9 billion), and 9 percent of development ($9.1 billion). In 1993 the federal government employed over 60,000 scientists and engineers in R&D activity (National Science Foundation, 1996b).

For the most part, national R&D statistics shed little light on the volume and character of R&D funding and performance by U.S. state and local governments. Nonfederal government entities collectively funded roughly 2 percent of all R&D performed in the United States in 1993 (National Science Board, 1996). Most state and local R&D monies are used to support applied research at doctorate-granting state universities. These funds come either directly in the form of research grants and contracts or indirectly in the form of general-purpose funds that end up being used for research by the recipient academic institutions. Collectively, state and local governments funded between 12 and 17 percent of U.S. academic research in 1995, 7.4 percent (or $1.6 billion) through research contracts and grants and an additional 5 to 10 percent ($0.9 to $2.0 billion) through general purpose funds (National Science Board, 1996).

INDUSTRY

Since 1980, industry has been both the primary source of R&D funds and the largest R&D performer in the United States, financing 59.4 percent ($101.7 billion) and performing roughly 71 percent ($121.4 billion) of all U.S. R&D in 1995 (Figure 2.3). Industry performs the overwhelming majority of the research that it funds, $99.3 billion in 1995, with the remainder, $2.4 billion, going to support research in colleges and universities and other nonprofit research institutions. Industry performed an additional $20.3 billion worth of R&D supported by federal funds in 1995; most of this was defense-related development work financed by the Department of Defense (DOD).[4] In addition to R&D performed directly by private firms, federal agencies also funded about $1.8 billion of R&D at industry-administered FFRDCs that year.

In 1995, industry funded 70.4 percent of all development work, 56.8 percent of all applied research, and 25.3 percent of all basic research performed in the United States. In turn, industry performed about 86 percent ($87.6 billion) of all development work, 67 percent ($26.7 billion) of all applied research, and 24.2 percent of basic research ($7.2 billion) that year. Roughly 764,500 scientists and engineers were engaged in R&D in U.S. industry in 1993.

COLLEGES AND UNIVERSITIES

Colleges and universities (both private and public) performed 12.6 percent, or $21.6 billion worth, of R&D in 1995. That year, university-administered FFRDCs performed an additional $5.3 billion of R&D. Institutions of higher

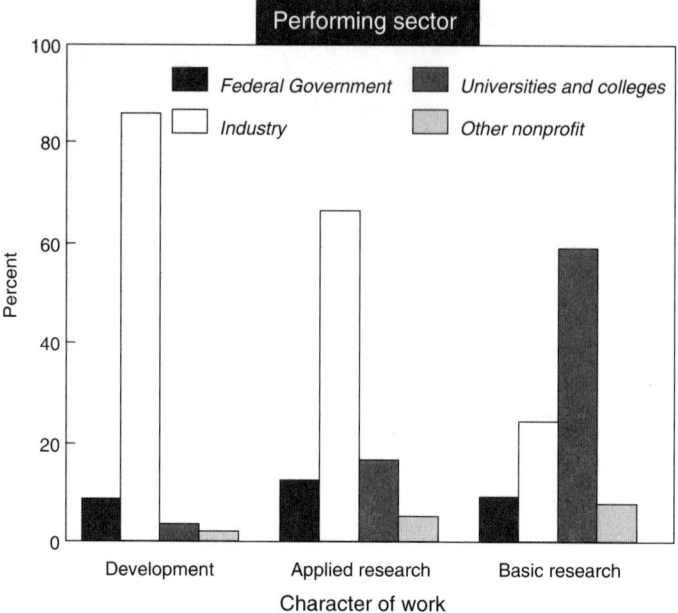
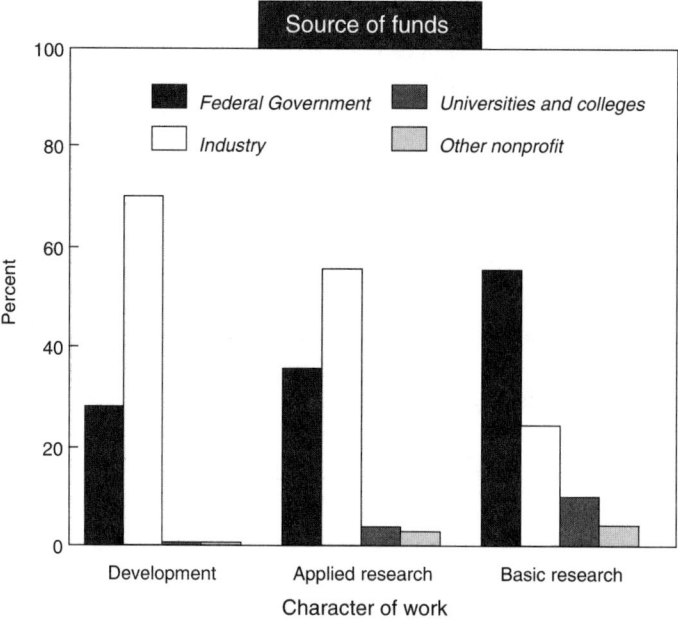

FIGURE 2.4 National R&D expenditures, by performing sector, source of funds, and character of work, 1995. SOURCE: National Science Board (1996).

education are the primary locus of basic research in the United States, accounting for roughly 49 percent of all basic research performed in 1995. Colleges and universities performed about 14 percent of all applied research and only 1.6 percent of total U.S. development work that year. Although 875 institutions of higher education reported performing R&D in 1995, the 100 largest of these accounted for 80 percent of all academic research conducted in the United States (National Science Board, 1996).

The federal government has long been the primary source of academic R&D dollars (Table 2.2). Although the federal share has fallen significantly since the early 1970s, when it accounted for more than 70 percent of academic research funds, federal agencies still financed over 60 percent of all academic research in 1995. Private universities, which represent an important, highly productive part of the nation's academic research enterprise, depend much more heavily on federal R&D funds than do public universities, which receive both targeted research funding and general-purpose appropriations from state governments. In 1993, federal agencies funded roughly 56 percent of all research at public universities and 74 percent of research performed at U.S. private universities (National Science Board, 1996). The second largest source of research funding for colleges and universities is their own institutional funds. This collection of general purpose state or local government appropriations, general purpose grants from outside sources, tuitions and fees, endowment income, and unrestricted gifts totaled roughly $3.9 billion in 1995. The share of academic research supported by institutional funds increased from 13.8 percent in 1980 to 18.1 percent in 1995. In addition to their indirect investment in academic research through general purpose appropriations to colleges and universities, state and local government directly funded 7.4 percent of U.S. academic research in 1995. Other nonprofit institutions funded an additional 7.4 percent of the total. Industry accounted for the smallest share (6.9 percent) of academic research support in 1995. However, since the mid-1980s, industry support has increased more rapidly than any other source of academic R&D funding.

TABLE 2.2 Support for U.S. Academic R&D, Percent Shares by Sector

	1970	1980	1990	1995 (est.)
Federal government	70.5	67.5	59.2	60.2
State and local government	9.4	8.2	8.1	7.4
Industry	2.6	3.9	6.9	6.9
Academic institutions	10.4	13.8	18.5	18.1
All other sources	7.1	6.6	7.3	7.4
TOTAL	100.0	100.0	100.0	100.0

SOURCE: National Science Board (1996).

In 1993, universities and colleges employed nearly 150,000 doctoral scientists and engineers, 10,500 individuals with professional degrees, 5,500 scientists and engineers with degrees at the master's and bachelor's level, and roughly 90,000 graduate students in R&D activity (National Science Board, 1996).

OTHER NONPROFITS

The least well-documented and well-measured sector of the U.S. R&D enterprise is that comprising a diverse population of "other nonprofit" R&D funders and performers. Led by private nonprofit foundations, such as the Howard Hughes Medical Institute, "other nonprofit" organizations funded 1.8 percent of total U.S. R&D in 1995 (National Science Board, 1996). That year, these organizations funded 5.5 percent of all basic research, 2.6 percent of all applied research, and less than 0.5 percent of development work in the United States. In 1995, private nonprofit foundations, independent R&D institutes, private research hospitals, independent medical research centers, consortia, and their affiliates (more than 2,300 institutions altogether) performed about 3.5 percent of all U.S. R&D. In 1995, other nonprofits conducted 7.5 percent of basic research ($2.2 billion), 4.8 percent of applied research ($1.9 billion), and less than 2 percent of development ($1.9 billion). Other nonprofit research institutions are particularly prevalent in medical- and health-related research. In 1994, more than 45 percent of all R&D funded by other nonprofit institutions was in the area of health, as was nearly 42 percent of R&D performed by other nonprofit institutions. Other nonprofit institutions employed 10,200 scientists and engineers in R&D activities in 1993 (National Science Foundation, 1996b).

Distribution of Publicly Funded R&D

Since the 1940s, the federal government has focused its support of the nation's technology enterprise on mobilizing technical resources to further specific national missions. These missions, championed by various federal agencies, have included national security, the cure of disease, space exploration, food production, and world leadership in basic science. National economic development and international competitiveness have rarely been explicit objectives of federal technology policies and investments.

THE DEFENSE IMPERATIVE

A defining feature of the U.S. government's R&D portfolio has long been its heavy commitment to the needs of national security. In 1955, during the height of the Cold War, defense-related R&D claimed over 85 percent of all federal R&D dollars. During the 1980s, national security accounted for nearly two-thirds of federal R&D spending and one-third of total national (public and private) R&D

expenditures. In spite of a significant decline in defense spending during the past 5 years, defense-related R&D still accounted for 55 percent of federal R&D spending, or roughly one-quarter of all R&D spending in 1995 (Table 2.3).[5] Although federally funded R&D as a share of total industrial R&D has declined rapidly since the late 1980s, from 33 percent in 1988 to 17 percent in 1995, DOD remains the source of over 80 percent of all federal R&D dollars spent by private industry.

During the past 4 decades, defense-related R&D and procurement have fostered the development of important "dual-use" technologies (technologies having both civilian and defense applications) and provided a powerful stimulus to innovation in a select number of high-tech civilian industries such as microelectronics, software, and aerospace.[6] As of 1994, federal R&D dollars (predominantly DOD funds) still accounted for 61 percent of industrial R&D in the aerospace sector (Table 2.3). Nevertheless, the overwhelming majority of this defense-related R&D (an estimated 90 percent as of the early 1990s) has been for the

TABLE 2.3 U.S. Defense-Related R&D, Various Comparisons

	1955	1960	1970	1980	1990	1995
Share of federal R&D that is defense related	85	80	58	51	63	55
Share of total U.S. R&D that is defense related	48	52	33	24	26	23
Share of federal support of academic engineering research that is defense related	*	*	45[a]	55	44	45[b]
Share of all government-funded R&D in U.S. industry that is defense related[c]	*	81	68	63	83	80
Federal share of total R&D funds in aerospace industry	88[d]	89	77	72	76	61[e]
Federal share of total R&D funds in electrical machinery and communications	66[d]	65	52	41	38	14[e]

*Data not available.
[a] 1971 data.
[b] 1993–1995 average federal academic research obligations.
[c] Department of Defense only, data for 1962, 1981, and 1989.
[d] 1957 data.
[e] 1994 data.

SOURCES: National Science Foundation (1990; 1991; 1992a,b; 1994; 1996b).

"development, testing, and evaluation" of weapons and other systems that have no markets other than the military[7] (Alic et al., 1992).

National security has also long been the focus of government support for engineering R&D in U.S. universities and government laboratories. Although DOD accounted for only 12.2 percent of federal funding for all fields of academic R&D in 1995, the agency remains a major funder of university-based engineering research. As of 1994, DOD accounted for over 49 percent of all federal obligations for academic research in math, computer sciences, and all fields of engineering combined. This included 60 percent of federal funds for academic electronics and electrical engineering research, 54 percent for metallurgy and materials research, 52 percent for aerospace engineering research, 41 percent for mechanical, 47 percent for civil, and 4 percent for chemical engineering research[8] (National Science Foundation, 1997).

Finally, the demands of national defense have largely determined the structure and objectives of the government's system of federal laboratories, particularly in the physical sciences and engineering research. In 1995, DOD accounted for nearly half of all obligated expenditures of federal laboratories and, as of 1993, employed more than half of all federal laboratory R&D scientists and engineers[9] (National Science Board, 1996; National Science Foundation, 1995a).

GOVERNMENT CIVILIAN R&D PRIORITIES

Between 1987 and 1994, the share of federal R&D funds dedicated to civilian or nondefense-related agency missions increased from 31 percent to 45 percent (Table 2.4). In 1994, over 60 percent of the federal civilian R&D portfolio was allocated to the missions of health and civilian space exploration. The shares of federal civilian R&D funds dedicated to the missions of health, energy, the "advancement of research," and agriculture all declined slightly between 1987 and 1994. These declines were offset by increases in the shares allocated for research related to civilian space, infrastructure, environmental protection, and industrial development. Of these four mission areas, industrial development R&D has grown most rapidly since the late 1980s, albeit from a very small base.

More than two-thirds of federal civilian R&D funds went for basic and applied research in 1995. In contrast, 90 percent of federal defense-related R&D went for exploratory development (Figure 2.5). The vast majority of federal support for basic research flows from a few civilian agencies. The Department of Health and Human Services (DHHS), more specifically its National Institutes of Health, is overwhelmingly the largest funder of basic research—DHHS obligations in 1995 were $6.3 billion, three or more times those of the National Science Foundation (NSF) ($2.0 billion), NASA ($1.8 billion), and the Department of Energy ($1.7 billion). By way of comparison, DOD's obligations for basic research were $1.2 billion in 1995 (National Science Board, 1996). Likewise, that year, civilian agencies accounted for over 78 percent of all federal obligations for

TABLE 2.4 Distribution of Government R&D Appropriations by Socioeconomic Objective in the United States, 1987 and 1994

	Percent of Public R&D Funds				
	Total Funds		Civilian Funds		
Objective	1987	1994	1987	1994	Percent change in share 1987–1994
Agriculture	2.3	2.5	7.3	5.6	−23
Industrial development	0.2	0.6	.6	1.3	117
Energy	3.6	4.2	11.5	9.4	−18
Infrastructure	1.8	2.9	5.7	6.5	14
Environmental protection	0.5	0.8	1.6	1.8	13
Health	11.9	16.5	37.9	36.9	−3
Civilian space	6.0	10.9	19.1	24.4	28
Defense	68.6	55.3	—	—	—
Advancement of research	3.6	4.0	11.5	8.9	−23
General university funds[a]	—	—	—	—	—
Not elsewhere classified	—	2.3		5.1	

SOURCES: National Science Board (1989, 1996).

applied research. Nearly half of all federal obligations for basic research and a third of those for applied research went to support research in the life sciences (biological, agricultural and medical sciences) in 1994 (Figure 2.6).

Until recently, direct support by states of applied research projects (predominantly at academic research institutions) appears to have been concentrated in a relatively small number of fields or mission areas, including health, agriculture, and transportation. Although data regarding the distribution of state and local government R&D funds are fragmentary, it is estimated that between 60 and 75 percent of all of research supported with nonfederal government dollars in 1994 was health related. During the past decade, however, some states have broadened their R&D portfolios to support industrial and economic development more explicitly and aggressively (Coburn, 1995).

NEW FEDERAL INDUSTRIAL R&D INITIATIVES

The small claim of industrial development on the federal R&D budget testifies to the weak commitment of the federal government to economic development as an explicit mission of public R&D and technology policy. Indeed, the federal government and the private sector have long maintained a stark division of roles with regard to the funding of research versus the funding of development and deployment of technology for most sectors of the nation's economy.[10] Basic research and the development and application of technology relevant to accepted

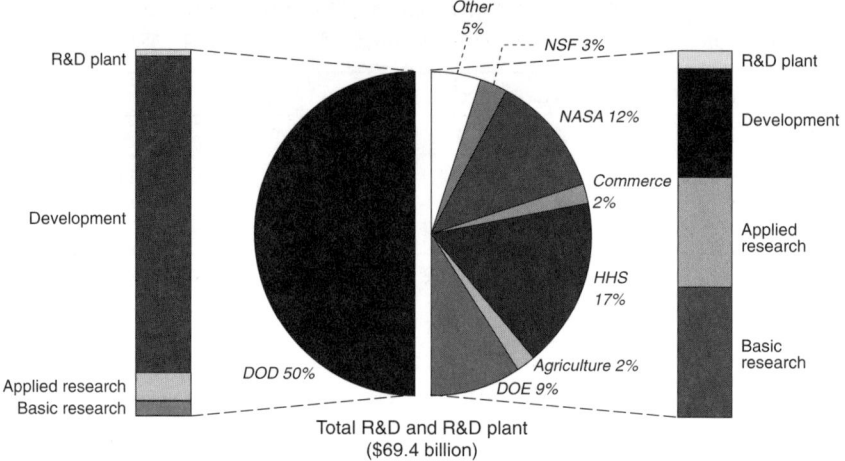

FIGURE 2.5 Federal obligations, by agency and type of activity, 1995. SOURCE: National Science Board (1996).

federal agency missions (though conducted principally by private-sector actors) have been regarded as legitimate activities for funding by the public sector. The identification, development, and adoption of technology for commercial products and services not directly associated with public missions has been seen as the preserve of the private sector. Until relatively recently, the only notable exception to this division of labor was the technical support (standards, testing, and evaluation) provided by the Department of Commerce's National Institute of Standards and Technology (NIST, formerly the National Bureau of Standards).

During the 1970s and 1980s, growing concerns regarding the health and competitive performance of the U.S. commercial technology enterprise prepared the way for a number of new initiatives by the federal government that would engage it explicitly, however tentatively, in support of civilian technology for national economic development. A series of laws were passed to promote government-industry partnerships and to foster technology transfer and collaborative R&D between and within sectors of the nation's technology enterprise. The 1980 Patent and Trademark Amendments (P.L. 96-517), known as the Bayh-Dole Act, permitted recipients of federal grants and contracts to retain title to inventions developed with government funds. Bayh-Dole provided a major impetus for universities and colleges in particular to get into the business of patenting and licensing technologies developed on their campuses.[11] Similarly, the Stevenson-Wydler Technology Innovation Act of 1980 (P.L. 96-480) and subsequent amendments to it during the ensuing decade were directed at engaging federal laboratories more extensively in the transfer of technologies to private firms as well as fostering cooperative research among federal laboratories, state and local governments, universities, and private firms. The Federal Technology Transfer Act of 1986

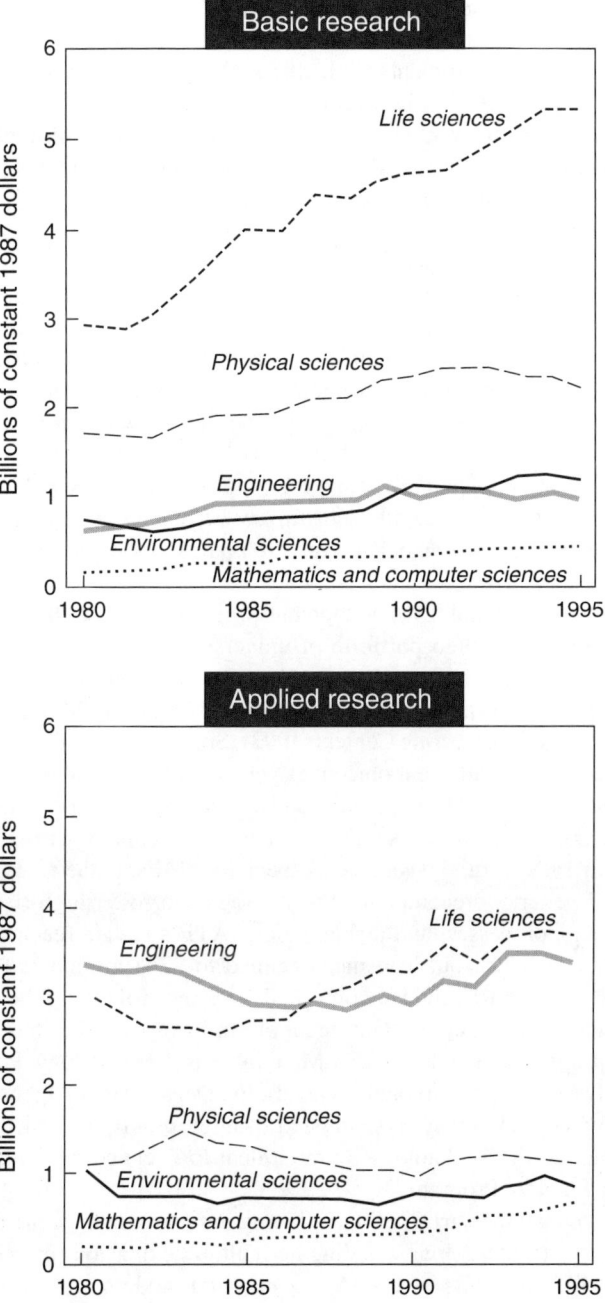

FIGURE 2.6 Federal obligations for basic and applied research, by field. SOURCE: National Science Board (1996).

(P.L. 99-502) amended Stevenson-Wydler to authorize cooperative research and development agreements (CRADAs) between federal laboratories and other entities, including state governments.[12] In 1984, Congress passed the National Cooperative Research Act (P.L.98-462), which fostered the proliferation of industrial R&D consortia and joint ventures by removing the threat of treble damages under U.S. antitrust law for firms that filed with the Department of Justice information concerning their involvement in such activities.[13]

In addition to these changes in law, Congress funded a range of programs designed to foster industrial technology development and technology transfer during the 1980s and early 1990s. Coburn (1995) identifies five basic types of federal and state cooperative technology programs that were established during the past 10 to 15 years. These include programs directed at *technology development* (i.e., those that support the development and application of new or enhanced industrial products or processes); programs focused on *industrial problem solving*, particularly for small business, through the diffusion of technology and best-practice applications; *technology financing* programs involving public capital or facilitated access to private capital; *start-up assistance,* primarily through public support of technology incubators and research parks; and *teaming*, or assistance in forming partnerships.

Major federal technology development programs include the National Science Foundation–sponsored portfolio of university-industry research centers—Industry/University Cooperative Research Centers (begun in 1973), Engineering Research Centers (1985), Science and Technology Centers (1987), Materials Research Science and Engineering Centers (1993), Supercomputer Centers (1986)—each designed to serve different objectives yet sharing a commitment to facilitate university-industry research cooperation and technology transfer.[14] Another technology development initiative is NIST's Advanced Technology Program (ATP), established in 1989 to fund businesses, especially SMEs, in the research and development of generic, precompetitive technologies to foster high-risk, high-potential products, processes, and technologies. ATP's budget reached a high of $340.5 million in 1995, but has since declined to $225 million in fiscal 1997. With the advent of a Republican-controlled Congress following the election of 1994, ATP has been under constant threat of elimination. Other technology development programs include DOD's Manufacturing Technology Program and SEMATECH (now totally privately funded), the Department of Transportation's Intelligent Vehicle Highway Systems and Maglev programs, and multiagency initiatives such as the Technology Reinvestment Project and the Small Business Technology Transfer Program.[15]

Primary federal industrial problem-solving initiatives include the Department of Agriculture's (USDA) long-standing agricultural extension service, set up in the 1914 to diffuse results of USDA research and modern farming technology and methods, and NIST's Manufacturing Extension Partnership (MEP). MEP was conceived initially in 1988/89 as a system of manufacturing technology centers designed to transfer advanced production technology from NIST's research facilities and other federal laboratories to small and medium-sized enterprises (SMEs). By the early 1990s, however, the focus of the program shifted to the provision of technical extension/industrial modernization services for SMEs through what would become a nationwide network of extension centers and agents. MEP centers assist SMEs with adoption of improved manufacturing technologies, training, management, and networking. MEP's budget in fiscal 1997 was $95 million. (For further information on MEP, see Annex II, pp. 207–209.)

Although no federal programs provide direct general-purpose financing of technology-based companies, there are several federal grant programs that help finance the development and commercial application of technologies relevant to federal agency missions. Most notably, in 1982, the Small Business Innovation Research (SBIR) program was created to direct a small share (initially not less than 1.25 percent, now 2.5 percent) of each major mission agency's total annual R&D budget to fund R&D at small and medium-sized firms and to stimulate the commercialization of new products and services (National Science Board, 1996). Other programs that help finance the commercialization of technology by private companies include NASA's Aerospace Industry Technology Program, and the Environmental Protection Agency's Environmental Technology Initiative.

Four federal agencies, the Department of Commerce, DOD, the Department of Labor, and NASA, sponsor "teaming or network-building" programs that provide assistance to industry through information dissemination, networking, and databases. Start-up assistance in the form of technology incubators and research parks remains the exclusive preserve of state and local governments.

Coburn (1995) estimates that the federal investment for all cooperative technology programs grew from $1.7 billion in fiscal 1992 to $2.7 billion in fiscal 1994. Forty percent of federal spending on cooperative technology programs in fiscal 1994 was for technology development initiatives, 28 percent for technology financing, and 25 percent for industrial problem solving (Table 2.5).

Although the proliferation of federal cooperative technology programs during the past decade has been impressive, collectively these programs amounted to less than 4 percent of the fiscal 1994 federal R&D budget. Furthermore, since the Congressional elections of 1994, several major federal cooperative technology programs have ended (e.g., TRP and DOD funding for SEMATECH[16]), or are on the verge of being eliminated (ATP) by a more skeptical Republican controlled Congress. In other words, at the federal level at least the role of government in direct financial support of industrially relevant civilian R&D and technology transfer is still seeking potential legitimacy.

STATE INDUSTRIAL TECHNOLOGY PROGRAMS

In contrast with the federal government, state governments traditionally have had few political reservations about using public funds to actively promote indus-

TABLE 2.5 Federal and State Government Investment in Cooperative Technology Activities, by Type of Program, FY 1994

Federal Agency	Technology Development		Industrial Problem Solving		Technology Financing[a]		Start-Up Assistance[b]		Teaming	
	\$ millions (number of programs)									
Department of Agriculture	0.0	(0)	434.6	(1)	16.1	(2)	0.0	(0)	0.0	(0)
Department of Commerce	199.5	(1)	30.2	(1)	3.7	(4)	0.0	(0)	0.0	(1)[c]
Department of Defense (FY93)	374.5	(6)	205.7	(2)	346.0	(1)	0.0	(0)	48.2	(2)
Department of Energy	1.7	(1)	9.0	(3)	61.3	(3)	0.0	(0)	0.0	(0)
Department of Health and Human Services	4.1	(1)	0.0	(0)	129.0	(1)	0.0	(0)	0.0	(0)
Department of Labor	0.0	(0)	1.4	(4)	0.0	(0)	0.0	(0)	0.2	(1)
Department of Transportation	345.4	(4)	0.0	(0)	4.2	(1)	0.0	(0)	0.0	(0)
Environmental Protection Agency	0.0	(0)	0.0	(0)	56.8	(4)	0.0	(0)	0.0	(0)
National Aeronautics and Space Administration	3.6	(1)	8.1	(1)	149.2	(3)	0.0	(0)	7.7	(2)
National Science Foundation	161.2	(6)	0.0	(0)	29.8	(2)	0.0	(0)	0.0	(0)
Total federal support by type	\$1,090.0	(20)	\$689.0	(12)	\$761.1	(20)	\$0.0	(0)	\$56.1	(6)
Total state support by type	127.5	(69)	59.5	(97)	101.8	(108)	7.2	(28)	5.9	(34)
Grand Total, federal and state support by type	\$1,217.5		\$748.5		\$862.9		\$7.2		\$62.0	

[a] Dollar amounts for the ATP are included in the Technology Development totals.
[b] While there are no formal federal programs in this category, there is ad hoc activity in some agencies.
[c] The Department of Commerce teaming program is the Clearinghouse for State and Local Initiatives, which has received no funding so far.

SOURCE: Coburn (1995).

trial and economic growth within their borders. Most of the 50 states sponsor economic development programs, the earliest of which date from the late 18th century (Coburn, 1995). The use of internal technology resources to foster economic development was first promoted by North Carolina in the early 1960s, an effort that led to the creation of the Research Triangle Park complex. Other states have followed suit, and now all 50 states have technology-based development programs of one sort or another.

Coburn (1995) estimates that state governments spent just over $384 million in fiscal 1994 on cooperative technology programs. Of this, approximately one-third was used for technology development, mostly matching support for university-industry technology centers funded primarily by federal initiatives (Box 1). One-quarter went to support technology financing, about 60 percent of which went to projects and 30 percent to companies. Fifteen percent of state cooperative technology funds were used for industrial problem solving, predominantly for technology extension and deployment programs (state initiatives as well as MEP matching funding). Twenty-one percent of state funds went to educational programs at institutions of higher education that sponsor the development, diffusion, and use of technology and improved practices to benefit specific companies. As of 1994, 42 states had some form of industrial problem solving[17] and technology-financing program in place; 31 had technology development programs; 18 supported start-up assistance incubators or industrial technology parks; 21 funded teaming (Coburn, 1995).

North Carolina, one of the pioneers, invests the most of any state in cooperative technology programs ($37 million in 1994), followed by Pennsylvania ($34 million), Texas ($30 million), Georgia ($30 million), Connecticut ($27 million), Ohio ($27 million), New York ($23 million), New Jersey ($20 million), Michigan ($14 million), and Maryland ($13 million). The highest per-capita investments are made by Arkansas, Connecticut, Nebraska, North Carolina, and South Dakota (Coburn, 1995).

The Industrial R&D Enterprise

The U.S. industrial R&D enterprise is distinguished by its large size, both in terms of R&D volume and the number of firms involved; its dynamism as reflected in the changing sectoral distribution of R&D activity over time; and its capacity for spawning new technology-based products and industries.

In 1995, over 18,000 manufacturing and 23,000 nonmanufacturing companies reported performing a total of $102 billion of R&D in the United States.[18] Collectively, these firms employed 764,500 scientists and engineers in R&D activity in 1993 (National Science Foundation, 1996b). The vast majority of industrial R&D spending is concentrated in a small number of firms. In 1993, for example, the 20 largest R&D spending companies accounted for one-third of all industrial R&D expenditures; the 200 largest firms accounted for 71 percent

> **BOX 1**
> **State Spending, by Category,**
> **on Cooperative Technology Programs, 1994**
>
> | | ($000s) |
> |---|---:|
> | Technology Development (34 percent) | |
> | University-Industry Technology Centers | $104,606 |
> | University-Industry Research Partnerships | 12,118 |
> | Government-Industry Consortia | 4,810 |
> | Equipment and Facility Access Programs | 5,965 |
> | Technology Financing (26 percent) | |
> | Project Financing | $62,172 |
> | Company Financing | 30,861 |
> | Small Business Innovation Research | 3,185 |
> | Technology Reinvestment Program/ Advanced Technology Program | 5,593 |
> | Related Educational Initiatives (21 percent) and Other | $82,635 |
> | Industry Problem Solving (15 percent) | |
> | Technology Enterprise Divisions | $54,851 |
> | Federal Technology Application Programs | 3,805 |
> | IMPs | 850 |
> | Start-Ups (2 percent) | |
> | Incubators | $7,238 |
> | Research Parks | not available |
> | Teaming (2 percent) | |
> | Networks | $4,376 |
> | Databases | 1,531 |
> | **TOTAL** | **$385,000** |
>
> SOURCE: Coburn (1995).

(National Science Board, 1996). In 1994, 19 companies each spent more than $1 billion on R&D, and another 49 spent more than $200 million.

SHIFTING SECTORAL DISTRIBUTION OF R&D ACTIVITY

There has been significant change in the sectoral distribution of U.S. industrial R&D in recent decades, reflecting changes in the composition of the nation's industrial base, changes in the relative R&D intensity of different industries over time, and changes in the way U.S. industrial R&D activity is measured (Figure 2.7). The most notable change over the past 10 years has been the rapid increase in the share of R&D claimed by nonmanufacturing (predominantly service) industries. Until fairly recently, nonmanufacturing industries were believed to account for less than 5 percent of all industrial R&D spending. Since the early 1980s, however, their share of the total has increased rapidly, from 5.1 percent in 1983 to 26.7 percent in 1993. Much of the increase in nonmanufacturing R&D over this period can be attributed to changes in NSF's survey of industrial R&D in 1991, which changed and greatly expanded the sample of companies surveyed, thereby incorporating more accurate information on the R&D performance of smaller firms and firms classified in the nonmanufacturing sector. According to NSF, these changes resulted in an upward revision of total nonmanufacturing R&D in 1991 from roughly $10 billion previously reported to $21 billion. That year, an additional $7 billion of R&D was reclassified from manufacturing to nonmanufacturing categories. Much of this latter shift is believed to accurately

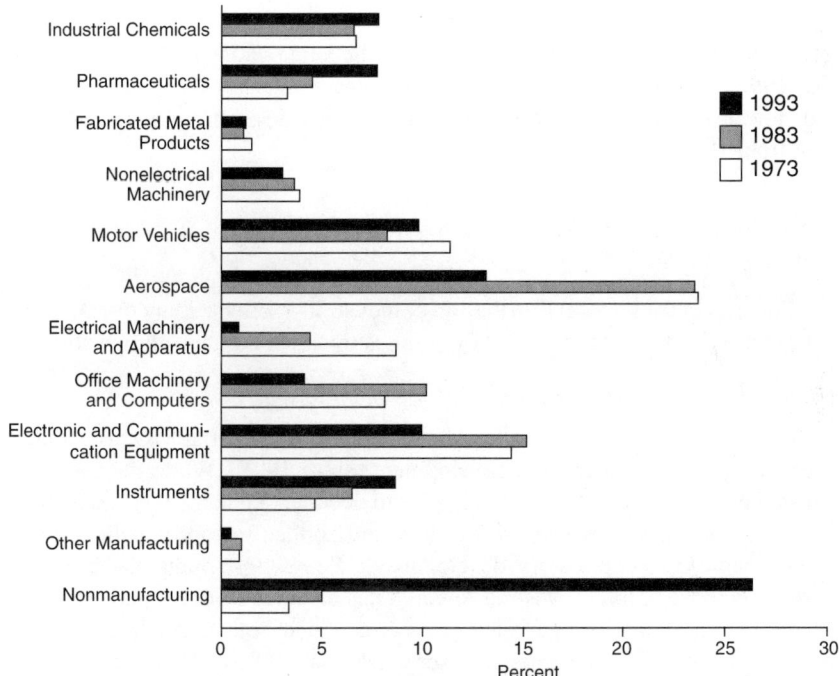

FIGURE 2.7   U.S. industrial R&D spending, by sector, 1973, 1983, and 1993. SOURCE: Organization for Economic Cooperation and Development (1996a).

reflect changes in the output mix of companies formerly classified in manufacturing industries (National Science Foundation, 1995b).

Three industries account for the majority of nonmanufacturing R&D: computer programming and related services, including software (8.1 percent of total R&D in 1993); communications services (4.4 percent); and research, development, and testing services (1.5 percent) (National Science Board, 1996).[19]

Two manufacturing industries—pharmaceuticals and professional and scientific instruments—have also significantly increased their share of total industrial R&D expenditures during the past 2 decades. Notably, the growth of these industries' share of industrial R&D tracks the growth in the share of federal R&D dedicated to health-related research as well as the associated growth of the nonindustrial research base in the life and medical sciences during the period. (See Figure 2.6, p. 75.)

Four manufacturing industries—aerospace, electronics and communications equipment, office machinery and computers, and electrical machinery—have seen their shares of total industrial R&D contract dramatically during the past 10 to 20 years. Historically, these industries, particularly aerospace, have been the beneficiaries of DOD R&D and procurement, which has declined dramatically during the past decade. As of 1988, the aerospace industry absorbed more than 60 percent of all federal R&D funds for industry. However, by 1994, industrial aerospace R&D amounted to about 39 percent of the total (National Science Foundation, 1996a). Moreover, data from the Aerospace Industry Association (1994) also indicate a 25-percent decrease in revenues from sales of military-related hardware from 1990 and 1993.

CHANGES IN THE COMPOSITION AND ORGANIZATION OF INDUSTRIAL R&D

Changes in the sectoral distribution of industrial R&D spending over the past 2 decades have been accompanied by compositional and organizational shifts.

Relative Decline in Industrial Basic and Applied Research

First, there has been a change in the character of industrial R&D (i.e., basic research, applied research, and development) since 1991. While the inflation-adjusted industrial R&D expenditures overall declined 5.9 percent between 1991 and 1995, industrial performance of basic and applied research declined more than did industrial exploratory development. Since 1991, industrial basic research as a share of total industrial research has declined from 6.7 percent to 5.9 percent,[20] that of industrial applied research declined from 23.5 to 22.0 percent, while that of industrial development increased from 69.8 to 72.2 percent. These shifts are explained, in part, by the dismantling of several companies' large central research facilities and a general movement in several industries away from long-term fundamental research toward more short-term applied research and development in order to meet intensifying international competition (National Science Board, 1996). Accompanying this latter trend have been an increased emphasis on R&D as a tool for scanning for and exploiting knowledge generated or applied beyond national boundaries, as well as closer integration of R&D with activities farther downstream in the value-added process (i.e., changes designed to leverage scarce R&D dollars and speed commercialization of new technology) (National Academy of Engineering, 1993, 1996b).

Increased Cooperative R&D and R&D Outsourcing

Second, there has been an increase in both cooperative R&D and R&D outsourcing among firms as well as between firms and nonindustrial R&D performers during the past decade. This is in part explained by the rapid growth in the number of R&D consortia, joint ventures, and other forms of strategic alliances in R&D at the hands of U.S.-based companies during the past decade.[21] Though by no means a measure of all U.S. R&D consortia and joint venture activity, the number of "joint research ventures" (JRVs) registered each year with the Department of Justice (DOJ) has grown significantly since passage of the 1984 National Cooperative Research Act (P.L. 98-462). As of 1995, more than 565 JRVs had been registered with the DOJ (Vonortas, 1996). Likewise, Hagedoorn (1995) has documented a marked increase in the level of U.S.-firm participation in international strategic technology alliances since the early 1980s (Figure 2.8).

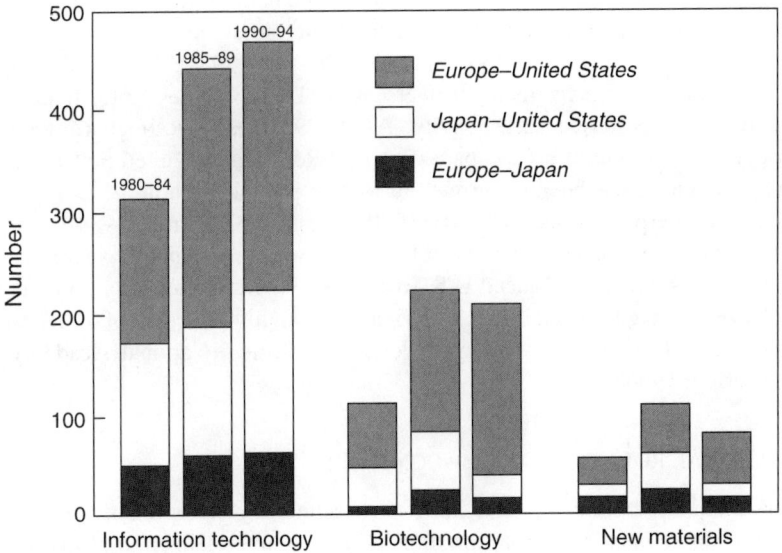

FIGURE 2.8 Number of new strategic technology alliances, by industry and region. SOURCE: National Science Board (1996).

Between 1985 and 1995, industry funding of R&D at universities and colleges (in inflation-adjusted dollars) nearly doubled, and industry research funding at other nonprofit organizations grew nearly 65 percent. During the same period, company-financed R&D performed within industry grew less than 27 percent in constant 1987 dollars. Consistent with these trends has been a significant increase in the level of industrial involvement in collaborative research with academic researchers via university-industry research centers as well as rapid growth (albeit from a very small base) in the volume of technology licensed by industry from academic research institutions during the past decade (Association of University Technology Managers, 1996; Cohen et al., 1994).[22] Similarly, there has been rapid growth in the number of CRADAs between companies and federal laboratories since the mid-1980s.[23]

Yet another indicator of the growth of research collaboration between industry and nonindustrial research institutions is the rapid increase in the share of scientific and technical articles that are coauthored by individuals in industry and researchers based at nonindustrial research institutions. Between 1981 and 1993, the share of scientific and technical articles that had industry-based authors grew from 27.3 percent to 47 percent (National Science Board, 1996). Most of this increase was accounted for by growth in the volume of academic-industry coauthored literature.

Internationalization of U.S. Industrial R&D

Third, the past 2 decades have witnessed a growing internationalization of U.S. industrial R&D activity, predominantly at the hand of foreign direct investment (multinational companies) and international strategic alliances (National Academy of Engineering, 1996b). Between 1985 and 1993, U.S.-owned companies increased their investment in overseas R&D three times faster than their investment in U.S.-based R&D activity. As of 1994, these investments amounted to roughly 10 percent of all company-financed R&D in the United States. Even more pronounced has been the growth of foreign participation in the U.S. industrial R&D enterprise since the early 1980s (Figure 2.9). From 1984 to 1994, R&D spending by the U.S. affiliates of foreign-owned companies[24] increased as a share of all company-financed U.S. R&D from 9 percent to nearly 16 percent. As of 1994, foreign-owned companies financed roughly 2 percent of all research conducted at U.S. universities and federal laboratories (National Academy of Engineering, 1996b).

THE SPECIAL ROLE OF START-UP COMPANIES

A unique feature of the U.S. industrial technology enterprise is the critical role start-up companies play in the transfer and commercialization of fast-moving, science-based technologies. This happens generally via movement, or "spinout," of researchers and technology from universities, large established compa-

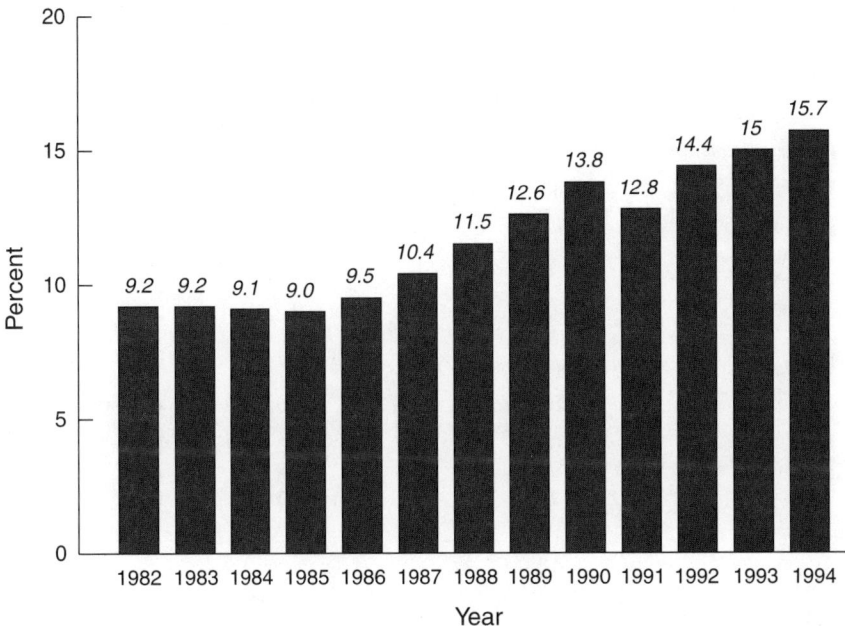

FIGURE 2.9 R&D spending by U.S. affiliates of foreign-owned firms as a percentage of all privately funded U.S. R&D, 1982–1994. SOURCES: National Science Board (1996) and U.S. Department of Commerce (1996a).

nies, and government laboratories. U.S.-based high-tech start-ups are credited with commercializing the technologies that launched the new biotechnology and computer software industries. Although growth in the number of new U.S. high-tech companies established during the past decade is considerably slower than that from the mid-1970s to the mid-1980s, nearly half of all U.S. high-tech companies operating in 1994 were established during the past 15 years (Table 2.6) (National Science Board, 1996). More than one-quarter of all new businesses started since 1980 (and operating in 1994) were software companies, and software continues to create more new start-ups than any other technology field.[25] Similarly, the rate of formation of new firms dedicated to the exploitation of one or another aspect of recent advances in biotechnology has been phenomenal: 800 new enterprises were founded in the 1980s, and the industry currently includes more than 1,200 firms. A few of these firms have become large, successful operating companies (e.g., Amgen), however, the vast majority are still small, investor-funded ventures. From 1980 to 1994, the shares of start-ups in computer hardware, advanced materials, photonics, optics, and telecommunications also increased.[26]

High-tech start-ups have played important roles in the U.S. technology enter-

TABLE 2.6 High-Tech Companies Formed in the United States, 1960–1994

Period Formed	All High-Tech Fields	Auto-mation	Biotech-nology	Computer Hardware	Advanced Materials
Number of Companies					
1960–1994	29,358	1,939	735	2,845	1,045
1980–94	16,660	917	546	1,907	487
1980–84	7,727	483	213	842	212
1985–89	6,510	331	225	756	194
1990–94	2,423	103	108	309	81
Percentage of all high-tech companies formed during each period					
1960–1994	100.0	6.6	2.5	9.7	3.6
1980–94	100.0	5.5	3.3	11.4	2.9
1980–84	100.0	6.3	2.8	10.9	2.7
1985–89	100.0	5.1	3.5	11.6	3.0
1990–94	100.0	4.3	4.5	12.8	3.3
Percentage of all U.S. high-tech companies					
1960–1994	100.0	100.0	100.0	100.0	100.0
1980–94	56.7	47.3	74.3	67.0	46.6
1980–84	26.3	24.9	29.0	29.6	20.3
1985–89	22.2	17.1	30.6	26.6	18.6
1990–94	8.3	5.3	14.7	10.9	7.8

[a]Other fields are chemicals, defense related, energy, environmental, manufacturing equipment, medical, pharmaceuticals, test and measurement, and transportation.

SOURCE: National Science Board (1996).

prise because they can accept a level and type of risk that larger companies usually cannot. Able to serve highly dynamic niche markets, start-ups often serve as a "test-bed" for new products and services, a few of which might develop into large-volume businesses (National Academy of Engineering, 1995c). Furthermore, start-ups are considered particularly adept at drawing effectively upon new product ideas of customers, suppliers, universities, research laboratories and others as well as at rapidly commercializing innovations.[27]

Many factors have enabled high-tech start-up companies to perform their unique roles in the U.S. innovation system.[28] The following are among the most important:

- the existence of sophisticated financial markets, particularly access to a large volume of venture capital and highly developed public equity markets;

Photonics and Optics	Software	Electronic Components	Telecommunications	Other Fields[a]
977	7,661	2,923	1,556	9,677
507	5,196	1,293	933	4,874
221	2,467	629	408	2,252
191	1,962	508	370	1,973
95	767	156	155	649
3.3	26.1	10.0	5.3	33.0
3.0	31.2	7.8	5.6	29.3
2.9	31.9	8.1	5.3	29.1
2.9	30.1	7.8	5.7	30.3
3.9	31.7	6.4	6.4	26.8
100.0	100.0	100.0	100.0	100.0
51.9	67.8	44.2	60.0	50.4
22.6	32.2	21.5	26.2	23.3
19.5	25.6	17.4	23.8	20.4
9.7	10.0	5.3	10.0	6.7

- the large scale and technological intensity of relatively homogeneous segments of the U.S. domestic market;
- the large size, high mobility, accessibility, and entrepreneurial orientation of the U.S. technical workforce;
- the sheer scale and accessibility of U.S. publicly funded nonproprietary research, particularly university-based research;
- the scale of federal procurement combined with explicit preferences or set-asides for small and medium-sized vendors and suppliers;
- a history of regulatory and other public policy commitments conducive to high-tech start-up companies, including the competition-oriented or technology diffusion–oriented enforcement of intellectual property rights and antitrust law (competition policy), as well as relatively risk-friendly system of company law, particularly bankruptcy law; and
- a highly individualistic, entrepreneurial culture nurtured in industry and many U.S. research universities by private practices, public policies, and various institutional mechanisms such as technology business incubators and venture capital firms that encourage risk taking.

COMPARATIVE STRENGTHS AND WEAKNESSES OF THE INDUSTRIAL R&D ENTERPRISE

The U.S. technology enterprise excels in the development and exploitation of new commercial technologies. Major shifts in the sectoral composition of U.S. industrial R&D and industrial production during the past 2 decades, as well as the large and rapidly expanding population of U.S. high-tech start-up companies, attest to this fact. Further evidence of the dynamism and future growth orientation of the U.S. industrial technology base is offered by the U.S. patent and export statistics.

The patent activity of U.S. companies encompasses a broad spectrum of technologies and new product areas. However, recent patenting by U.S. companies demonstrates a strong emphasis on technologies or fields—medical and surgical devices, telecommunications, aeronautics, electricity transmission, advanced materials, biotechnology—that are expected to serve as engines of future economic growth[29] as well as technologies associated with the extraction and use of the nation's abundant natural resources (Table 2.7). Not surprisingly, these areas of patent emphasis reflect the competitive strength of U.S. industry in global high-technology product markets. In 1994, 25 percent of U.S. manufacturing exports were high-tech manufactured goods, and 3 of the 10 classifications of high-technology products accounted for nearly 85 percent of these technology exports: information technology (computers, software, and communications) (35.5 percent), aerospace (29.0 percent), and electronics (21.3 percent) (National Science Board, 1996). By way of comparison, U.S. patent activity by German companies in 1993 indicates an emphasis on technology areas associated with heavy manufacturing industries (motor vehicles, printing, power generation, and new chemistry and materials) that have long been a source of German comparative industrial strength in world markets.[30]

In contrast to the relative strength of the U.S. industrial R&D enterprise and its supporting nonindustrial R&D infrastructure in opening up new technological frontiers and launching new industries, the U.S. enterprise appears to be less effective than some of its trading partners at serving the R&D and technology transfer/diffusion needs of technologically mature industries.

In particular, U.S. companies in many technologically mature manufacturing industries appear to operate increasingly on the periphery of the nation's nonindustrial R&D system. The R&D portfolios of U.S. research universities, federal laboratories, and most nonprofit research institutes have not overlapped much with the process and product R&D needs of firms, particularly small and medium-sized firms, in these industries. Many observers have noted gaps in the R&D portfolios of major technologically mature industries (Competitiveness Policy Council, 1993; National Academy of Engineering, 1993). Of particular concern have been perceived emerging gaps in these industries' "infrastructural" R&D portfolios—R&D directed at the discovery and development of low-technical-risk, difficult-to-appropriate technologies that have the potential to enhance

TABLE 2.7 Top 20 Most-Emphasized U.S. Patent Classes for Inventors from the United States and Germany, 1993

Ranking of class	United States	Germany
1	Wells	Fluid-pressure brake and analogous systems
2	Mineral oils; processes and products	Plant protecting and regulating compositions
3	Surgery, patent class 604	Printing
4	Surgery, patent class 606	Internal combustion engines
5	Chemistry, hydrocarbons	Organic compounds[a]
6	Special receptacle or package	Synthetic resins or natural rubbers[b]
7	Surgery: light, thermal, and electrical applications	Organic compounds[a]
8	Chemistry: analytical and immunological testing	Conveyors: power-driven
9	Fluid handling	Organic compounds[b]
10	Liquid purification or separation	Winding and reeling
11	Error detection/correction and fault detection	Organic compounds[a]
12	Illumination	Land vehicles
13	Chemistry: natural resins or derivatives	Plastic articles
14	Receptacles	Organic compounds[a]
15	Amusement devices: games	Synthetic resins or natural rubbers[b]
16	Communications: directive radio wave systems and devices	Organic compounds[a]
17	Information processing system organization	Fluid sprinkling, spraying, and diffusing
18	Surgery	Organic compounds[a]
19	Hydraulic and earth engineering	Compositions: coating or plastic
20	Supports	Material or article handling

[a]Part of the class 532–570 series.
[b]Part of the class 520 series.

SOURCE: National Science Board (1996).

the performance of a broad spectrum of firms within an industry or related industries. Also of concern are gaps in these industries' "pathbreaking" R&D base—R&D aimed at discovering and developing high-technical-risk technologies with the potential for transforming existing industries (Alic et al., 1992).

Factors that have helped weaken the connection between firms in many industries and the nation's nonindustrial research enterprise include the highly concentrated (by industry and technology field) and mission-driven nature of federal R&D funding; the fragmented structure and low levels of industrial self-organization of many technologically mature U.S. industries; and changes in the industrial composition of the U.S. economy (i.e., the increasing shares of total U.S. output accounted for by service and high-tech manufacturing industries). Numerous federal industrial technology initiatives of the past decade have sought to strengthen government-university-industry R&D cooperation as well as foster industrial consortia in selected industries (e.g., semiconductors, automotive). However, the volume of federal R&D dollars devoted to these initiatives has been small, and it is not yet clear whether these programs have been effective at forging tighter linkages between industrial and nonindustrial R&D performers in established technologically mature industries.

Another relative weakness of the U.S. industrial R&D/technology transfer enterprise is its limited capacity for diffusing new technology and know-how, particularly manufacturing or production technology, within technologically mature industries and SMEs in particular (National Academy of Engineering, 1993). In recent years, there has been a concerted effort at both the federal and state levels to develop a more far-reaching network of private- and public-sector providers of technical extension/industrial modernization services to SMEs. Examples of this are NIST's manufacturing extension partnership and related state initiatives. (See Part II, pp.76–79, and Annex II, pp. 205–209.) There are indications that a growing percentage of U.S.-based manufacturers are adopting advanced manufacturing technologies more rapidly (National Science Board, 1996). Nevertheless, compared with its German counterpart, the U.S. infrastructure for diffusion of production technology and other technologies to established industries is much more uneven and fragmented.

Technology Transfer to U.S. Industry in Context

In order to begin to place technology transfer from nonindustrial R&D performers to U.S. industry in context, it is important to recognize that the volume of technology transfer that takes place internally among divisions of large private firms and externally between firms is by far the largest segment of U.S. technology transfer. This activity occurs through formal measures (such as mergers and acquisitions, and licensing of patents, software, and trade secrets) as well as through less formal mechanisms (such as sharing technical know-how, exchanges of personnel, and technical and marketing assistance). Data collected by the U.S. Internal Revenue Service show that in 1992, corporate royalty income in the U.S. manufacturing sector alone was almost $33 billion, roughly 100 times the royalty income of all of U.S. universities and federal laboratories combined. Indeed, that year several large technology-intensive firms reported royalty incomes of over $1 billion (e.g., IBM, Texas Instruments, and Bellcore).

Several recent surveys of R&D-intensive companies shed light on the perceived relative importance of industrial and nonindustrial sources of commercializable ideas and technology. A 1992 survey by Roessner (1993) of member companies of the Industrial Research Institute (mostly large, research-intensive firms) found that respondents considered other companies (U.S. and foreign) to be the most significant sources of external technology, with universities second, private databases third, and federal laboratories fourth. Similarly, a 1994 pilot study of U.S. industrial innovation by the NSF and the U.S. Bureau of the Census found that the three most important sources of information leading to the development and commercial introduction of new products (according to the "innovating firms"[31] that responded to the survey) were internal sources, clients and customers, and suppliers of materials and components (National Science Board, 1996). This study found that the least important sources of such information were government laboratories, technical institutes, and consulting firms.

The NSF/Census study also revealed that the channels used most frequently by innovating firms to access new technology were hiring skilled employees, purchasing equipment, and using consultants. Likewise, the channels used most often by innovating firms to transfer new technologies to other organizations included communication with other companies, mobility of skilled employees, and R&D performed for others (National Science Board, 1996).

The following sections explore in greater detail the organization and dynamic of technology transfer to U.S. industry within the three major sectors of the nation's nonindustrial R&D enterprise: research universities and colleges, federal government laboratories, and the diverse population of privately held, nonacademic, mostly nonprofit organizations (e.g., independent and affiliated R&D institutes, consortia, incubators and research parks, and technical and professional associations).

TECHNOLOGY TRANSFER FROM HIGHER EDUCATION TO INDUSTRY

There are over 3,600 publicly and privately funded colleges and universities as well as 6,900 vocational and technical institutions offering post-secondary education in the United States. Only about 875 public and private universities and colleges conduct science and/or engineering research, and of these, the 100 largest account for 80 percent of all academic R&D (National Science Board, 1996). It is this latter, highly diverse subset of 100 public and private institutions that constitute the heart of the U.S. basic research enterprise and the main object of analysis in this chapter.

To understand the structure and dynamic of technology transfer from these institutions of higher education to industry, it is useful to review briefly several major distinguishing characteristics of the U.S. academic research enterprise as well as an overview and the history of university-industry technology transfer in the United States.

Distinguishing Characteristics of the Enterprise

SCALE

One major distinguishing feature of the U.S. academic research enterprise is its size. In 1995, U.S. universities and colleges performed $21.6 billion worth of research and development,[32] or 12.6 percent of all R&D conducted in the United States that year. This expenditure was roughly the same as that by all federal laboratories and FFRDCs ($25 billion in 1995) and was nearly half of total German R&D spending in 1994. Academic institutions performed 49 percent of all basic research, 14 percent of all applied research, and less than 2 percent of all development work performed in the United States in 1995. In 1993, U.S. universities and colleges employed over 149,800 doctoral scientists and engineers (S&E), 10,500 individuals with professional degrees, and 5,500 S&Es with S&E degrees at the masters and bachelors levels in R&D activities. In addition, nearly 90,000 full-time graduate students (27 percent of total full-time enrollment) relied on research assistantships as their primary source of support (National Science Board, 1996).

U.S. universities and colleges graduate roughly 24,000 Ph.D. scientists and engineers each year. In 1993, these institutions received nearly 6,600 invention disclosures and applied for over 3,000 patents (including roughly 2,000 new patents). In 1993, U.S. academic researchers authored nearly 100,000 articles in professional journals, representing 25 percent of the world's scientific and technical literature.[33]

DIVERSITY

A second distinguishing feature of U.S. research colleges and universities is their diversity. There is no U.S. university "system" in the formal sense of the term. Rather, the academic research enterprise is a heterogeneous, highly autonomous population of research colleges and universities, each of which was established and has evolved in response to a unique combination of local, regional (state), and national needs. Some are public, state-owned institutions; others are privately owned. Although all institutions that receive federal funding must comply with common federal rules and regulations, each institution, or state-run system of institutions, has a distinct governing body, administration, accounting practices, and mission statement.

U.S. academic research institutions differ greatly in size and research focus. Some institutions perform significant amounts of industry-sponsored research, while others do very little (Table 2.8). The distribution of R&D spending by science and engineering field of the top 20 research universities illustrates how diverse their research portfolios are (Table 2.9). (These 20 institutions conducted roughly a third of all U.S. academic research in 1993.) Some universities maintain research portfolios that are more national or international in scope and reputation. Others conduct research that is more heavily weighted to the needs of local industries or their region's or state's economy. Some remain focused almost exclusively on their traditional missions of education and research, while others have become deeply involved in a broad spectrum of technology transfer and outreach activities.

TABLE 2.8 Industry-Sponsored Research as a Share of Total Academic Research Expenditures at the Top 20 Research Universities, Fiscal Year 1994

Institution and Ranking	Total Research Expenditures (thousands of $)	Industry Sponsored Research (thousands of $)	Industry Sponsored as Percentage of Total Research Expenditures
Johns Hopkins University	784,043	10,418	1.33
University of Michigan	430,778	26,732	6.21
University of Wisconsin-Madison	392,718	13,729	3.50
Massachusetts Institute of Technology	363,918	55,500	15.25
Texas A&M University	355,750	28,576	8.03
University of Washington	343,910	33,199	9.65
University of California-San Diego	331,901	9,764	2.94
Stanford University	318,561	14,714	4.62
University of Minnesota	317,865	23,726	7.46
Cornell University	312,683	17,199	5.50
University of California-San Francisco	312,393	10,977	3.51
Pennsylvania State University	302,997	45,408	14.99
University of California-Berkeley	289,632	12,547	4.33
University of California-Los Angeles	279,869	13,394	4.79
Harvard University	289,459[a]	10,228	3.53
University of Arizona	269,939	15,053	5.58
University of Texas-Austin	260,602	4,268	1.64
University of Pennsylvania	251,461	12,107	4.81
University of Illinois-Urbana	245,407	13,527	5.51
Columbia University	236,417	1,632	0.69
TOTAL	6,679,303	372,698	5.58

NOTE: Because of rounding, figures may not add to the totals shown.

[a]Estimated

SOURCE: National Science Foundation (1996a).

SPONSORED RESEARCH

A third distinguishing feature of U.S. academic research is the way in which it is funded. The vast majority of U.S. academic research in science and engineering is sponsored directly via grants or contracts from federal mission agencies. In other words, it is not supported by public "general university" or "base institutional" funds as is the case in Germany, Japan, and other advanced industrialized countries. In 1995, federal government agencies funded 60.2 percent of

TABLE 2.9 R&D Expenditures at Universities and Colleges, by Science and Engineering Field, Fiscal Year 1994 (dollars in thousands)

Institution and Ranking	Total	Engineering	Physical Sciences	Environmental Sciences
Johns Hopkins University	784,043	210,522	117,188	40,593
University of Michigan	430,778	88,837	22,972	20,823
University of Wisconsin-Madison	392,718	55,021	39,838	21,898
Massachusetts Institute of Technology	363,918	153,530	95,154	16,094
Texas A&M University	355,750	82,565	21,890	80,878
University of Washington	343,910	20,332	19,375	57,912
University of California-San Diego	331,901	15,806	35,450	102,266
Stanford University	318,561	92,946	44,030	6,192
University of Minnesota	317,865	30,625	15,802	11,560
Cornell University	312,683	41,416	45,211	4,389
University of California-San Francisco	312,393	0	0	0
Pennsylvania State University	302,997	129,313	22,486	21,360
University of California-Berkeley	289,632	61,654	59,996	4,466
University of California-Los Angeles	279,869	29,544	24,069	14,130
Harvard University	278,459[a]	6,027[a]	31,718[a]	9,714[a]
University of Arizona	269,939	20,659	91,765	20,861
University of Texas-Austin	260,602	106,743	64,108	25,826
University of Pennsylvania	251,461	11,918	23,245	801
University of Illinois-Urbana	245,407	51,634	38,500	27,052
Columbia University	236,417	14,407	21,433	39,786
TOTAL	6,679,303	1,223,499	834,230	526,601

NOTE: Because of rounding, figures may not add to the totals shown.

[a]Estimated

SOURCE: National Science Foundation (1996a).

U.S. academic R&D, state and local governments 7.4 percent, industry 6.9 percent, individuals and nonprofit institutions 7.4 percent, with the remaining 18.1 percent coming directly from academic institutions themselves.[34] Most federal funds for academic research are awarded on a competitive basis to individual investigators or to research teams. Researchers submit project proposals that are then peer reviewed according to "best-science" principles. This approach demands that principal investigators invest a great deal of time in grant management (i.e., non-research-related) activities, both as grant applicants and "volunteer" reviewers of the grant proposals of other researchers. However, it also fosters intensive and valuable competition among ideas and rapid exploitation of new research directions and concepts within the academic research community.

Math & Computer Sciences	Life Sciences	Psychology	Social Sciences	Other Sciences
119,297	270,314	1,021	9,784	15,324
19,186	212,198	9,098	51,094	6,570
10,031	222,482	11,540	31,028	880
18,514	37,690	8,503	8,179	26,254
6,963	141,130	1,570	17,547	3,207
6,516	218,998	7,321	10,675	2,781
13,542	156,724	3,998	4,115	0
14,513	152,104	3,710	5,066	0
218	219,241	6,970	11,852	0
23,614	184,425	3,670	9,958	0
0	312,393	0	0	0
3,518	96,520	6,393	10,409	12,998
4,836	122,182	6,617	24,830	5,051
8,291	178,014	7,514	18,307	0
4,169[a]	168,143[a]	3,117[a]	46,480[a]	9,091[a]
7,296	116,202	2,546	8,666	1,944
15,897	23,584	3,961	16,183	4,300
8,408	183,502	2,296	21,291	0
15,395	55,519	6,305	14,096	36,906
4,637	148,100	2,386	5,668	0
326,438	3,219,465	98,536	325,228	125,306

Most research performed by U.S. universities and colleges is basic or long-term applied in nature. Basic research accounted for 67 percent of total academic R&D in 1995, applied research 25 percent, and development only 8 percent. Nevertheless, because of the way it is funded, U.S. academic research (even so-called basic research) in many fields is shaped largely by the applied needs of federal agency missions.

The distribution of U.S. academic research expenditures by field shows a heavy emphasis on the life sciences, particularly the medical sciences (Table 2.10). In 1993, the medical and biological sciences consumed 45 percent of all academic research dollars. All engineering disciplines together accounted for less than 16 percent of the total.

Despite the fact that U.S. funding of academic research has not kept pace with the financial demands of a growing population of academic researchers, U.S. academic research expenditures grew faster than those of any other major

TABLE 2.10 R&D Expenditures at Universities and Colleges, Percent Share by Major Science and Engineering Field, Fiscal Year 1994

Source and Field	1994
Engineering, total	15.77
Aeronautical and Astronautical	1.03
Chemical	1.31
Civil	1.86
Electrical	3.44
Mechanical	2.34
Metallurgical and materials	1.51
Other, n.e.c.	4.27
All sciences, total	84.23
Physical sciences	10.30
Environmental sciences	6.76
Mathematical sciences	1.32
Computer sciences	3.13
Life sciences	54.65
Psychology	1.70
Social sciences	4.51
Other sciences, n.e.c.	1.86

NOTE: Because of rounding, figures may not add to the totals shown. n.e.c. = not elsewhere classified.

SOURCE: National Science Foundation (1996a).

R&D performing sector during the 1984–1994 period. During this period, academic research grew at an average annual rate of 5.8 percent, compared with 2.8 percent for FFRDCs and other nonprofit laboratories, 1.4 percent for industrial laboratories, and 0.7 percent for all federal laboratories (National Science Board, 1996).

History of University-Industry Relations

The history of U.S. university-industry interaction with respect to research and development and technology transfer can be divided roughly into three periods: from the mid-1800s to the eve of World War II; from the early 1940s through the mid-1970s; and from the late-1970s to the present.

During the first of these periods, the development of U.S. higher education and research was influenced heavily by the more immediate, practice-oriented training and technical problem-solving needs of U.S. agriculture and industry. Although this era witnessed the emergence of a small number of elite research universities whose faculties engaged in basic research, it was during this period that U.S. colleges and universities made their greatest strides in the applied sciences and engineering disciplines, largely in response to the demands of local or regional industries.

Government at both the state and federal levels had a strong hand in shaping the practical, regional economic orientation of higher education and research during the period. Indeed, many public universities were founded by state governments with an explicit mandate to support the technical needs of the regional economy. In 1936, state governments funded 14 percent of all U.S. academic research. Throughout this time, federal government support of academic research, education, and extension activities was concentrated in areas critical to the technological development of large sectors of the U.S. economy that lacked a privately funded R&D base, in particular agriculture, forestry, and mining.[35] University-based agricultural research and extension activity alone claimed about 40 percent of federal research funds during the mid-1930s (Matkin, 1990; Mowery and Rosenberg, 1993).

By the eve of World War II, the federal government accounted for no more than one-quarter of total academic research funding. Private foundations funded the majority of academic R&D during this second period. The R&D-intensive industries of the day, such as electrical manufacturing and chemicals, helped to develop the research and training capabilities of select U.S. universities, but mainly as a complement to the extensive in-house R&D efforts of the companies themselves (Matkin, 1990).

World War II represented a watershed in the relationship between U.S. research universities and the federal government. Academic research was enlisted very effectively in service of the war effort and was instrumental in the development of new technologies such as atomic energy and radar, and new fields like aeronautics. This greatly enhanced the public reputation of academic research institutions and engendered a new appreciation for the importance of basic and long-term applied research for U.S. military security and economic prosperity, as well as other national interests. Accordingly, academic research assumed a central role in the new federal science policy articulated during the mid-1940s—a policy based on a new "social contract" that explicitly harnessed the academic science community in service of national objectives through greatly increased federal support for academic research and its associated infrastructure (Bush, 1945).

By the early 1950s, agencies of the federal government, led by the Department of Defense, had become the principal patrons of U.S. academic research, sponsoring 60 percent of all academic R&D in 1955. In the decades to follow, the academic research community would be enlisted in support of a broad range of federal agency missions, including national defense, energy independence, the cure of disease, space exploration, as well as the broader goal of achieving U.S. preeminence in virtually all fields of science and engineering.

With the shift in the funding base of U.S. academic research came a corresponding shift in the orientation of much academic research and graduate education in science and engineering. Rather than focusing on the more immediate practical and applied R&D needs of private industry, academic research became more concerned with the basic and long-term applied research agendas of the federal agencies.[36] A majority of academic research funds were now allocated by federal agencies through a system of peer-review evaluation, which was guided by "best-science" principles. This new funding environment fostered a more pronounced division of labor between universities and industry with regard to basic and applied research, and reinforced differences between the two sectors' research cultures.[37] Academia rewarded research faculty primarily for the originality of their research; the quality, number, and timeliness of their research publications; and their success in competing for research funding from government agencies and nonprofit foundations. Accordingly, the academic research community placed a premium on the openness, free exchange, and rapid dissemination of new knowledge and ideas. By contrast, industry-based researchers continued to be rewarded according to the standards of the marketplace (e.g., the number and value of patents received, the successful commercialization of technologies). In short, private industry concerned itself with capturing and protecting the economic value embodied in new ideas through intellectual property and trade secrets.

Throughout this second period, the transfer of technology from academic research institutions to industry was treated generally as an ancillary activity by most major research universities. These institutions considered their primary contributions to the technological capabilities of American industry to be well-trained graduates, published research results, and faculty consultants.

The third and current phase of university-industry interaction dates from the late 1970s and is characterized by a renewed interest in collaborative research and technology transfer between the two sectors. This changing dynamic is the result of several factors. First, the 1970s heralded the commercial take-off of industries with strong technological roots in academic research, including microelectronics, software, and biotechnology. These successes generated a new wave of industrial interest in particular areas of academic research and expertise. Second, the emergence of major new challenges to the competitiveness of many U.S. technology-intensive industries during the 1970s prompted federal and state efforts to harness the capabilities and outputs of the U.S. academic research enterprise to serve the R&D and technology needs of American industry more effectively. Finally, although federal funding of academic research has grown rapidly in absolute terms throughout the period, the increased cost of research and an expanding population of academic researchers have made competition for federal support tighter than ever. These trends have encouraged university-based researchers to look increasingly to the private sector for sources of research support.

At the federal level, two changes in policy fostered the shift to a more collaborative era in U.S. university-industry relations. First, in 1980, Congress passed the Bayh-Dole Act, which made it possible for universities, other nonprofit organizations, and small businesses to retain rights to most of their federally funded inventions. Under the terms of the act, academic research institutions are granted considerable autonomy in licensing or otherwise commercializing intellectual property they develop with public funds, as long as they (a) give preference to businesses located in the United States, particularly small companies, when licensing such intellectual property; and (b) grant exclusive rights or sell this intellectual property to companies willing and able to manufacture substantially in the United States products embodying the invention or produced through application of the invention (U.S. General Accounting Office, 1992).[38]

The federal government has also sought to promote greater university-industry collaboration by funding university-based research centers that engage academic and industrial researchers in collaborative, often multidisciplinary, research. Most prominent among these are the National Science Foundation's Industry-University Cooperative Research Centers (begun in 1973), Science and Technology Centers (1987), Engineering Research Centers (1985), and Materials Research Science and Engineering Centers (1993).[39] Recent federal industrial technology initiatives such as the Advanced Technology Program of the National Institute of Standards and Technology or the multiagency Technology Reinvestment Project have also included provisions supportive of university-industry collaborative research.[40]

State governments, too, have tried to promote closer ties between public universities and their host region's economies and industrial base. The 1980s witnessed a shift to increasingly science-and-technology-driven economic development strategies among most of the 50 states. Public universities stand at the center of many of these new initiatives, as state governments seek to recreate the success of Route 128, the high-tech corridor around Boston said to have been spawned and nurtured by the technical capabilities of MIT (Etzkowitz, 1988; Feller, 1990).

Technology Transfer by Research Universities and Colleges

Recent surveys of R&D-performing companies attest to the fact that the most valued output of U.S. research universities from the perspective of corporate America is the human capital they generate in the form of well-trained scientists and engineers.[41] For the most part, the value of science and engineering graduates to a firm (or the economy at large) is defined by the research and learning skills these individuals have acquired through their academic training, rather than by the volume of specific (and often rapidly outdated) knowledge they have amassed during their course of studies.

Researchers based at universities and colleges account for over 70 percent of all U.S. scientific and technical articles (see Figure 2.10). In certain fields the

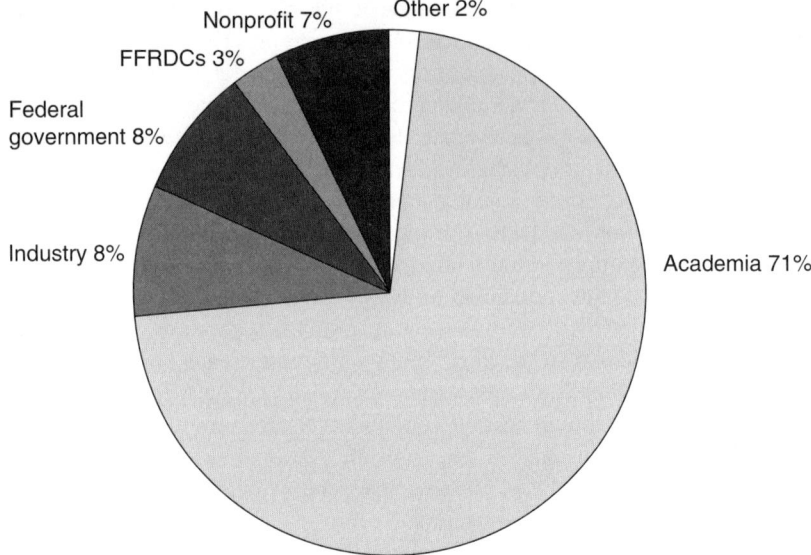

FIGURE 2.10 Distribution of U.S. scientific and technical articles, by sector, 1993. FFRDC = federally funded research and development center. SOURCE: National Science Board (1996).

research literature represents an important source of highly specialized knowledge of direct relevance and value to the technology strategies of companies in some industries. In recent years, citations of research literature on the first page of U.S. patent applications (an indication of the potential contribution of published research to patentable inventions) have risen rapidly. About half of all publications cited were papers from academic institutions (National Science Board, 1996). In most industry sectors, the most valuable contribution of fundamental academic research is its role in helping companies understand existing technologies better and in exposing promising paths for and enhancing the productivity of industrial applied research and development (David et al., 1992; Pavitt, 1991). Indeed, university research is usually more useful for improving on inventions already made than for making them (i.e., one has to thoroughly understand how and why an invention works before one can have a strategy, other than pure trial and error, for improving on it).

The U.S. panel accepts that the production of graduates and new knowledge remain the primary contribution of American higher education to the technical needs of U.S. industry. It also acknowledges the important role academic research publications play in the transfer of highly specialized knowledge in a number of industries. However, in this report, the panel focuses primarily on those activities that, though related to the missions of education and research, involve the intentional or "directed" transfer of intellectual property or specific knowledge (i.e., "proto-technology") from universities and colleges to industry.

Even within this narrower definition, university technology transfer encompasses a wide range of transfer mechanisms. Some can be defined and measured relatively easily (e.g., the transfer of codified technology or proto-technology via patents, copyrights, and research publications). Others are little more than proxies for actual technology transfer and are very difficult, if not impossible, to quantify. These mechanisms include faculty consulting; the movement of graduates and faculty from academia to industry; university investments in the transfer and commercialization of technology; industry-sponsored or collaborative academic-industrial R&D; and a range of other market-making activities by industry and academia directed at the commercially valuable outputs of academic research.

TECHNOLOGY TRANSFER MECHANISMS

There are three types of mechanisms for technology transfer from academia to industry in the United States.[42] The first includes such things as faculty consulting and the transfer of university intellectual property and proto-technology embodied in graduates and faculty who are hired by private companies. These mechanisms, closely related to the education and research missions of universities and colleges, were the predominant modes of technology transfer prior to the mid-to-late 1970s. The second type, also linked to the traditional missions of universities, has only seen extensive use or significant growth in use since the late 1970s (the third phase of university-industry relations). These mechanisms include patent licensing, university acquisition of private-sector licensees, and various approaches for enhancing industry access to and sponsorship of university-based research. The third type includes activities, such as technical assistance programs and technology business incubators, associated with commercializing research or improving university-industry relations more generally. These mechanisms, which have also seen significant growth since the late 1970s, are more ancillary to the traditional missions of the research university.

The following sections review each of these mechanisms separately. It is well to remember, however, that universities and individual academic researchers employ many of these mechanisms in concert in order to take advantage of the synergies and complementarities among them.

Faculty Consulting

No aggregate data exist on the number of U.S. academic research faculty involved in consulting with private industry or the number of scientist or engineer man-hours academic researchers devote to consulting with industry each year. Nevertheless, panel members estimate that more than half of the academic engineering faculty at the top 20 U.S. research universities spend 10 to 15 percent of their time consulting with industry.[43] Each consultant might work for 1 to 10 clients; the type of work relationship varies widely. Academic consultants are generally paid hourly or daily fees for their services. Annual retainer fees are uncommon. Perhaps the best measure of the effectiveness of a consultancy arrangement is whether it is terminated or continued by the client firm.

Academic researchers and industry are attracted to consultancy for different reasons. For university faculty, consultancies offer important learning opportunities, additional sources of support for their research (both material and intellectual), as well as opportunities for placing their students with client organizations. This latter benefit enables faculty to attract the best students and ensure ongoing bi-directional technology transfer with the client firms. Industry, in turn, receives solutions to specific technical problems and enhanced access to academic research results and highly trained graduates. The fact that U.S. university faculty are salaried for only 9 months out of the year and rely heavily on external sources of funding for their research also provides a strong incentive for them to engage in consultant work.

Movement of University-Based Researchers to Industry

The movement of academic researchers—graduates, postdoctoral fellows, and faculty—to private industry is an important transfer mechanism for technology, proto-technology, and highly specialized knowledge and skills. It is extremely difficult, however, to come up with useful measures of this type of technology transfer.

Proxies such as the number of newly minted science and engineering Ph.D.'s that are hired by private companies each year (roughly two-thirds of the total) or the number of Ph.D. scientists and engineers that move from academic to industrial employment (more than 12,000 between 1988 and 1993) shed some light on the importance of this mechanism (National Research Council, 1993b). Moreover, leading U.S. research universities often temporarily exchange research personnel with private industry in the context of collaborative research projects.

Data on the number of start-up companies founded by university graduates or research staff do not exist.[44] However, it is fair to assume that a respectable share of many high-tech start-ups in science-based industries, such as biotechnology, have been built directly on the intellectual capital of university-based researchers.[45] Numerous case studies, including several prepared by the panel, demonstrate the many ways in which university graduates and research staff have brought technology or proto-technology to new or established companies (Box 2).

Patent Licensing

Prior to the early 1970s, patent licensing was a fairly limited tool of technology transfer for American universities. In 1965, only 96 U.S. patents were granted to 28 U.S. universities or related institutions. However, the commercial success of new science-based industries in the fields of microelectronics, information technology, biotechnology, and advanced materials, along with passage of the Bayh-Dole Act in 1980, fueled rapid growth of university patenting during the following 2 decades.

By 1995, more than 127 U.S. universities had patent portfolios and were aggressively involved in the business of technology licensing, according to a survey by the Association of University Technology Managers (1996). These 127 institutions collectively employed 618 full-time equivalent (FTE) professional staff in licensing university intellectual property and in technology transfer activities. This represented roughly a 27 percent increase in professional FTEs over 1992. In 1995, these institutions received nearly 7,427 invention disclosures, applied for 5,100 patents (including 2,373 new patents), and executed 2,142 license options. Gross annual royalty receipts for the 127 universities were roughly $274 million in 1995, over ten times those of federal laboratories but only one-hundredth those of industry.[46]

Many universities have established in-house offices of technology transfer or technology licensing, whose primary activities focus on locating, patenting, and licensing university-developed intellectual property and less frequently on spinning off inventions to start-up companies. Other universities have established semiautonomous technology transfer organizations to pursue some or all of the university's patenting, licensing, and technology transfer functions. These organizations are usually established in the nonprofit sector, although some are profit making. Some examples are ARCH (for Argonne-Chicago) which manages inventions from the University of Chicago and the Argonne National Laboratory, which Chicago manages for the Department of Energy, and WARF (Wisconsin Alumni Research Foundation).

Along with establishing technology transfer offices, many universities have developed financial incentive programs to encourage their research faculty to innovate. At Stanford, for example, 15 percent of license revenues goes to support the technology licensing office. (Revenues in excess of the office's expenses go into a research incentive fund to assist researchers without sponsorship.) The remaining 85 percent or royalties are then divided among the inventors, their department, and the school of medicine.[47] A similar policy is in effect at MIT and the University of California at Berkeley.

There is great diversity among U.S. research universities with respect to their approach to patenting and technology licensing. Some universities, public institutions in particular, lay claim to all research output generated in their labs; others are more flexible in negotiating the disposition of intellectual property resulting from research on their campuses. Likewise, some institutions look to their technology licensing offices to generate revenue, and others see these units as instruments for building long-term relationships with private companies as research patrons or partners (Box 3). To date, however, only a small number of institutions can claim success meeting any of these objectives. Many research universi-

> **BOX 2**
>
> **Cree Research, Inc.: From a Ph.D. Thesis to a World-Class Company in 10 Years**
>
> Until relatively recently, scientists and electrical engineers could not create silicon electronic devices that operated at elevated temperatures or, if used as light emitting diodes (LEDs), that produced blue light. Calculations showed that if semiconductor-grade silicon carbide (SiC) were available, it would overcome these limitations and open up additional opportunities in power and high-frequency electronics. Companies like Bell Labs, Hewlett-Packard, IBM, GE, or Motorola and universities such as Harvard and Stanford might be expected to discover the secret of how to make SiC devices, but that was not the case.
>
> Rather, a team at North Carolina State University headed by Professor Bob Davis in the Materials and Science Engineering Department began solving some of the tough problems associated with growing perfect crystals of SiC. In the early 1980s, Eric Hunter participated in this research program as a student. His brother Neal was also acquainted with the project, although he was studying mechanical engineering. On graduation, they both found jobs in conventional industries and forgot that SiC even existed.
>
> After a few years, however, they realized that their real desire was to start their own company. Meanwhile, Eric reestablished contact with the SiC program at NCSU, and when John Edmond, one of the stars of this program, announced that if no one was going to make a business out of SiC, he would take his Ph.D. and go elsewhere, the Hunter brothers, along with two other star members of the project, Calvin Carter and John Palmour, decided that the best new business opportunity was to use SiC to produce blue diodes.
>
> One would think venture capital firms found this opportunity attractive. None did. So, Neal and Eric pooled their own funds and, by selling stock at $0.18 a share, raised $20,000 from family members and friends. They

ties are still searching for effective ways to manage and grow their R&D and technology transfer activities with industry.

As of the late 1980s, drug and medical device patents accounted for about 35 percent of all university patents among five broad classes of technologies defined by Henderson et al. (1995) (Figure 2.11). Chemical patents accounted for 25 to 30 percent, electronic and related patents for 20 to 25 percent, mechanical patents for 10 to 15 percent, and all other patents for 5 percent. Since the early 1970s, university-based inventors have been much more focused on drugs and medical technologies and much less focused on mechanical technologies than their coun-

> BOX 2—Continued
>
> promised John Edmond at least 4 months of work if he would refrain from joining an established company. In September 1987, they successfully negotiated with the University Research Office of North Carolina State for the one SiC patent the university was planning to file as well as for all other SiC technology that the newly formed company, Cree Research, felt more excited about patenting than did the university. In exchange, the university received $10,000 plus repayment of their patent expenses and 5 percent of the stock of the new company.
>
> Benefiting from the entrepreneurial spirit invading the North Carolina Research Triangle, the company was able to raise $400,000 from four private investors, after having been turned down by professional venture capital groups. By this time, the firm had an after-market value of $6 million even though it had not yet made its first blue diode. Over the next 6 months, with help from Professor Davis, the young research team achieved a very faint blue diode. In March 1988, with this proof that the technology worked, private investors put in $3 million. A year later, General Instrument and Polaroid agreed to purchase $1 million worth of the diodes. By the summer of 1990, the company had raised another $3 million and, when that ran out, it raised another $5 million. Then, in February 1993, the company went public and raised $11,000,000 at $4.12 a share, giving the business a market value of $45,000,000.
>
> In recent years, the company has established partnerships with major companies around the world that are excited about working with Cree's SiC wafers in a wide range of electronic applications. Government agencies have contracted for over $20 million worth of research. In the last 5 years, annual sales of SiC wafers and blue diodes have grown to $6 million. Meanwhile, the stock has gone as high as $31.00 a share, giving the company a price-to-earnings ratio of infinity, and a price-to-sales ratio of over 50 to 1.
>
> SOURCE: Walter Robb, Vantage Management Services.

terparts from other sectors of the U.S. R&D enterprise (compare Figure 2.11 to Figure 2.12).

Despite rapid growth over the past 20 years in the number of universities involved in patenting, university patent activities remain highly concentrated. Although patents were awarded to over 150 universities and related institutions in 1991, the top 20 institutions accounted for about 70 percent of all patents granted, with MIT alone receiving 8 percent of the total (Henderson et al., 1995).

University royalty income is distributed very unevenly. In fiscal 1995, only six institutions received more that $10 million in gross royalties—the University

> **BOX 3**
> **Computer-Aided Design for Microelectronics**
>
> Well into the 1970s, designers created the complex geometric patterns needed to manufacture microelectronic chips using manual or computer-based drafting tools. As chip complexity increased, it became nearly impossible to complete error-free designs in one attempt. Designers sought computer aids to enforce rules linking the functional and electrical specifications required for a chip, and the geometric mask patterns used for its manufacture. Engineers at many semiconductor manufacturers, and researchers at a few universities, understood this problem. Better computer-aided design (CAD) tools for microelectronics became a necessity.
>
> Incremental improvements in existing computer-based drafting tools proved to be an inadequate approach. Several firms developed proprietary software for chip design and verification based on mainframe computers. These individual efforts were costly, however, and each proprietary CAD package had its particular strengths and weaknesses. A critical problem was adapting the design tools rapidly enough to match the rapid advances in semiconductor technology.
>
> To meet needs for research and instruction at the University of California at Berkeley, faculty members and graduate students developed several generations of software for design tasks including circuit analysis, chip layout, design-rule verification, and pattern generation. Students were the "guinea pigs" who used the prototype software as a part of class assignments. A vision gradually emerged of modular design software. The university team adopted the UNIX software-development environment because it enabled rapid iterative refinements in the design software.
>
> During the 1970s, progress in CAD software development accelerated due to close working relations between faculty members and CAD engineers at several leading electronics companies, many of whom were graduates of the Berkeley program. After several stages of software refinement by university scientists, colleagues in industry agreed to evaluate the software. The university received valuable feedback from several industrial laboratories.
>
> In the early stages of these collaborations, disagreements arose often concerning intellectual property rights. Faculty members believed that restrictions on intellectual property would inhibit the open exchange of ideas and prototype software. The university team adopted a policy of making source code available to others and of placing its work in the

BOX 3—Continued

public domain. Experiences like this indicate that, apart from copyrights, protections on intellectual property rarely are important to successful software development.

Leading firms in the U.S. semiconductor industry established the Semiconductor Research Corp. (SRC) in 1982. Soon thereafter, the federal government became an SRC sponsor. The goal of SRC is to foster graduate education and research in fields relevant to the semiconductor industry. The UC Berkeley CAD program received one of the first major SRC grants. Additional research support came from the Advanced Research Projects Agency. With these new resources, research and prototyping of new, improved CAD tools accelerated. A parallel industry initiative, the Computer Aided Design/Computer Aided Manufacturing Consortium, provided $18 million in cash and computers to construct and equip a large new research facility on the UC Berkeley campus.

Direct design synthesis of chips from formal specifications became an additional goal. Berkeley continued to distribute software, including source code, to sponsoring firms. Feedback from many users contributed importantly to the evolution of improved tools.

No one expected that the university could be the long-term provider of support, documentation, training, and service for industrial software. Semiconductor manufacturers recognized it would be wasteful for every user of CAD software to create their own software development and support capability. Vendors of earlier computer-based drafting software did not aggressively pursue the new generation of design software. So, about 1985, entrepreneurs including several graduates of the Berkeley CAD program established a successful new business supplying CAD software and support. Several other similar firms subsequently entered the market. Even today, many of the commercial CAD software modules have roots in the early Berkeley prototypes.

Berkeley's research and graduate program in electronic CAD continues. Technical goals have evolved to include process and device modeling, multichip assemblies, boards, and miniaturized interconnection technologies. Other focus areas are performance-driven design and very-low-power design for portable equipment. The patterns of sponsorship and interaction with industry continue much as they have in the past. The graduates of this program are leaders and major technical contributors to the world's top CAD vendors.

SOURCE: David Hodges and Donald Pederson, University of California at Berkeley.

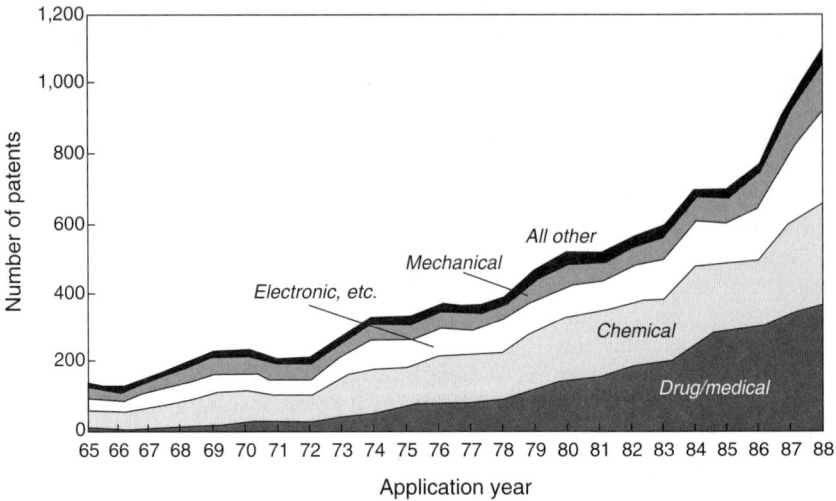

FIGURE 2.11   University patents by broad fields. SOURCE: Henderson et al. (1995).

of California system ($57 million), Stanford University ($39 million), Columbia University ($34 million), Michigan State University ($15 million), and University of Wisconsin-Madison/Wisconsin Alumni Research Foundation ([WARF] $12 million). Yet these six institutions accounted for over 56 percent of total gross royalties received by U.S. universities. Only 25 of the 117 universities that reported gross royalty receipts to AUTM in 1995 received more than $2 million in royalties, whereas 82 institutions reported less than $1 million in royalties. Universities with "home-run" inventions often have order-of-magnitude higher royalty income streams than universities that lack such blockbusters. For example, as of 1993, WARF received $99 million in license royalties for vitamin D and related technologies; the University of California system and Stanford shared $97 million in royalties on the Cohen-Boyer gene splicing technique;[48] Michigan State earned $86 million in royalties on cisplatin; the University of Florida brought in $33 million in royalties related to Gatorade; and Iowa State received $27 million in licensing fees for fax technology. Some universities that encourage the formation of new companies and spin-offs often take equity in these new ventures in lieu of some or all of the royalties to which they would be entitled from license fees for a patented process or product. When these equities are eventually sold, universities receive additional income, sometimes years after the original invention.

Equity Ownership in Start-Up Companies

It is estimated that academic licensing has contributed to the establishment of 1,633 new companies since 1980, 464 (or 28 percent) of these were established

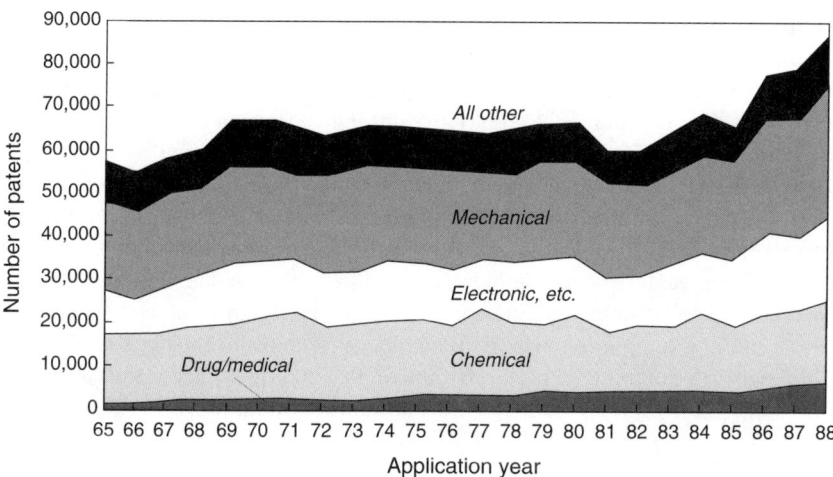

FIGURE 2.12   All U.S. patents by broad fields. SOURCE: Henderson et al. (1995).

in 1994 and 1995. Although a small number of universities have a long history of taking equity positions in companies engaged in the commercialization of new technology, it is only recently that significant numbers of universities have engaged in this type of technology transfer activity. As of 1995, over 50 universities had reported negotiating more that 560 licenses with equity, 99 of these in 1995 alone (Association of University Technology Managers, 1996).

There are many reasons why universities have chosen in recent years to enter into the venture capital business.[49] First, although not without substantial risks, acquiring equity in start-up companies founded to exploit university-generated intellectual property holds the promise of a much larger financial return than could be earned from licensing alone. Second, acquiring equity in companies can be a way to hedge against the risk of having university-owned patents infringed upon or rendered obsolete. Third, by accepting stock in licensee companies in lieu of royalties, universities are able to negotiate mutually beneficial deals with cash-strapped start-ups. Fourth, some universities view their venture fund activities as a way to attract and retain high-powered faculty (this is said to be particularly important for medical schools). Fifth, taking equity in companies often provides universities with increased opportunity for sharing research instruments and facilities. And last, but by no means least, by acquiring equity stakes in local start-up companies, universities are able to make a highly visible commitment to the local or regional economy, thereby generating good will with current or potential future patrons within state administrations or legislatures.

There are two main avenues by which universities invest in start-up companies: through portfolio investment of the university's endowment and through administratively separate or independent organizations established specifically for this purpose. The first route, wherein a university's treasurer makes investments solely according to standard investment criteria, is fairly "arms length" in nature. MIT, for example, is said to invest roughly 10 percent of its endowment in venture capital projects. The second route, the establishment of administratively separate or independent organizations, provides a mechanism that (a) allows the university to bring in outside venture-capital expertise unfettered by university policies, (b) offers an effective structure within which participants in new business ventures can communicate and negotiate, and (c) helps shield the university from commercial concerns (financial risks, perceived conflicts of interest, etc.). Examples of successful university ventures of this type include the University of Rochester's nonprofit University Ventures, and Johns Hopkins University's for-profit Triad Investors Corporation (Matkin, 1990).[50]

Industry-Sponsored Research

Between 1980 and 1995, private industry's share of funding for research at American universities and colleges increased from 3.9 percent to 6.9 percent. As of 1994, of the 200-plus universities and colleges that reported conducting some amount of industry-sponsored research and development, 39 institutions received $10 million or more of industry support. As noted in Table 2.8, there is significant variation in the extent of industry-sponsored research among different universities. While the average industry share of total sponsored academic research was 6.9 percent, MIT and Pennsylvania State University both received roughly 15 percent of their total research budgets from private industry in 1994. Meanwhile, total research funding at other top-20 research universities, including the University of Wisconsin at Madison, the University of California at San Francisco, the University of California at San Diego, and the University of Texas at Austin, averaged industry shares of less than 4 percent.

Company-sponsored research at U.S. universities frequently is carried out via contracts or grants. The distinction between the two instruments is subtle and varies among institutions. In general, research contracts, more than research grants, obligate university-based researchers to provide their corporate sponsor with more-frequent and more-formal reports on their progress. Contracts also usually specify particular deliverables, whereas grants are generally more open ended. National statistics on the sponsorship of academic research do not distinguish between contracts and grants because of the definitional vagaries and reporting inconsistencies among institutions. However, at several top-ranked institutions, including MIT and the University of California at Berkeley, the vast majority of industry-sponsored research is in the form of grants (National Academy of Engineering, 1996b).

Research grants may demand more of a quid pro quo from university-based researchers than the term "grant" implies. For example, companies providing research grants to university-based researchers may receive favorable consideration in licensing negotiations, even though they do not receive royaltyfree or exclusive rights. For example, at the University of California at Berkeley and MIT, some engineering departments have agreed to accept visiting fellows from major industrial donors (National Academy of Engineering, 1996b).

In addition to contracts or grants with individual academic researchers or research teams, industry sponsorship of university research can involve the establishment of formal university-industry research centers; research consortia involving other universities/departments, multiple firms, government laboratories, and other nonprofit research organizations; and "support-for-research-access" initiatives such as industrial liaison or affiliate programs. Each of these is discussed below.

Formal University-Industry Research Centers[51]

A 1994 study by Cohen et al. defined university-industry research centers (UIRCs) as university-affiliated research centers, institutes, laboratories, facilities, stations, or other organizations that conducted research and development in science and engineering fields with a total budget (1990 dollars) of at least $100,000 and with part of that budget consisting of industry-sponsored funds.

More than 1,000 centers located at more than 200 universities and colleges throughout the United States are thought to have met those criteria in 1990. More than half of these centers had been established since 1980 (Figure 2.13). In 1990, UIRCs spent $2.53 billion on R&D involving approximately 12,000 faculty, 22,300 Ph.D.-level researchers, and 16,000 graduate students. That same year,

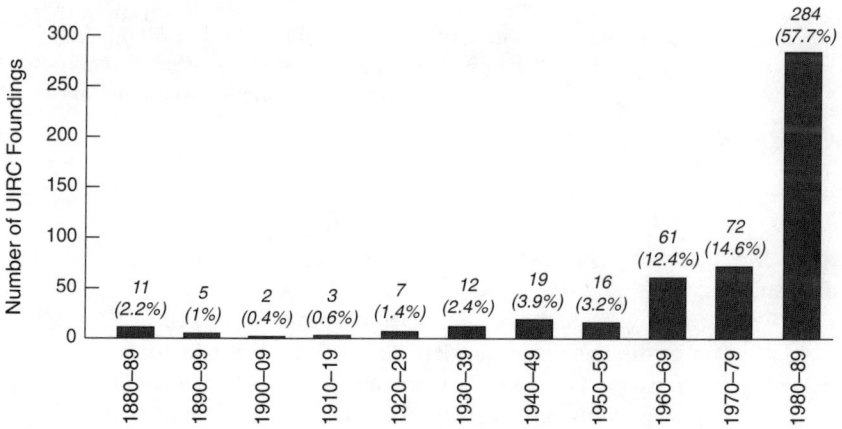

FIGURE 2.13 UIRC foundings by decade, 1880–1989, for UIRCs existing in 1990. SOURCE: Cohen et al. (1994).

> **BOX 4**
> **The Engineering Research Center in Data Storage Systems at Carnegie Mellon University**
>
> The Data Storage Systems Center (DSSC) traces its beginnings to a 1982 workshop organized by Carnegie Mellon University (CMU) professor Mark Kryder. Attending the workshop were a dozen key technical managers from various firms in the U.S. data storage industry and a similar number of faculty from CMU who had experience in magnetics technologies. At the time, outside of CMU, there were only a few academic researchers in the United States who worked in magnetics—even though the magnetic recording industry, which relied heavily upon advances in magnetic materials and devices, was comparable in size to the semiconductor industry. The goal of the workshop was to identify topics suitable for Ph.D. thesis research.
>
> Based upon the list of suggested topics, Professor Kryder wrote a proposal for a university-based Magnetics Technology Center, which would conduct research on magnetic storage technologies, including magnetic recording, magneto-optic recording and magnetic bubble memories. The privileges of membership in the center would vary according to the amount a firm contributed. Intellectual property was to be owned by the center and provided royalty free to associate members paying $250,000 per year, while affiliate members paying $50,000 per year were to be given the right to license intellectual property for a reasonable fee. This arrangement would make it possible for the center to pursue patents and copyright protection for intellectual property, and provide that benefit to its industrial sponsors, without requiring the segregation of the research projects for individual sponsors. Thus, all sponsors would gain access to the research in the center in proportion to their contributions.
>
> The CMU administration was highly supportive of the effort and committed to build a clean room for the center. Throughout the remainder of

the average number of companies participating in each center was 17.6; the median number was 6.

UIRCs vary significantly in size, whether measured in terms of overall research budget or the number of academic researchers or industrial partners involved. Large centers, such as the Engineering Research Center for Data Storage Systems at Carnegie Mellon University (Box 4) and Stanford University's Center for Integrated Systems, involve dozens of firms as sponsors, operate with budgets in excess of $10 million per year, and support 50 or more faculty researchers across multiple departments. Nearly 23 percent of all centers, however, had bud-

> *BOX 4—Continued*
>
> 1982 and into 1983, Professor Kryder, Angel Jordan (then dean of engineering at CMU), and Richard Cyert (then president of CMU) worked together to solicit industry support. In May 1993, IBM and 3M joined at the associate member level, committing to provide $750,000 each over a 3-year time frame. A number of other corporations joined at the affiliate and associate member levels. By April 1984, the Center had over $3 million per year in funding, most of it coming from industry and most committed for 3 years.
>
> Professor Kryder used this funding to seed research efforts by CMU faculty who had expertise relevant to magnetic data storage technologies. Some of the faculty had a background in magnetics research, but the majority had never worked on magnetic-storage technologies before. Most learned the requirements of magnetic storage technologies very quickly and have since become experts in the field. By 1988, the center had an annual budget of over $5 million, most of it from U.S. industry. In 1990, CMU obtained funding from the NSF for an Engineering Research Center (ERC) in data storage systems. The NSF award has amounted to between $2 million and $3 million per year.
>
> Following the initial NSF award, the industrial sponsors of the center formed the National Storage Industry Consortium (NSIC), with the goal of providing leveraged funding for research on data storage technologies. Professor Kryder worked with NSIC to obtain several Advanced Technology Program awards and an ARPA grant for work on advanced data storage technologies (magnetic disk, magnetic tape, and optical disk). As a result of the collaboration with NSIC, funding for the center has risen to over $10 million per year, with over 40 percent of this coming from industry.
>
> SOURCE: Mark H. Kryder, Carnegie Mellon University.

gets of less than $500,000 in 1990, and roughly 45 percent of all centers involved less than 6 companies as participants (Cohen et al., 1994).

Forty percent of the research conducted by UIRCs is basic research, 40 percent is applied research, and 20 percent is development work. In other words, UIRCs perform a significantly higher proportion of applied research and development than do universities. On average, UIRCs devoted two-thirds of their effort to R&D and one-fifth to education and training.

As a group, UIRCs receive 46 percent of their funding from public sources (34 percent from federal government and 12 percent from state governments), 31

TABLE 2.11 UIRC Research by Discipline, 1990

Discipline	Number of UIRCs	Percent of UIRCs
Basic science:		
Chemistry	192	38.6
Biology	169	34.0
Physics	120	24.1
Geology and earth sciences	98	19.7
Mathematics	54	10.7
Engineering:		
Materials	171	34.4
Electrical	159	32.0
Mechanical	155	31.2
Chemical	137	27.6
Civil	103	20.7
Industrial	87	17.5
Aeronautical and astronautical	58	11.7
Applied science:		
Materials	145	29.2
Computer science	130	26.2
Agricultural	106	21.3
Medical sciences	93	18.7
Applied math and operations research	57	11.5
Atmospheric	45	9.1
Oceanography	27	5.4
Astronomy	6	1.2

NOTE: Total number of UIRCs reporting was 497. Many of the centers had more than one disciplinary focus.

SOURCE: Cohen et al. (1994).

percent from private industry, and 18 percent from universities themselves. Some 70 percent of all industry support for academic R&D was channeled through UIRCs in 1990. The vast majority of public and private support for research at UIRCs comes in the form of grants. Most industrial support of UIRCs appears to be directed at more basic and long-term applied research. In addition to direct funding, industry contributions to individual centers also include equipment, instrumentation, and internship opportunities for students.

The goals and missions of individual centers vary considerably, as do their disciplines (Table 2.11), technology (Table 2.12), and industry orientation, and their organizational form. Collectively, these centers engage a broad range of traditional and high-technology industries in their research (Table 2.13). Some centers are more focused on industry's immediate needs, for example product and process improvements. Other centers are focused on more traditional academic objectives, such as education and the advancement of knowledge (Table 2.14).

TABLE 2.12 UIRC Research by Technology Area, 1990

Technology Area	Number of UIRCs	Percent of UIRCs
Environmental technology and waste management	147	29.8
Advanced materials	135	27.3
Computer software	129	26.1
Biotechnology	109	22.0
Biomedical	108	21.9
Energy	100	20.2
Manufacturing (industrial, automotive, and robotics)	98	19.8
Agriculture and food	89	18.0
Chemicals	77	15.6
Scientific instruments	67	13.6
Semiconductor electronics	64	13.0
Aerospace	61	12.3
Pharmaceuticals	61	12.3
Computer hardware	50	10.1
Telecommunications	48	9.7
Transportation	37	7.5

NOTE: Total number of UIRCs reporting was 494. Many of the centers had more than one technology focus.

SOURCE: Cohen et al. (1994).

The primary impetus for establishing nearly three-quarters of all UIRCs in existence in 1990 came from university-based researchers themselves. Government and industry each took the initiative in 11 percent of all centers established. The most aggressive federal sponsor of UIRCs during the 1980s was the NSF, which helped establish a raft of university-based centers, including Engineering Research Centers, Science and Technology Centers, Industry-University Cooperative Research Centers, Materials Research Centers, and Supercomputer Centers. NSF provided seed money for these centers with the expectation that the host institutions would raise matching funds from industry, state and local governments, and internally. While the objectives of these centers' programs vary in many respects (research focus, relative emphasis on research, education, and technology transfer, etc.), all share a commitment to facilitate industry access to university research results, engage industry in the definition of a research portfolio, and otherwise promote technology transfer to participating firms.

Recent assessments of the NSF centers indicate that, on the whole, they are effective mechanisms for forging university-industry research partnerships.[52] In aggregate, UIRCs graduated an average of four to five Ph.D.'s and seven to eight master's recipients per year (Table 2.15). On average, roughly 6 students from each UIRC found permanent employment with a participating company during the 2-year period 1989–1990. UIRCs accounted for 211, or about 20 percent, of

TABLE 2.13 UIRC Research by Industry, 1990

Industry	Number of UIRCs	Percent of UIRCs
Chemical/Pharmaceutical	213	41.7
Computer	179	35.0
Electronic equipment	148	29.0
Petroleum and coal	144	28.2
Software and computer services	133	26.0
Food products	110	21.5
Fabricated metals	107	20.9
Agriculture	102	20.0
Utilities	100	19.6
Rubber and plastics	88	17.2
Transportation	86	16.8
Transportation equipment	79	15.5
Mining	78	15.3
Communications	78	15.3
Industrial/Commercial machinery	78	15.3
Lumber and wood	77	15.0
Primary metals	76	14.9
Paper and allied products	75	14.7

NOTE: Total number of UIRCs reporting was 511. Many of the centers engaged more than one industry in cooperative research.

SOURCE: Cohen et al. (1994).

the 1,174 patents granted to universities in 1990. The nature and level of UIRC performance varies by technical field and funding source and is heavily influenced by the mission orientation of the particular center. Moreover, the scope and type of UIRC outputs is influenced heavily by the area of technology specialization (Cohen et al., 1995). For example, UIRCs focused in the fields of biotechnology and advanced materials lead in the production of patents. UIRCs emphasizing biotechnology develop the most new products, whereas those specializing in software lead in the development of new processes.

Nevertheless, some observers have expressed concern that the benefits resulting from deepening academic ties with industry through UIRCs and other mechanisms may come at a cost to core comparative strengths of the U.S. academic research enterprise—in particular, its capacity for basic research and its relative openness—that is unacceptable (Dasgupta and David, 1994; Rosenberg and Nelson, 1994) In fact, recent empirical studies indicate that university faculty receiving support from industry tend to conduct research that is more applied on average and to accept restrictions on the dissemination of their research findings (Blumenthal et al., 1986a,b; Cohen et al., 1994; Morgan et al., 1994a,b). While these documented changes appear to offer benefits to firms directly involved in UIRC

TABLE 2.14 Distribution of UIRCs by Importance of Selected Goals

	Number and Percentage [in brackets] of UIRCs Scoring Goals as:				Mean Score[a]
	Not Important	Somewhat Important	Important	Very Important	
To advance technological or scientific knowledge (N=497)	5 [1.0]	20 [4.0]	88 [17.7]	384 [77.3]	3.71
Education and training (N=499)	14 [2.8]	56 [11.0]	149 [29.9]	281 [56.3]	3.40
To demonstrate the feasibility of new technology (N=486)	44 [9.1]	118 [24.3]	160 [32.9]	164 [33.7]	2.91
To transfer technology to industry (N=496)	55 [11.1]	127 [25.6]	185 [37.3]	129 [26.0]	2.78
To improve industry's products or processes (N=491)	68 [13.8]	133 [27.1]	164 [33.4]	126 [25.7]	2.71
To create new business (N=483)	203 [42.0]	163 [33.7]	76 [15.7]	41 [8.5]	1.91
To create new jobs (N=481)	199 [41.4]	157 [32.6]	76 [15.8]	49 [10.2]	1.95
To attract new industry to the local area or state (N=474)	201 [42.4]	144 [30.4]	85 [17.9]	44 [9.3]	1.94

[a]Mean computed where 1 = not important; 2 = somewhat important; 3 = important; and 4 = very important.

SOURCE: Cohen et al. (1994).

TABLE 2.15 Output per UIRC, 1990

	Mean[a] (N=425)	Mean[b] (N)	Median[a]	Median[b]
Research papers	42.47	43.60 (414)	20	20
Invention disclosures	1.60	2.11 (321)	0	1
Copyrights	1.09	1.73 (268)	0	0
Prototypes	1.00	1.49 (286)	0	1
New products invented	0.69	1.06 (277)	0	0
New processes invented	0.92	1.39 (281)	0	0
Patent applications	1.08	1.39 (330)	0	0
Patents issued	0.50	0.68 (311)	0	0
Licenses	0.38	0.53 (301)	0	0
Ph.D.'s	4.38[c]	4.60 (410)	2	2
Master's degrees	7.03[c]	7.53 (402)	3	3

[a]Computed assuming blank responses signify zero, as long as there is a response to at least one of the category items.
[b]Composed assuming blank responses are missing values.
[c]N = 431
SOURCE: Cohen et al. (1994).

collaborative research, they may weaken channels of communication and redirect resources away from areas of basic research that benefit firms more broadly.

Industrial Liaison Programs

Industrial liaison programs (ILPs) charge membership fees to companies in return for providing them with facilitated access to the results of university research, to researchers, and to laboratories in specified fields. ILP members are generally entitled to receive research publications (some prepublications) from university-based researchers; to attend workshops, lectures, and conferences on research topics of interest; and to participate in an annual conference at which faculty and student research is formally presented and summarized. Some ILPs are universitywide in scope (i.e., a corporate member receives facilitated access to a broad range of university research for a fee that is added to the university's unrestricted funds). Most ILPs, however, are focused on a narrowly defined research area involving individual academic departments or research clusters, or, in some cases, individual UIRCs.[53] These more typical ILPs involve closer interaction between academic researchers and technical staff from industry and a higher level of faculty engagement overall in their management. Accordingly, corporate membership fees go to the sponsoring academic department or UIRC.

As part of its 1992 survey of 35 leading U.S. research universities, the U.S. General Accounting Office (GAO) (1992) gathered information on the growth of industrial liaison programs. Thirty of these institutions had at least one ILP.

Carnegie Mellon University alone accounted for 59 of 278 such programs that were identified. Eighteen of the universities surveyed provide liaison program members, domestic or foreign, with access to the results of federally funded research before those results are made generally available, while the other 12 institutions do not.

Research Consortia

Research consortia involve a university, academic research department, or UIRC with multiple corporate sponsors, and often state and federal government funding agencies, in the sponsorship of a specific field of academic research. Examples of such consortia include the Biotechnology Process Engineering Center Consortium at MIT (Box 5) and the Computer Aided Design/Computer Aided Manufacturing Consortium at the University of California at Berkeley. (See Box 3, pp. 106–107.) As in the case of formal UIRCs, consortia partners from industry and government are involved directly in helping define the research agenda of the academic research performer. Moreover, research consortia, like UIRCs, may also encompass targeted industrial liaison programs.

Technical Assistance Programs

Technical assistance programs are designed to serve small and medium-sized enterprises (SMEs) within a defined geographic region by providing them with technical advice and problem-solving capabilities usually related to manufacturing and production issues. Technical assistance programs may have a permanent staff of assistance providers or merely serve a broker function by putting companies in contact with expert consultants, including university faculty.

Most technical assistance programs are associated with universities. As of 1992, all but 8 of 75 members of the National Association of Management and Technical Assistance Centers were associated with college or universities. Included among the population of university-affiliated programs are the several hundred small-business development centers in community colleges established by the U.S. Small Business Association, the various technical and management assistance centers in universities funded by the Department of Commerce (such as the Manufacturing Extension Partnership and the Manufacturing Technology Centers), as well as many of the 42 centers funded by the U.S. Department of Transportation that provide technical advice to state departments of transportation.

As one observer has noted, these technical assistance programs "are public service activities and rarely have strong alliances with teaching or fundamental research. They require heavy subsidies and therefore must be attentive to the purposes and requirements of funding agencies. . . . [and they] exist on the periphery of the university, uncertain of their place and often unsupported by the administration" (Matkin, 1990). Whether such activities are worth the diversion of effort from the core missions of the university is an open question. Nevertheless, as in the case of equity investments in start-up companies, these activities may

BOX 5
The MIT Biotechnology Process Engineering Center

The Biotechnology Process Engineering Center (BPEC) at MIT is a pioneering program in education and research for the biotechnology industry (Biotechnology Process Engineering Center, 1995). BPEC takes an innovative, cross-disciplinary approach to biotechnology, integrating life sciences and bioprocess engineering with the goal of producing advanced manufacturing technologies. Established at MIT in 1985 by the National Science Foundation, the BPEC maintains active collaborative ties with the biotechnology industry.

A team of 14 faculty members with complementary areas of expertise lead the research and educational programs of BPEC. The faculty are from the MIT departments of chemical engineering, biology, chemistry, electrical engineering and computer science, and the Harvard University department of chemistry. Undergraduate and graduate students, postdoctoral fellows, visiting scientists, and industrial associates are integral participants in the center's activities. The center's vision is to establish, through research and education, the advanced manufacturing concepts and processes that will ensure the competitiveness of the U.S. biotechnology industry.

The research thrust of BPEC is the production of complex therapeutic proteins, specifically in areas of generic needs expressed by the industrial manufacturing sectors. One particular goal is to develop proteins in high concentration (quantity) and with high productivity (rate). A second major goal is to ensure the stability, formulation, and delivery of the therapeutic protein during processing and delivery.

The Biotechnology Process Engineering Center Consortium offers industry the opportunity to exchange information and personnel, share equipment and facilities, and perform collaborative research with the BPEC or with other consortium members. Consortium members keep in contact with BPEC faculty and students and receive advance notice of new technologies developed in the center's laboratories. Presently, nearly 60 companies from the chemical, pharmaceutical, and biotechnology industries are members of the consortium. The Consortium program puts on workshops for the purposes of information and technology exchange. Technology and information transfer also are accomplished via an annual symposium, publications, seminars, theses from center students, and consortium workshops. Direct industrial collaborations between industry and the center's students, research staff, and faculty have also been quite active.

SOURCE: Arthur Humphrey, Pennsylvania State University.

help buy sponsoring universities continued political/financial support within state legislatures. More importantly, as underscored by Armstrong (1997), such programs have the potential for exposing basic researchers in academia to other institutional cultures in the technological innovation system, to the benefit of all parties involved.

TECHNOLOGY BUSINESS INCUBATORS

The purpose of university-based technology business incubators is the care and feeding of start-up ventures through their early phases of development. Generally, incubators provide laboratory or building space at below-market rental rates, as well as a variety of technical and general business services. The incubators' principal service is to provide clients with access to academic researchers, including faculty, postdocs, and graduate students. In early 1997, there were more than 100 technology business incubators operating in the United States. Roughly half of these were affiliated with research universities (Association of University-Related Research Parks, 1997; National Business Incubators Association, 1997).[54]

ASSESSING TECHNOLOGY TRANSFER FROM UNIVERSITIES AND COLLEGES

The preceding review of the major technology transfer mechanisms of U.S. universities and colleges testifies to the dynamism, flexibility, and innovativeness of the nation's academic research enterprise in this area. Since the early 1980s there have been strong fiscal and public-policy-related incentives for academia to engage industry more intensively as a research partner and client. In this context, the highly diverse and autonomous population of U.S. research colleges and universities and their research faculties have had great latitude to experiment with new institutional arrangements to this end. Responding to the economic development challenge, academic research institutions have expanded their portfolio of technology transfer activities to encompass collaborative research centers, consortia, proactive technology licensing offices, venture capital funds, and technical extension programs.

While it is difficult to assess the aggregate impact of or attribute specific causality to these experiments, the past 10 to 15 years have witnessed a number of significant readily documented changes in university-industry research interaction that are at least consistent with the logic of these initiatives. Industrial support for academic research has grown more rapidly than funding by any other sector since 1980. The number of academic research publications cited in U.S. patent applications has increased markedly in the last 5 years. University licensing revenues have grown rapidly in the past decade, albeit from a small base. Although most academic researchers involved in collaborative work with industry still view the advancement of knowledge as their primary research objective, the more entrepreneurial among them are now faced with greater opportunities (and incentives) to become involved directly in the commercialization of technologies developed or seeded within the academy through start-up companies or other mechanisms. Through more intense research collaboration, firms in a number of industries have gained enhanced access to academic researchers—faculty, postdocs, and graduate students—with highly specialized knowledge.

With respect to the impact of academic research and technology transfer on industrial performance there are clearly significant inter-industry variations in experience. As the survey of UIRCs suggests, the relative importance of different technology transfer mechanisms varies widely according to the nature of the technology being transferred and the industry being served. The extent and nature of a given research university's contribution to the technology needs of a particular industry or company depends largely on the specific characteristics of that industry's key technologies (e.g., whether they are highly science-based or not, whether they are relatively new and dynamic or more mature and stable, whether intellectual property rights are central or tangential to their successful commercialization, etc.). For example, patent licensing is a critical instrument of technology transfer in biotechnology, where control of intellectual property rights is essential for the long and expensive development/commercialization cycle of human therapeutic compounds. Yet patents are much less important in software or microelectronics, where the pace of technology life cycles is much shorter.

Research universities, which constitute the locus of most basic research in molecular biology and computer sciences in the United States, are considered the most important nonindustrial source of external technology for the relatively new, highly science-based biotechnology and software industries (see Annex II). Yet aside from their critical contribution of well-trained, learning-equipped science and engineering graduates, U.S. research universities have not figured prominently as a source of new technology or proto-technology for more technologically mature or established industries (e.g., automobiles, machine tools).

Surveys of industrial researchers by Nelson and Levin (1986) and related research by Mansfield (1995) have shown that there are only a few industries where technology transfer from universities in the form of codified intellectual property, or the direct contribution of academic research to the commercializable products and processes are perceived to be important. Here again, software and biotechnology (i.e., new technologies where the step from basic research to application is direct) are the only two areas where corporate managers see universities as major sources of "invention." From the perspective of most other technology-intensive industries, academic research mainly stimulates and enhances the power of R&D performed by private companies. Those who produce nonbiotech pharmaceuticals assert that they look to academic research primarily to improve their understanding of technologies, particularly new technologies, yet only rarely for new products. Likewise, electronics manufacturers view academic research as an important source of radically new designs and concepts, but as a relatively insignificant contributor to incremental technological advance in their industry (Rosenberg and Nelson, 1994). Yet even in less "science-based" industries, better understanding of technologies, illuminated by academic research, may enable industrial researchers to search more efficiently for incremental changes. In other words, academic research helps identify a much wider range and variety of options for incremental improvement, but the selection among these options for further pursuit can be better done by industrial researchers more intimately familiar with all the surrounding constraints and requirements (many of them nontechnical).

Our understanding (both quantitative and qualitative) of the current nature and dynamics of university-industry partnerships in individual industries and research fields remains very limited. However, the large degree of variation in company practices, in the demands of technology in different industries, and in the nature and practices of universities documented in these and other case histories makes it clear that no single set of approaches will fit all situations.

From a U.S. perspective, an effective system of collaboration among universities and industry is a keystone of technology policy for economic growth. It is clear that companies and universities are good at different aspects of research, development, demonstration, and commercial innovation and that the process of allocation of effort and resources should reflect those differing capabilities. It is not clear, however, that either companies or universities know how to be good partners. In many partnerships, the missions, cultures, norms, and concerns of the two organizations could not be farther apart. Corporate technology strategies call for justifiable R&D expenditures and focus on speeding the contribution of new technology to commercial success. University mission statements and culture value contributions to education, learning, and long-horizon fundamental research. Because of these differences, partnerships can be strained, with neither party being particularly satisfied. Indeed, increased emphasis on applied research at universities and growing limitations on the disclosure of academic research results, both fueled by deepening university-industry research ties, may be undermining core strengths of the academic research enterprise and its capacity for serving the less proprietary, more long-term knowledge/research needs of industry.

Amidst rising public enthusiasm for and expectations of university-industry partnerships, companies, universities, and public policymakers are faced with a number of critical questions. For companies, there are a host of operational questions as to what can and cannot be accomplished working with universities and which practices work best. For universities, there is an equally complex set of operational questions—about how best to serve companies as clients—made even more difficult by the educational mission of universities and a long-standing historical remove of many universities from commercial concerns. For example, what intellectual property policies guide successful collaborations in different industries and fields of research? And how can conflicts of interest and exploitation be avoided? For policymakers at the state and federal levels there are important questions regarding, among others, the opportunity costs of diverting resources and effort from traditional university missions to strengthen industrial outreach and research collaboration for economic growth, the structure and effectiveness of programs designed to foster such university-industry collaboration, the allocation of public research monies more generally, and the disposition of intellectual property generated with public funds.

These questions have given rise to a substantial body of research focused on measuring the rate of return of academic research to specific industries, evaluating the performance of particular institutional modes of university industry collaboration, or extracting generalizable lessons concerning effective strategies and practices for university-industry collaboration from multi-industry, multidisciplinary surveys, and patent data.[55] Nevertheless, the pace of cross-institutional learning remains slow. Many leading U.S. research universities appear to have developed effective policies, practices, and institutional frameworks for engaging private companies in mutually beneficial cooperative research. There is, however, considerable evidence that a great many more U.S. research universities are still struggling to put effective policies and practices in place.

U.S. FEDERAL LABORATORIES AND TECHNOLOGY TRANSFER TO INDUSTRY*

Overview

The U.S. federal government maintains over 720 laboratories, encompassing more than 1,500 separate R&D facilities. These facilities were established and developed to support the public missions of federal agencies, such as national security, energy independence, the cure of disease, food production, or science and engineering research. Federal laboratories and research facilities are the second largest segment of U.S. R&D enterprise, performing nearly $25 billion worth, or 14.4 percent, of all U.S. R&D in 1994. These institutions perform roughly 18 percent of all basic research, 16 percent of all applied research, and 13 percent of all technology development in the United States. Collectively, they employ roughly 100,000 scientists and engineers nationwide[56] (National Science Board, 1996).

Federal laboratories vary widely in their size, mission, organization, and management. Although the total number of federal laboratories is large, most federal laboratories are either very small, or have a very narrow technical mission. Fewer than 100 federal laboratories have the technologies and resources to

---

*This section draws extensively on a background paper prepared by Robert K. Carr (1995) for the U.S. delegation to the binational panel.

engage in significant technology transfer activities. Included among these are all of the large multiprogram laboratories[57] of the Department of Energy, many of the Defense Department's laboratories, most field centers of the National Aeronautics and Space Administration (NASA), as well as facilities of the USDA, the Public Health Service, including the NIH, and NIST.

As agents of technology transfer to industry, federal laboratories rank a distant third behind universities and private companies as measured by licensing revenues. The roughly $19 million in royalties and fees received by federal laboratories in 1993 represented less than 8 percent of the total collected by U.S. universities ($250 million) (Association of University Technology Managers, 1995) and less than 0.1 percent of licensing royalties and fees earned by private companies in 1992 ($33 billion) (Internal Revenue Service, 1993). Most federal labs designate only a small percentage, if any, of their total R&D budget to technology transfer and related activities. A recent survey of technology outsourcing by large U.S. firms found that federal laboratories rank fourth as sources of external technologies after other firms, universities, and private-sector research databases (Roessner, 1993).

THREE TYPES OF FEDERAL LABORATORY MANAGEMENT

Most federal laboratories are governmental-owned, government-operated facilities, referred to often by the acronym GOGO. Their land and facilities are usually owned by the federal government, and their employees and managers are career civil servants. A second category of federal research facility, the federally funded research and development centers (FFRDCs), contractor-operated and mostly contractor-owned research facilities established at the request of federal agencies with congressional authorization, draw over 70 percent of their funding from the federal government. FFRDC employees and managers are not civil servants. An important subset of FFRDCs is a small group of large government-owned, contractor-operated laboratories, or GOCOs. Their land and facilities are usually owned or leased directly by the federal government, but the labs themselves are operated for the government by private contractors, including companies, universities, and nonprofit institutions. Most GOCO laboratories are administered by the Department of Energy.

Because GOCO laboratories and FFRDCs operate largely outside the government's personnel and contracting systems, they are freed from many of the regulatory and administrative requirements by which GOGO labs must abide. Because of their special status, FFRDCs are often used by federal agencies to execute new federally funded R&D programs that require rapid start-up.[58]

FEDERAL LABORATORY R&D EXPENDITURES

Federal laboratories spend fully one-third of all federal dollars devoted to research and development (Figure 2.14). In 1994, GOGO labs (and other intra-

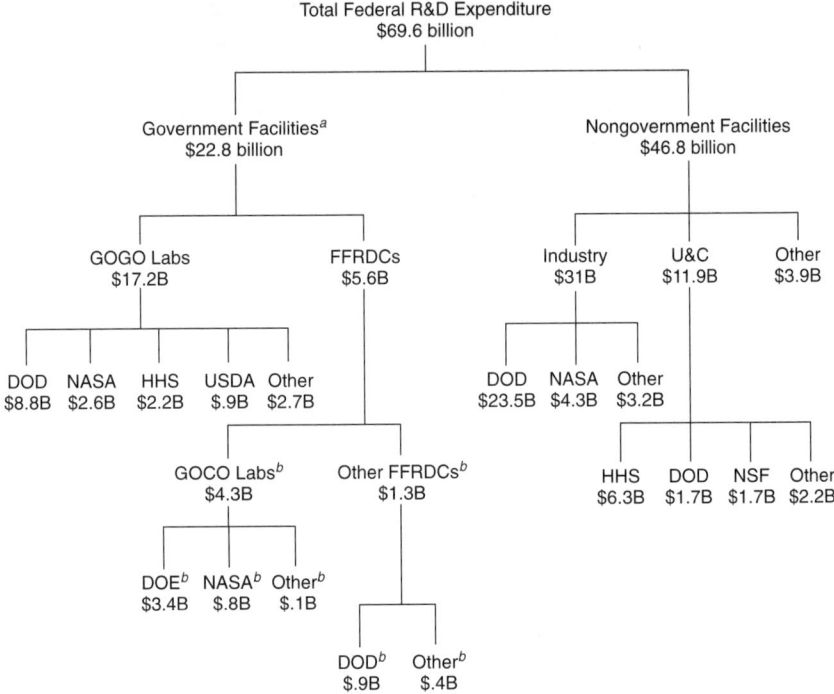

FIGURE 2.14 Federal R&D funds by selected categories of performers, estimated values for FY 1994. [a]Includes costs associated with the administration of intramural and extramural programs by federal personnel as well as actual intramural performance. [b]Author's estimates based on NSF data for all FFRDCs. NOTE: U&C = universities and colleges; DOD = Department of Defense; DOE = Department of Energy; HHS = Health and Human Services (primarily the National Institutes of Health); NASA = National Aeronautics and Space Administration; NSF = National Science Foundation; and USDA = Department of Agriculture. SOURCE: Carr (1995).

mural facilities) received $17.2 billion in federal R&D support, and FFRDCs an additional $5.6 billion. Of this latter figure, GOCO labs received an estimated $4.3 billion, and other FFRDCs approximately $1.3 billion.

Federal Laboratories by Major Mission Area

DEFENSE LABORATORIES

In this chapter, "defense laboratories" refers to the laboratories of the military services and the Department of Energy's (DOE's) three nuclear weapons laboratories (Los Alamos, Lawrence Livermore, and Sandia).

Department of Defense Laboratories

DOD and the Departments of the Army, Navy and Air Force collectively own and operate 81 GOGOs, many of which are grouped for command and management purposes into larger entities. In fiscal 1994, DOD and the service agencies funded $35.6 billion worth of R&D, yet only $8.8 billion, or 25 percent of this total, was performed by the 81 intramural facilities. The vast majority of DOD-funded R&D is performed extramurally by private companies, universities, and FFRDCs. DOD-funded basic research is done primarily in universities, and most DOD technology development is performed by private defense firms. DOD intramural laboratories perform mainly exploratory development.

Since the end of the Cold War, the U.S. defense laboratory system has come under increasing pressure to downsize. In recent years, the Air Force and Army have consolidated their laboratory structure, forming a smaller number of "super labs." However, few facilities were closed and few positions eliminated. In 1994, the Defense Science Board (Defense Science Board, 1994) recommended that DOD laboratory personnel be reduced by 20 percent and that "vigorous programming of outsourcing of defense laboratory activities" be pursued. Yet, the Base Realignment and Closure Commission has recommended closing only a few R&D facilities. Although rationalization of the defense laboratory system has been slow, the system is certain to involve fewer, smaller laboratories in the future.

DOE Laboratories

The three DOE defense laboratories, Los Alamos and Sandia National Laboratories in New Mexico, and Lawrence Livermore National Laboratory in California, are the largest federal laboratories. Although their annual budgets are declining, the three labs still spend nearly $1 billion each on R&D, and each employs many thousands of scientists and engineers. Los Alamos and Livermore National Laboratories are both managed by the University of California. Their early work focused almost exclusively on nuclear weapons design. However, in the past 50 years, their missions have broadened to include many diverse technology areas. R&D at the two labs is divided roughly equally among weapons R&D, nondefense nuclear work, nonnuclear defense work, and nondefense, nonnuclear R&D.

Sandia National Laboratories, operated by Lockheed Martin, are engineering laboratories whose mission is to "weaponize" the nuclear weapons designs created at Los Alamos and Livermore. As is the case with the other defense labs, Sandia has also developed a broad array of technology capabilities in addition to its weapons-related functions. Sandia is perhaps the most industry-oriented of the three nuclear labs.

Like the DOD R&D system, DOE's defense laboratories are also faced with excess capacity. In recent years, these facilities have experienced a steady reduction in their nuclear weapons design work and other programs, with a corresponding reduction in personnel. However, they seem less likely to be subject to the same percentage reductions as the DOD laboratories, particularly given their new mission for "science-based stewardship" of the U.S. nuclear stockpile.

CIVILIAN LABORATORIES

Department of Energy

In addition to its defense laboratories, the DOE operates a number of other multiprogram laboratories, most of these under the aegis of the department's Energy Research Program. The largest of these laboratories (all GOCOs) are described below.

Argonne National Laboratory in Chicago, Illinois, encompasses engineering research (advanced batteries, fuel cells, and advanced fission reactor); physical research (materials science, physics, chemistry, high-energy physics, mathematics and computer science); the Advanced Photon Source, (the nation's most brilliant X-ray beam); and energy and environmental science and technology.

Brookhaven National Laboratory, located on Long Island, New York, maintains user facilities for investigation in a multitude of scientific disciplines, including experimental and theoretical physics, medicine, chemistry, biology, environmental research, engineering and many other fields.

The Idaho National Engineering Laboratory, located near Idaho Falls, specializes in natural resource processing and environmental management, spent nuclear-fuel management, environmental technology development, mixed-waste characterization and treatment, non- and counterproliferation, advanced manufacturing, alternate energy supply and energy efficiency, and transportation technologies.

Lawrence Berkeley Laboratory in Berkeley, California, which conducts research in advanced materials, biosciences, energy efficiency, detectors, and accelerators, focuses on national needs in technology and the environment.

The National Renewable Energy Laboratory in Denver, Colorado, is the nation's primary federal laboratory for renewable energy research. It focuses on alternative fuels, analytic studies, basic sciences, buildings and energy systems, industrial technologies, photovoltaics, and wind technology.

Oak Ridge National Laboratory in Oak Ridge, Tennessee, has major programs in energy conservation, materials development, magnetic-fusion energy, nuclear safety, robotics and computing, biomedical and environmental sciences, medical radioisotope development, and basic chemistry and physics.

Pacific Northwest Laboratory in Richland, Washington, focuses on resolving environmental issues, such as waste cleanup and global climate change. Other areas of research activity include molecular science, advanced processing technology, biotechnology, global environmental change, and energy technology development.

The National Institute of Standards and Technology

NIST, formerly the National Bureau of Standards, was established by Congress "to assist industry in the development of technology . . . needed to improve product quality, to modernize manufacturing processes, to ensure product reliability . . . and to facilitate rapid commercialization . . . of products based on new scientific discoveries" (National Institute of Standards and Technology, 1997). An agency of the U.S. Department of Commerce, NIST's primary mission is to develop and apply technology, measurements, and standards and to promote U.S. economic growth. It carries out this mission through work in four areas:

- research planned and implemented in cooperation with industry and focused on measurements, standards, evaluated data, and test methods;
- the Malcolm Baldrige National Quality Award and an associated quality-outreach program;
- the Advanced Technology Program, which provides cost-shared grants to industry for the development of high-risk technologies with significant commercial potential; and
- the Manufacturing Extension Partnership, which helps small and medium-sized companies adopt new technologies.

As a GOGO laboratory, NIST's staff of more than 3,200 scientists, engineers, technicians, and support personnel are federal employees (National Institute of Standards and Technology, 1997). In addition, some 1,200 visiting researchers work at NIST each year. In fiscal 1994, about 46 percent of NIST's R&D budget went for intramural work. Nearly all of the remainder (52 percent) of the budget supported R&D in private industry. NIST's intramural R&D and related activities are performed at principally two sites, one in Gaithersburg, Maryland, and the other in Boulder, Colorado. The main areas of research pursued by these laboratories are electronics and electrical engineering, manufacturing engineering, chemical science and technology, physics, materials science and engineering, building and fire-prevention research, computer systems, and computing and applied mathematics.

National Aeronautics and Space Administration

NASA's in-house research and development is carried on in a number of field centers, all of them GOGOs with the important exception of the Jet Propulsion Laboratory, a GOCO operated by the California Institute of Technology. The field centers conduct about one-third of all R&D funded by the agency. In addition to in-house research, an important function of the centers is the management of NASA contractors, principally aerospace firms, that perform roughly half of all NASA-funded R&D. NASA's principal research centers and their missions are described below.

- Ames Research Center in Moffett Field, California, focuses on fluid dynamics, life sciences, earth and atmospheric sciences, information, communications, and intelligent systems and human factors.
- Dryden Flight Research Center in Edwards, California, specializes in aero-

dynamics, aeronautics, flight testing, thermal testing, and integrated systems and validation.
- Goddard Space Flight Center in Greenbelt, Maryland, is NASA's center for earth and planetary sciences missions, LIDAR, cryogenic systems, tracking, telemetry and command.
- Jet Propulsion Laboratory in Pasadena, California, conducts work in near- and deep-space mission engineering, microspacecraft, space communications, information systems, remote sensing, and robotics.
- Johnson Space Center in Houston, Texas, has technological strengths in artificial intelligence and human computer interface, life sciences, human space flight operations, avionics, sensors, and communications.
- Kennedy Space Center, near Cape Canaveral, Florida, is the principal NASA launch site and also specializes in emissions and contamination monitoring, sensors, corrosion protection, and biosciences.
- Langley Research Center in Hampton, Virginia, focuses on aerodynamics, flight systems, materials, structures, sensors, measurements, and information sciences.
- Lewis Research Center in Cleveland, Ohio, specializes in aeropropulsion, communications, energy technology, and high temperature materials research.
- Marshall Space Flight Center in Huntsville, Alabama, has strengths in materials, manufacturing, nondestructive evaluation, biotechnology, space propulsion, controls and dynamics, structures, and microgravity processing.
- Stennis Space Center in Hancock County, Mississippi, specializes in space propulsion systems, testing and monitoring, remote sensing, and nonintrusive instrumentation.

NASA has recently reviewed the activities of its research centers with the goal of reducing technology overlap and bringing more focus to their activities.

The National Institutes of Health

Begun as the Laboratory of Hygiene in 1887, the NIH is one of eight health agencies of the Public Health Service, which, in turn, is part of the U.S. Department of Health and Human Services. NIH is made up of 24 separate institutes, centers, and divisions, with a total R&D budget of more than $10 billion in 1994. As of 1992, NIH funded roughly 80 percent of all biotechnology-related R&D supported by the U.S. federal government (National Research Council, 1992d).

Eighty-one percent of NIH-funded R&D is performed by extramural institutions, three-fourths of this by universities and colleges. Only about 11 percent of the NIH R&D budget supports research within in its own laboratories. NIH on-campus research facilities include:
- the Research Hospital and its laboratory complex, containing a 470-bed facility where patients participate in clinical studies;

- the Outpatient Clinic and the Ambulatory Care Research Facility and laboratories, supporting the NIH Clinical Center's outpatient programs;
- the Mary Woodard Lasker Center for Health Research and Education, the location for the NIH–Howard Hughes Medical Institute research program for medical students; and
- the National Library of Medicine, the world's largest medical library, with a collection of 5 million items and its computerized index, MEDLINE.

The agency's off-campus facilities include:

- the National Institute of Environmental Health Sciences, located in Research Triangle Park, North Carolina;
- the NIH Animal Center in Poolesville, Maryland;
- the National Institute on Aging's Gerontology Research Center in Baltimore, Maryland;
- the Addiction Research Center of the National Institute on Drug Abuse, in Baltimore, Maryland; and
- the National Institute of Allergy and Infectious Diseases' Rocky Mountain Laboratories in Hamilton, Montana.

In addition to its own research, NIH maintains active and long-standing partnerships with universities, independent research institutions, private industry, and voluntary and professional health organizations through which research programs and product development activities based on federally funded research are transferred.

Environmental Protection Agency

In 1994, the Environmental Protection Agency (EPA) had an R&D budget of $557 million. At that time, work within EPA's intramural laboratories accounted for only one-fifth of the agency's total R&D spending. Academic institutions and state and local governments each received another 20 percent of the R&D budget, while the remaining 40 percent was awarded to private firms. EPA has recently consolidated 12 laboratories and 7 field centers into 3 national laboratories and 2 national centers, all GOGOs, employing a total of 800 scientists and engineers. These reorganized facilities will focus on a redefined science mission based on the National Academy of Science's risk assessment/risk management model (National Research Council, 1983, 1994). In a complementary move, EPA is seeking to increase the role of the extramural science community in environmental research.

The laboratories and centers in EPA's new R&D structure are

- the National Risk Management Research Laboratory in Cincinnati, Ohio, which is the principal EPA research laboratory responsible for environmental risk management;
- the National Health and Environmental Effects Research Laboratory in Research Triangle Park, North Carolina, which is the EPA focal point for

toxicological, clinical, epidemiological, ecological, and biogeographic research;
- the National Exposure Research Laboratory, also in Research Triangle Park, which has the task of reducing and quantifying the uncertainty in the EPA's exposure and risk assessments for all environmental stressors;
- the National Center for Extramural Research and Quality Assurance located near Washington, D.C., which is responsible for managing the agency's extramural research grant programs, the Environmental Technology Initiative, small business innovation research grants, quality assurance policy and oversight, and conduct of peer reviews; and
- the National Center for Environmental Assessment located near Washington, D.C., which is responsible for risk assessment research, methods, and guidelines; health and ecological assessments; development, maintenance, and transfer of risk assessment information and training; and setting research priorities.

Department of Agriculture

In fiscal 1994, the USDA spent $1.4 billion on R&D, including about $900 million for work performed intramurally and $450 million for research in U.S. universities and colleges. The Agricultural Research Service (ARS), which conducts more than two-thirds of all USDA in-house R&D, oversees some of the oldest federal research facilities in the United States. The ARS is treated as a single entity for the purpose of technology transfer, although it operates a number of small and a few large research facilities. The total number of ARS R&D sites is around 110; some of these facilities are being closed in response to budget pressures.

With a staff of 350 scientists, the Beltsville Agricultural Research Center in Beltsville, Maryland, is the largest ARS laboratory. ARS has five regional laboratories that are focused on finding new uses for agricultural commodities. There are also two large animal-disease research centers and a series of smaller laboratories, each with a narrow focus such as a particular crop. In addition, nearly every land grant college has an ARS facility, usually integrated into the research facilities of the institution's academic departments.

ARS research falls into six categories:

- soil, water, and air (conservation, management, reduction of agricultural environmental impacts, and efficiency of use)
- plant productivity (crop productivity and quality)
- animal productivity (productivity and health of farm animals)
- commodity conversion and delivery (converting raw agricultural commodities into food, textiles, industrial materials, and other products)
- human nutrition and well-being (nutrients in foods, how nutrients work in humans, what nutrients are needed by humans, and what nutrients are provided by foods)

- systems integration (integrating scientific knowledge into systems that improve the efficiency of resource use and enable technology transfer from laboratory to farm).

Federal Laboratories and Technology Transfer: History and Legislation

The current era of federal technology transfer began in 1980 with passage of the Bayh-Dole (P.L. 96-517) and Stevenson-Wydler (P.L. 96-480) Acts. Although these legislative endeavors marked new directions in federal policy, they were based on nearly a century of federal cooperation with the private sector and on earlier legislation that encouraged technology transfer. These precursors of current federal technology cooperation include still-ongoing programs to support important sectors of the U.S. economy.

In 1862, Congress passed the Morrill Act, which provided resources to the states to develop colleges offering practical instruction in agriculture and the mechanical arts. Twenty-five years later, the 1887 Hatch Act created a system of state agricultural experiment stations under the auspices of the land grant colleges and universities. In 1914, the Smith-Lever Act created the Cooperative Agricultural Extension Service, a partnership among federal, state, and county governments to deliver the practical benefits of research to citizens through an extension service. The Smith-Lever Act represents the first U.S. law to promote intentionally technology transfer from federally funded research activities. As late as 1940, research by USDA and state agricultural extension programs accounted for almost 40 percent of the federal R&D budget. Reflecting the changed position of agriculture in the U.S. economy, USDA's share of the federal R&D budget has fallen to 2 percent in recent years. However, the Department of Agriculture remains the only federal agency that explicitly allocates a large share (roughly half) of its overall R&D budget to the dissemination and transfer of technology to the private sector.

The U.S. Geological Survey (USGS), established by Congress in 1879, provided technical support critical to the development of the nation's natural resource industries. Placed within the federal Department of the Interior, the USGS was charged with the "classification of the public lands, and examinations of the geological structure, mineral resources, and products of the national domain." During the late 1800s and early 1900s, the Survey's mining geology research program served as a primary research base for U.S. minerals industries, and was a major factor in the development of economic geology as a distinct field in geology (Rabbitt, 1997).

The National Advisory Committee on Aeronautics (NACA), the predecessor to NASA, was created in 1915, the result of an early competitiveness concern. NACA's charge was "to supervise and direct the scientific study of the problems of flight, with a view to their practical solutions" (Bilstein, 1989). NACA research in the post–WWI era was focused on civil aviation and was closely coordinated with the U.S. aircraft industry, to which much of the resulting technology was transferred. The 1958 Space Act (P.L. 85-568), which created NASA, incorporated NACA and its operations into the new structure. It specifically required that NASA engage in technology transfer, and for many years, NASA had the most active federal technology transfer program outside the USDA.

During the 1960s and 1970s, the responsibilities of the federal laboratory system grew to include the construction and operation of major user facilities, such as particle and photon accelerators, environmental research parks, and materials laboratories. These new facilities opened the laboratory system increasingly to U.S. and foreign researchers from industry and academe. Throughout this period, federal agencies relied heavily on the R&D capabilities of academic institutions and private companies to advance public missions, contracting or collaborating with a large number of private-sector R&D-performing institutions. In the process, a number of federal labs transferred significant amounts of know-how and other uncodified technology to private firms. Nevertheless, relatively little in the way of codified government-owned intellectual property was commercialized by private companies prior to the 1980s.

Through the end of the 1970s, the philosophy behind the dissemination of federally funded research was that if the public paid for the research, the resulting intellectual property should be made equally available to all interested parties. While universal access is a normal feature of most government programs, it is not a typical feature of the business world. Hence, few businesses were willing to risk substantial sums to develop government technologies into commercial products when competitors also had free access to the same intellectual property.

In 1980, Congress changed that philosophy in the belief that more could be done to increase the contribution of federal research to national competitiveness. The Bayh-Dole and Stevenson-Wydler Acts provided federal laboratories flexibility in granting individual companies varying degrees of exclusive access to federal intellectual property. The laws have been subsequently amended and supplemented with new legislation to support additional federal laboratory technology transfer activities. A brief review of the four principal technology transfer laws now on the books follows.

The Bayh-Dole Act of 1980 gave nonprofit organizations such as universities, as well as small businesses, the right to take title on inventions they developed with federal support; granted GOGO laboratories the authority to grant exclusive licenses to inventions that they patented; and protected inventions from public dissemination under the Freedom of Information Act, to allow for patent applications to be filed. Although Bayh-Dole did not originally apply to any of the DOE contractors responsible for laboratory management and operations, the law was subsequently amended to include them.

The Stevenson-Wydler Technology Innovation Act of 1980 mandated that federal laboratories actively seek to conduct cooperative research with state and local governments, academia, nonprofit organizations, or private industry and disseminate information about their activities and research. It established the Center for the Utilization of Federal Technology (CUFT) at the National Technical Information Service and required each federal laboratory to set up an Office of Research and Technology Applications (ORTA). These offices received a set-aside equal to 0.5 percent of each laboratory's budget to fund technology transfer activities. This act also established the National Medal of Technology.

The Federal Technology Transfer Act of 1986 (P.L. 99-502) amended Stevenson-Wydler to accelerate technology transfer by requiring that personnel evaluations of federal laboratory scientists and engineers include information about their support for technology transfer activities and that GOGO laboratories pay inventors a minimum of a 15-percent share of any royalties generated by the licensing of their inventions. It gave directors of GOGO laboratories authority to enter into Cooperative Research and Development Agreements (CRADAs), to license inventions that might result from CRADAs, to exchange laboratory personnel, services, and equipment with research partners, and to waive their rights to inventions and intellectual property developed in their labs under CRADAs. The act allows federal employees to participate in commercial development with private firms, if there is no conflict of interest, and it created a charter and funding for the Federal Laboratory Consortium (FLC), a 20-year-old grouping of federal laboratories.

The National Competitiveness Technology Transfer Act of 1989 (P.L. 101-189) further amended the Stevenson-Wydler Act to allow for the protection against disclosure of information, inventions, and innovations contained in CRADAs for a period of 5 years. It also established a technology transfer mission for the nuclear weapons laboratories and clarified that GOCO laboratories could execute CRADAs and enter into other technology transfer activities.

The Federal Laboratories and Technology Transfer: Mechanisms

The federal laboratories fall into two general categories with respect to the technologies they develop. The first group includes federal laboratories that develop technologies that will ultimately be used in the private sector. The best examples are the ARS, NIH, NIST, and the DOE's National Renewable Energy Laboratory. Technology transfer from these labs is sometimes described as "vertical" (developed as a direct result of the principal mission), and almost all activities of these laboratories are potentially fertile areas for technology transfer to and cooperative R&D activities with the private sector. The second category of laboratories develops technologies that are more or less exclusively for government consumption (i.e. technologies not generally useful in the private sector). Commercially valuable technology transfer from these labs tends to be horizontal, that is, developed as a by-product, or spin-off, of the principal mission. The best examples of labs in this category are the military service laboratories and the DOE weapons laboratories. While the primary outputs of these laboratories—

defense-specific technologies—are not intended to contribute to the commercial technology base of the nation, commercially useful technology transfer and cooperative R&D occur in many technology areas that support the lab's primary defense missions.

MECHANISMS OF TECHNOLOGY TRANSFER

Analyses of technology transfer programs tend to focus on technology licensing and cooperative R&D. This is understandable, since these two activities generally involve the transfer of intellectual property and are the most formalized mechanisms. However, there are a number of other ways in which federal technology transfer benefits industry.

Licensing

Licensing is the traditional way technology is transferred from federal laboratories to industry. Licenses convey access to intellectual property arising from in-house laboratory research. Cooperative R&D programs also may generate licenses. Since passage of the Federal Technology Transfer Act (FTTA), there has been only a modest increase in the number of exclusive licenses issued by the

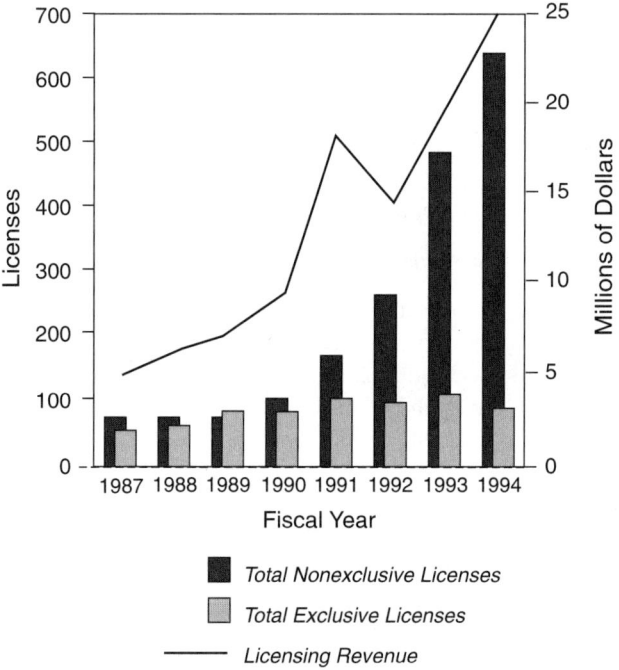

FIGURE 2.15 Federal laboratory licensing activity, 1987–1994. NOTE: 1991 "bump" represents a one-time AIDS-test payment. SOURCE: Carr (1995).

federal laboratories (Figure 2.15). Exclusive licensing is usually essential in cases in which the licensee will have to invest in considerable additional R&D to bring the technology to market. However, from the late 1980s onward, the number of nonexclusive licenses negotiated by federal laboratories and the revenue generated by these licenses have increased significantly, albeit from a relatively small base.

DOE accounted for nearly three-fourths of all nonexclusive licenses and one-half of all exclusive licenses granted by federal agencies and laboratories in fiscal 1994. The Department of Health and Human Services (HHS)/NIH accounted for another 20 percent of nonexclusive licenses. However, HHS/NIH received about 75 percent of all licensing income (royalties and fees) earned by federal agencies and laboratories that year.

Cooperative Research and Development and CRADAs

Federal laboratories have engaged in cooperative activities with the private sector through a variety of legal mechanisms long before the passage of the 1986 FTTA and the advent of CRADAs. These have included Space Act Agreements (authorized for NASA in the 1958 Space Act), various types of agreements authorized for the Department of Agriculture's ARS, and cooperative agreements under federal procurement legislation. In addition, three of the DOE laboratories have a joint cooperative R&D agreement with industry to support the Superconductivity Pilot Centers at Los Alamos, Argonne, and Oak Ridge National Laboratories. The agreements establishing these centers predate FTTA and have been described as proto-CRADAs. In addition, many earlier cooperative R&D agree-

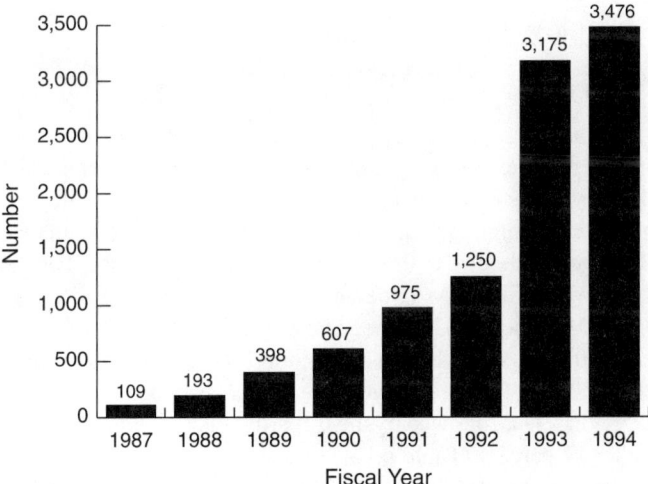

FIGURE 2.16 Active CRADAs at federal laboratories, 1987–1994. [a]Does not include NASA data. NOTE: 1994 data are estimated. SOURCE: Carr (1995).

TABLE 2.16 Active CRADAs by Federal Agencies and Laboratories, 1987–1994

Agency	1987	1988	1989	1990	1991	1992	1993	1994
Energy[a]	0	0	0	1	43	244	478	935
Defense:								
Air Force	0	2	7	13	26	50	126	176
Army	3	8	27	80	115	212	260	289
Navy	0	0	2	20	52	59	72	142
Commerce	0	9	44	82	115	194	311	414
Agriculture	9	51	98	128	177	237	273	276
Health and Human Services	22	28	89	110	144	149	184	209
Interior	0	0	1	12	11	38	56	87
Environmental Protection Agency	0	0	2	11	31	40	50	34
Transportation	0	0	0	1	9	18	26	33
Veterans Affairs	0	0	1	2	8	6	7	9
Housing and Urban Development	0	0	0	0	0	3	4	3
TOTALS	34	98	271	460	731	1,250	1,847	2,607
NASA[b]	75	95	127	147	244	N/A	328	869

[a]Most DOE laboratories are GOCOs, which were not covered under the Stevenson-Wydler Act until 1989.

[b]NASA has chosen to remain under the provisions of the Space Act with regard to its technology commercialization program. That Act permits NASA to enter into both reimbursable and nonreimbursable cooperative research and development agreements that are similar to those authorized under the Stevenson-Wydler Act.

These agreements are included separately because they do not fall under the terms of the Stevenson-Wydler Act.

NOTE: Data for 1994 are estimated.

SOURCE: U.S. Department of Commerce (1996b).

ments were negotiated between firms and federal laboratories without specific authority in the law. Several other contractual technology transfer mechanisms are used primarily in the health and medical sciences area. They include clinical trial agreements, screening agreements, and material transfer agreements.

As of 1992, only about one-third of cooperative agreements between firms and federal labs were CRADAs (Roessner, 1993). Nevertheless, growth in the number of CRADAs since 1987 has been impressive (Figure 2.16).[59] As of 1994, an estimated 3,500 CRADAs had been negotiated between federal laboratories and private companies. DOE accounted for the largest share, nearly a third, of all active CRADAs in 1994, followed by the Department of Commerce/NIST, which claimed another 12 percent (Table 2.16).

Limited data on the distribution of DOE CRADAs across technology/mission areas show that two categories, "manufacturing" and "advanced materials and instrumentation," account for the largest shares (18 percent in each case) of the total. These are followed by CRADAs in the field of "energy" (16 percent), "information and communication" (16 percent), "pollution minimization and remediation" (12 percent), "aerospace and transportation" (8 percent), "biotechnology and life sciences" (6 percent), and all other technical fields (8 percent).[60]

The CRADA has a number of advantages over other types of cooperative R&D agreements that were used prior to its inception. Foremost among these is the authority CRADAs give participating laboratories to protect any intellectual property relevant to the agreement from disclosure under the Freedom of Information Act. CRADAs constitute the only mechanism by which the federal government can define the disposition of intellectual property rights in advance. In addition, CRADAs provide authority for laboratories to contribute staff and equipment to a project undertaken with a private-sector partner. Participating firms can contribute staff, equipment, and funds (which can be transferred to a laboratory for CRADA-related activities), but laboratories cannot transfer CRADA funds to a private-sector partner.

CRADAs are intended to benefit the laboratory's own mission as well as the private-sector partner, and therefore they are generally funded from the laboratory's R&D budget. The DOE is the only federal agency that has received CRADA funding as an individual line item in its appropriations, an amount that has averaged $300 million per annum in recent years. Congress eliminated the line item from DOE's fiscal 1996 appropriations.

The federal laboratories have a mixed record implementing CRADAs. NIST, USDA, and some DOD laboratories have been able to execute CRADAs within a relatively short period of time, sometimes weeks. Initially, DOE had considerable difficulty executing CRADAs in a timely manner, with some agreements taking more than a year to be approved. In the past few years, DOE has significantly reduced the average processing time for CRADAs.[61]

Start-Up or Spin-Off Companies

Federal laboratories, led by DOE, have spun off at least 250 new technology-driven companies in the past 15 years.[62] Many of these companies have been started by former employees in some cases commercializing ideas or technologies developed in their laboratory research. Most of these spin-offs have involved no formal transfer of intellectual property—employee-entrepreneurs simply took ideas or know-how with them when they left the lab.[63] These employee start-ups usually operated on a shoestring budget. Many failed, but a few have succeeded and flourished. Among the better known are Amtech (spun out of Los Alamos National Laboratory) and EG&G-ORTEC (spun out of Oak Ridge National Laboratory). In general, federal laboratories have been slow to provide support to their spin-off companies, and some have taken a hands-off approach to start-ups to avoid problems related to conflict of interest and fairness of access to public technologies. This attitude has recently begun to change in some agencies, perhaps most notably within NASA.[64]

Information Dissemination

Large, comprehensive databases of federally developed technologies have been created during the last several decades. The largest, Federal Research in Progress (FEDRIP), is maintained by the Department of Commerce's National Technical Information Service (NTIS). FEDRIP provides information about more than 150,000 research projects under way and contains summaries of U.S. and foreign government-funded work in progress, as well as information about which agencies are providing funding. The database focuses on projects in the health and life sciences, physical sciences, agriculture, and engineering. Each FEDRIP entry describes a research project, its objectives and, when available, its funding and intermediate findings. Project descriptions include project title, starting date, principal investigator, performing and sponsoring organization, and a detailed abstract. Although FEDRIP's coverage is not complete, the scope is very broad.

Other databases or collections of databases that facilitate access to technologies developed by federal laboratories have been created and maintained by federally sponsored organizations and, to a lesser extent, by universities and the private sector. NASA operates the largest network of technology-referral and information organizations, consisting of the National Technology Transfer Center (NTTC), located in Wheeling, West Virginia, and six regional technology transfer centers (RTTCs), located across the United States. While the centers are funded by NASA, their technology focus spans the entire federal laboratory system. The NTTC's mission is to create a user-friendly system with a single point of contact that will permit business, government, and the general public to locate information on science and technology in the federal laboratory system and beyond. The NTTC has acquired and created a large set of databases, with more than a million records, that list the technologies and expertise available in federal laboratories and elsewhere. The center's technology agents use these databases to provide information to clients who reach the NTTC through a toll-free telephone number and, more recently, through the Internet and other electronic means.

Since the NTTC's Gateway service and its toll-free number (1-800-678-NTTC) opened for business in October 1992, the Center has received over 10,000 requests for assistance in accessing federal technologies; requests are currently coming in at the rate of over 4,000 per year. Just over 60 percent of the requests have been from small businesses (those with fewer than 100 employees), 9 percent have come from medium-sized business (with 100–499 employees), and the balance have come from large firms (500 or more employees). Requests span a range of technologies. The top five areas are materials science (16 percent); manufacturing technology (9 percent); computers, control and information (9 percent); electrotechnology (8 percent); and environmental pollution and control (7 percent).

The RTTCs were established by NASA in 1992, replacing an older network of industrial application centers. The RTTCs do not build databases as the NTTC does, but rather use existing sources of information to create contacts between federal labs (particularly NASA centers) and private firms. In addition, they engage in a number of other activities to "move technologies to the marketplace." To this end, they facilitate the marketing of technologies developed at NASA centers by helping to develop and then negotiate deals with commercial firms.

RTTCs reported handling over 6,000 requests for technical assistance (database searches, referrals, technical studies, technical problem solving, etc.) in a recent 12-month period. During that same time, RTTCs handled over 4,000 requests for commercial assistance (business plans, technology and market assessments, capital sourcing, and consortium building), assisted with over 100 technology-licensing deals, and played a role in over 200 technology-partnership agreements.

In addition to the NASA-funded technology transfer information system, the Federal Laboratory Consortium provides a similar function. The consortium operates a locator service, which is designed to link potential technology users in industry, government, or academia with federal laboratory capabilities. The FLC has built and maintains a database of federal laboratory expertise and uses a network of FLC regional and laboratory representatives to assist it in pinpointing expertise and technologies. During 1994, the locator system handled 832 requests, of which 296 sought advice about the technology transfer process and 531 sought technical information from labs. Small businesses were the principal user of the locator system (responsible for 81 percent of requests in 1994), followed by the federal government, including other federal labs (7.5 percent), and large business (5 percent). The consortium is supported by the federal government according to a formula that gives it a small percentage of federal funds appropriated for R&D.

Finally, a number of individual federal agencies and laboratories have established databases and retrieval systems for their laboratories' technologies. DOD, DOE, NASA, NIH, and USDA are the principal agencies publishing databases of their own technologies. The DOE has recently created the Technology Information Network, which makes its technology database available on an Internet home page, <http://www.dtin.doe.gov>.

Technical Assistance

Most technical assistance provided by federal laboratories is aimed at solving technical problems encountered by mostly small companies located in the same state or region as the laboratory. Technical assistance is often mediated by an intermediary, such as one of the many state technical extension programs or, at the national level, by the FLC or NTTC. NIST's Manufacturing Extension Program centers frequently use federal laboratories as resources to provide assistance to businesses in their region. DOE estimates that its labs are involved in several thousand technical assistance projects each year.

Exchange Programs

Exchange programs, as the term implies, promote the short- or long-term exchange of research personnel between federal laboratories and private firms.

Exchange programs have considerable technology transfer potential since they involve the most effective transfer agent, people. However, they are not much used. The number of researchers exchanged for significant periods between industry and federal laboratories is measured in the hundreds. Although there is an impetus to increase the number and size of exchange programs, there is a natural reluctance on the part of both management and staff to do so. Managers are reluctant to lose their best staff, and staff are fearful that a prolonged absence will hurt their chance of advancement in their home organization.

Work for Others and Use of Facilities

Reimbursable work for others consists of R&D services performed by a federal laboratory on behalf of a customer, in most cases another federal lab or agency. Other customers include state and local governments and private firms. Work for others is generally undertaken because the customer wants to take advantage of highly specialized scientific expertise and facilities in a federal laboratory that cannot be otherwise obtained.

The use of a federal research facility may or may not be reimbursed by the customer. A small but growing level of federal laboratory activity is devoted to cost-reimbursed projects. In some cases, the work is performed by the customer's staff, in others, by the federal laboratory scientists and engineers. In recent years, DOE's success with CRADAs has been attributable to a growing volume of industry-funded "work-for-others" business in its laboratories. Between 1992 and 1996, the volume of total "work-for-others" performed by DOE laboratories fluctuated between $1.1 and $2 billion. Yet over the 5-year period, the share of total "work-for-others" accounted for by non-federal entities (predominantly private companies) nearly doubled, from 6.5 percent to 12 percent.[65]

Consulting

Consulting is an important form of technology transfer between universities and industry. However, consulting is not as common between federal labs and industry. Most federal laboratory scientists are federal employees and are prohibited from working outside the government. To a lesser extent, the same is true for scientists working in GOCO laboratories. Some consulting does take place between firms and federal laboratories, but in these instances consulting staff continue to function as laboratory employees without extra compensation. Such arrangements are very similar to work-for-others arrangements. GOCO staff employees are generally allowed to engage in remunerated consulting activities with other public and private sector clients as long as they do it on their own time.

Collegial Interchange, Workshops, and Conferences

Collegial interchanges are important in fostering person-to-person contact, the primary channel through which technology transfer occurs. This mechanism is often said to be, in the aggregate, the most important form of technology transfer between federal labs and industry. These interactions involve informal contacts among researchers from the two sectors at a wide range of events, including technical workshops, laboratory tours, and conferences. The resulting exchange of information and technology can have considerable value for all sides, although that value is exceedingly difficult to describe, particularly with quantitative measures. Very often, these informal contacts lead to other, more formal, technology transfer activities.

The Special Case of Small and Medium-Sized Firms

Federal technology transfer legislation has been particularly supportive of increased interaction between labs and small and medium-sized enterprises (SMEs). The 1986 FTTA requires federal laboratory directors to give preference to small businesses when choosing CRADA partners or when licensing patents. The growth of technology transfer intermediaries, such as the National Technology Transfer Center and state economic development and technology assistance networks, has provided a new way for SMEs to find and access federal laboratory resources. Based on the recent experience of several agencies, it appears that these and other mechanisms have make it possible for some federal laboratories to serve SMEs effectively. As of 1994, 40 percent of NIST's 250 CRADAs were with small companies. DOE estimates that its national laboratories alone engage SMEs in several thousand technical assistance projects per year.[66]

LIMITS TO FEDERAL LABORATORY TECHNOLOGY TRANSFER

There are a number of factors that make technology transfer more difficult for federal laboratories than for private-sector organizations. Several of these are described below.

National Security

Particularly for defense laboratories, the classification of some technologies limits possibilities for technology transfer. Nevertheless, the procedures now in place for the commercialization of technologies from classified research programs, while lengthy, are reasonably well understood. Moreover, with the end of the Cold War, national-security concerns pose only a minor constraint to technology transfer for the defense laboratories.

Economic Performance and Reciprocity Requirements

The 1986 FTTA specifies that in the negotiation of CRADAs and licensing agreements, preference be given to business units located in the United States, and, when CRADAs are negotiated with foreign-owned companies, the home governments of these firms must permit American firms to participate in cooperative R&D programs in their country. The 1989 National Competitiveness Technology Transfer Act tightens these provisions and requires that any participating private firm, U.S. or foreign-owned, commit to undertake in the United States further design, development, and "substantial" manufacturing of products and processes embodying intellectual property resulting from the CRADA. If a firm is foreign-owned, additional reciprocity requirements are imposed. Many U.S. multinationals have voiced strenuous objections to the "substantial" U.S. manufacturing requirement. In response, DOE has relaxed these requirements somewhat in its CRADA agreements.[67]

Fairness of Opportunity

FTTA gives federal laboratories authority to negotiate terms and conditions of CRADAs. However, Congress also enacted a law requiring fair, or equal, access to federal technology transfer programs. Procedures for implementing this requirement have never been spelled out, and labs have adopted a wide range of practices to assure fairness. Although the selection of CRADA partners is not required by law to be competitive, particularly if a firm initiates the project, many laboratory technology transfer offices engage in extensive publicity and advertising prior to deciding on partners for licenses or CRADAs. While such publicity does increase awareness of federal laboratories' cooperative research activities, it adds additional delay to the CRADA negotiation process. Moreover, publicity sometimes discourages firms from engaging in partnerships, since they may not want their research aims to become public through a response to a public solicitation.

Conflict of Interest

In the federal government, conflict-of-interest laws and regulations developed over the past 20 years have placed limits on certain types of employee activities. While important, these limits have slowed the process of technology transfer, since administrators must focus more on conflict-of-interest issues at the expense of more productive activities, such as technology marketing.

Measuring the Performance of Federal Laboratory Technology Transfer

Many in government are asking whether increased technology transfer activities and increased spending on technology programs have had a significant impact on the economy and whether that impact was high relative to the federal dollars spent. The question is important but difficult, perhaps impossible, to answer fully.

MEASUREMENT OF ACTIVITIES UNDER THE STEVENSON-WYDLER
AND BAYH-DOLE ACTS

On the surface, there is evidence that Bayh-Dole and Stevenson-Wydler have produced the desired results. The number of CRADAs and license agreements, as well as royalties resulting from such agreements, are increasing steadily (Figures 2.15 and 2.16). Several studies support the conclusion of preliminary success, with one major survey of large R&D-intensive firms concluding that "the tech transfer legislation has 'worked' in the sense that companies are increasingly tapping the knowledge, expertise and facilities in federal labs" (Roessner, 1993). Nevertheless, aside from demonstrating an increase in the number of transactions between companies and federal laboratories, the activity data say little about the economic or social impact of federal technology transfer activity. To probe the economic impacts of technology transfer and other technology programs requires more sophisticated collection, organization, and interpretation of data.[68]

Over the past decade, a number of studies have attempted to find ways to detect economic benefits resulting from federally funded basic research. In addition, a few private firms have undertaken efforts to determine the impact of and returns from corporate R&D activities and internal technology transfer. While these two categories of investigation look at phenomena different from technology transfer, they nonetheless use techniques and produce a body of knowledge that are useful to the study of the economic value of federal laboratory technology transfer.

Anecdotal Evidence

Much of the early evidence of effectiveness of federal laboratories' technology transfer activities has been anecdotal. While anecdotes do not permit one to engage in cost/benefit or other systematic analyses, they do provide a form of "existence proof" that most people can relate to. Anecdotal evidence is particularly useful in a political context, especially if it can be focused to support the interests of individual politicians.

All of the early evidence for the benefits from federal technology transfer came in the form of anecdotes about spin-offs (primarily from work by NASA labs) (Doctors, 1971). Much of the conventional wisdom attributing specific space-age products (e.g., Teflon and Tang) to NASA is erroneous, but there were some significant outcomes in terms of individual products and, more importantly, the development of new technology areas. For example, the impact of NASA projects on electronics miniaturization (now widely diffused throughout the commercial sector) cannot be underestimated. In the past, anecdotes of technology transfer successes were common features of federal agencies' publicity about their technology transfer activities. However, recently this type of evidence has taken a back seat to the promise of more systematic analysis.

Formal Evaluations

In the past 3 years, federal agencies have begun to establish measurement and evaluation systems for their technology transfer programs. In addition, the Office of Management and Budget has been collecting basic data on technology transfer for several years as part of the federal budget process.[69]

From 1993 to 1995, a federal interagency working group chaired by the Department of Commerce met to define data elements describing technology transfer activities in most federal R&D agencies. However, the working group ceased its activities in mid-1995, and no data were ever collected using the draft data format it created. Such data could have lead to the creation of a government-wide technology transfer database that would have permitted government-wide identification and analysis of the activities under the Stevenson-Wydler and Bayh-Dole Acts. The working group's data elements were activity measures that would have provided only inferential evidence of economic impacts. However, the collection of activity measures constituted an essential beginning. Without comprehensive data on individual technology transfer claims or "events" as the working group called them, it is extremely difficult to design, collect, and interpret additional information that gauges impact.

A number of individual agencies and laboratories are collecting information about and evaluating the impacts of their technology transfer programs. As part of this effort, some agencies are beginning to survey their laboratories' private-sector R&D partners to gauge their satisfaction with the technology transfer programs. Very few of the results of these activities have been published, however.

These surveys also shed light on the relative importance R&D-intensive firms assign to the various technology transfer mechanisms employed by federal laboratories. Most survey respondents ranked cooperative R&D as the most likely means of achieving promising payoffs from interactions with federal labs. This was followed by workshops and seminars, visits to labs, technical consultation, contract research, and use of facilities, in that order. Licensing and employee exchange were ranked as least likely to have future payoffs (Roessner, 1993).

A 1995 study (Bozeman et al.) sponsored by NSF surveyed 219 private companies to get their views on the benefits and costs of working with federal laboratories. The responses covered a number of technology transfer mechanisms, including CRADAs, cooperative R&D other than CRADAs, technical assistance, and, to a lesser extent, licensing, use of facilities, and exchanges of research personnel.

Private-sector R&D partners reported mostly positive results from their experiences with federal laboratories. Twenty-two percent reported that the interaction had already led to a new product, 38 percent said that a product was under development, and 24 percent said that a product was improved. Overall satisfaction with the interactions was generally high, with 89 percent reporting that the interaction was a good use of company resources, and many responding companies were repeat customers.

Responding companies were asked to estimate in dollar terms the costs and benefits of their federal laboratory interaction. Firms reporting both costs and benefits received an average benefit of $1.8 million and experienced an average cost of $544,000, a three-to-one return on investment. These averages are highly skewed by a few big winners, however. Interestingly, the big winners were disproportionately basic research projects, demonstrating once again the high-risk/high-payoff nature of this type of inquiry. The survey found that the net impact of the interactions on job creation/retention was essentially zero, although most of the big winners, not surprisingly, reported job creation.

The Future of Federal Laboratory Technology Transfer

A number of agencies have ongoing processes to review and revise their technology transfer programs in response to stimulus from internal reviews, the administration, and Congress. This continuing analysis is likely to persist for some time, and the outcome of future reviews will depend heavily on the success of efforts to evaluate technology transfer. The most significant reassessment activities are reviewed below.

CONGRESS

The tenor of Congressional opinion about technology programs changed dramatically with the Republican victory in the November 1994 elections and their majority in the 104th Congress. Once enjoying modest bipartisan support, federal technology programs such as the Advanced Technology Program and the Pentagon's Technology Reinvestment Project came under attack. In spite of their public hostility toward the ATP and TRP, the Republican majority did not express significant opposition to federal laboratory technology transfer activities, including licensing and CRADAs.[70] Bipartisan legislation (H.R. 2196) sponsored by Rep. Connie Morella (R-Md.) and Sen. Jay Rockefeller (D-W.V.), which amended the Stevenson-Wydler Act to enhance the incentives for commercializing technologies developed at government labs, passed the 104th Congress and was praised by both Republican and Democratic leaders.

NIH and USDA

The technology transfer programs at NIH and USDA remain relatively stable. There have been some administrative changes at NIH to improve the processing of technology transfer agreements, and measured reductions, which have a minor impact on technology transfer, are occurring in the ARS laboratory structure.

DEPARTMENT OF DEFENSE

The Defense Authorization Act of 1992 (P.L. 102-484) required the secretary of defense to encourage technology transfer between the DOD laboratories and research centers and federal agencies, state and local governments, colleges and universities, and the private sector. That same legislation created the Office of Technology Transition in the Directorate of Defense Research and Engineering to monitor and encourage technology transfer from defense laboratories. The legislation also created the Federal Defense Laboratory Diversification Program to encourage greater cooperation between defense laboratories and private industry. In addition, an exhaustive study was conducted to assess the technology transfer potential of all defense laboratories and recommend ways each could promote additional transfer.

In June 1995, the secretary of defense issued a follow-up memorandum stating that domestic technology transfer and dual-use technology development were "integral elements of the Department's pursuit of its national security mission" (Perry, 1995) The memorandum required all R&D elements of the Defense Department to "make domestic technology transfer and dual-use technology development a priority element in the accomplishment of their science and technology missions" and gave the DOD's Office of Technology Transition increased oversight authorities. In a sense, these new DOD activities do little more than assure full implementation of the provisions of the 1986 and 1989 technology transfer acts by the military services and their labs. The very high level impetus provided by the secretary's memorandum should serve to accelerate the process, however.

DEPARTMENT OF ENERGY

The Department of Energy has recently shifted substantially the emphasis in its technology transfer program. The focus now is on activities that support the DOE mission, particularly those with potential for "spin on." In a September 1994 policy statement and review of technology transfer activities called "Our Commitment to Change," DOE focused almost exclusively on partnerships as a means to enhance U.S. industrial competitiveness (U.S. Department of Energy, 1994).

One year later, however, the DOE's statement of objectives and principles demonstrated a significant turnabout, noting that its first objective for technology partnerships was "to contribute to the Department's missions by leveraging the Department's resources with those of others" (U.S. Department of Energy, 1995). The statement also said that DOE would adhere to a series of principles, the first of which is that "All cost-shared DOE cooperative research and development partnerships will support DOE missions."[71] One catalyst for these changes was the so-called Galvin Report (Secretary of Energy Advisory Board, 1995), which recommended that the department no longer pursue industrial partnerships or the development of industrial technologies as a core mission.[72] Another was the election of a Republican-controlled Congress that viewed DOE's industrial competitiveness activities in a less favorable light than its Democratically controlled predecessor. In 1995, Congress eliminated DOE's line-item appropriation for CRADAs at DOE weapons labs, citing excessive bureaucratic rigidity in administering the funds from DOE headquarters (i.e., the objection was not to CRADAS per se). While there is little doubt that DOE, like all other federal R&D performing agencies, will continue to engage in CRADAs by drawing on regular laboratory program funds, the amount of funds available for DOE technology transfer has been reduced by Congress's action.

NASA

In response to a highly critical internal review (Creedon, 1992), NASA has begun to radically restructure its technology transfer programs. Among other things, NASA has committed to increase the commercialization of technologies developed by its private-sector contractors, who perform the majority of NASA's R&D.[73] Indeed, NASA now considers "the Commercial Technology Mission . . . a primary NASA mission, comparable in importance to those in aeronautics and space" (National Aeronautics and Space Administration, 1994). NASA continues to accelerate its efforts to work with industry in order to meet the Administration's goal of devoting 20 percent of its R&D budget to industrial partnerships. NASA is also taking steps to move its technologies to industry through more aggressive use of licensing. Finally, NASA has formed partnerships to establish three business incubators near NASA centers to provide assistance to entrepreneurs spinning off NASA technologies.

In summary, most federal agencies will continue to conduct technology transfer using mechanisms that have been employed since the 1980s and before. NIH and USDA, in particular, are likely to continue what are considered to be successful programs. Downsizing and reduced congressional appropriations are likely to limit technology transfer in the DOE weapons laboratories, and downsizing at the DOD laboratories may have the same effect, although the department is attempting to move the defense laboratories in the opposite direction (i.e., placing increased emphasis on dual-use research and technology development). Only NASA is involved in a major process of change in its technology transfer program. The program is ambitious and seeks to make NASA a leading player in federal technology transfer activities.

Conclusions

Until the 1980s, technology transfer from federal laboratories was tightly linked to the primary missions of the sponsoring federal agencies and, for the most part, involved closely related industries or regions. The Stevenson-Wydler Technology Transfer Act of 1980 and subsequent amendments throughout the decade provided a new technology transfer mandate. This new mandate encouraged licensing of technologies developed in government labs and allowed entrepreneurially minded lab directors (of some agencies) more discretion to enter into collaborative research with industry in fields less closely related to their agencies' traditional public missions.

For civilian agency laboratories that have long engaged in technology transfer to private industry as an integral part of their missions, the implications of these changes have not been as profound as they have been for defense laboratories. However, for some DOD and DOE defense laboratories faced with surplus capacity at the end of the Cold War, this new mandate seemed to offer a justification for significantly reorienting some of their activities and technological capabilities to serve commercial industries and thereby avoid downsizing. Questions surrounding the changing role of these latter institutions, particularly the three very large DOE defense laboratories, have fueled much of the debate over what the proper division of labor should be among federal laboratories, academic institutions, private industry, and other institutions in a U.S. innovation system widely perceived to be in a period of rapid and significant change.

Clearly, these laboratories are equipped with large numbers of highly trained personnel and in many cases are unique facilities housing valuable equipment. Some of these labs are at the forefront in areas of basic and applied research that are relevant to both public missions and private industry and already have successfully engaged in technology transfer to private companies. Many federal labs, however, have traditionally performed R&D that has little direct relevance to most civilian industries, and have had limited experience dealing with private companies as clients. For the most part (there are some notable exceptions), federal laboratories are more bureaucratically encumbered than other major R&D performers by virtue of their continuing public-mission focus and management structures (Secretary of Energy Advisory Board, 1995). At a time of increasingly constrained public R&D budgets and shifting national R&D needs, maintaining these laboratories at or near their present size may not be the most effective way to meet the nation's new public R&D priorities (i.e., other publicly or privately funded R&D performers may be better equipped to take these on). Given the large size of some of these facilities and their resulting economic and employment importance to their host states, however, the political impediments to their downsizing are likely to remain formidable. Ultimately, the nation will have to assess the relative strengths and weaknesses of its federal laboratories and other sectors of the U.S. R&D enterprise and seek to define the most productive role for each sector, while attending carefully to the interfaces between sectors and the potential for greater collaboration between them.[74]

At the same time, while there is no consensus concerning the appropriateness or cost-effectiveness of more federal laboratories getting into the business of technology transfer to private industry, all federal agencies with significant R&D portfolios are coming to rely more and more heavily on private-sector R&D capabilities and commercial components to advance their public missions. In many technology areas, commercial companies operating in competitive international markets are increasingly setting the pace of technological advance in areas critical to federal R&D missions. Yet, some federal agencies, particularly in the areas of defense and space, admit freely that they have much to learn about effectively harnessing private-sector technology and R&D to advance agency and national objectives.[75] Moreover, there is agreement in many quarters of the nation's R&D enterprise that many federal laboratories and their parent agencies have yet to develop effective mechanisms for ensuring appropriate industrial input into the formulation, execution, and evaluation of federal laboratory R&D and technology transfer activities (Secretary of Energy Advisory Board, 1995). In addition, the federal laboratories are often faulted for being unable to carry out technology transfer and partnership activities in a more businesslike and less bureaucratic manner.

Finally, from the standpoint of potential industry partners, the reliability of the federal laboratories as partners has been called into question by recent dramatic cuts in technology transfer funds for the Department of Energy laboratories. These cuts have been mostly limited to the three DOE weapons laboratories, forcing these labs to scale back and cancel cooperative R&D activities (including the high profile PNGV and Amtex partnerships). Nonetheless, there has been a strong spillover effect, strengthening the underlying perception that the entire federal laboratory system is a partner of uncertain reliability.

TECHNOLOGY TRANSFER BY PRIVATELY HELD, NONACADEMIC ORGANIZATIONS*

Overview

This section examines the scope and nature of technology transfer activities performed by a diverse population of privately held, nonacademic organizations (i.e., entities whose R&D and technology transfer activities fall outside those of the three major sectors of the U.S. technology transfer enterprise).[76] This "fourth sector" of the enterprise consists primarily of two types of institutions: those that transfer technology they have had a hand in developing and those that transfer or facilitate the transfer of technology developed by others. Included in the first group are independent and affiliated[77] R&D institutes and R&D consortia, predominantly nonprofit organizations. The second group includes providers of technology transfer referrals and information; technology business incubators and research parks; technology brokers, technology transfer consultants, law firms, and technology transfer conference organizers; and technical/professional associations, societies, and academies.

The R&D activities of privately held, nonacademic organizations, measured in dollar terms, are relatively small compared with the investments of other players in the R&D enterprise. These organizations perform somewhere between $8 billion and $12 billion worth of R&D annually, or about half the amount performed by academic institutions or federal laboratories and less than one-tenth of

---

*This section draws extensively on a background paper prepared by Robert K. Carr and Christopher T. Hill (1995) for the U.S. delegation to the binational panel.

the R&D conducted by U.S. industry.[78] As a group, fourth-sector institutions also account for significantly fewer patents and royalties than do any of the other three R&D-performing sectors.

However, aggregate quantitative measures understate the overall importance of fourth-sector organizations to U.S. technology transfer enterprise. First, many of these organizations perform significant R&D in several critical sectors, particularly in health and medical science. Second, many fourth-sector institutions perform strategic technology bridging and assistive technology transfer functions, often facilitating technology transfer among the three major R&D performing sectors. The panel believes that these services could become increasingly important to the nation's technology transfer enterprise as more and more U.S. firms are compelled to seek and use technology developed beyond their institutional boundaries in order to compete effectively in international markets.

Organizations That Create and Transfer Technology

The organizations in this category perform in-house R&D, contract for R&D, or perform contracted or cooperative R&D and transfer primarily technology that they have generated internally. These include independent R&D institutes, affiliated R&D institutes, and consortia or other private nonacademic organizations that conduct R&D and technology transfer. The National Science Foundation estimates that nonprofit institutes—more or less the same set as independent plus affiliated R&D institutes—performed $5.2 billion worth of R&D in 1994.[79] Of this, $2.9 billion came from the federal government, $1.5 from the nonprofit sector (mostly philanthropic foundations), and $800 million from industry. Over half of the 100 largest nonprofit institutes receiving federal funds are focused on research in the health and life sciences. Defense-related research also figures prominently among these institutions.

As of 1992, the 10 largest nonprofit recipients of federal funds were the Universities Research Association, Inc.[80] ($468 million) and SEMATECH ($98 million), both research consortia; Massachusetts General Hospital ($91 million), ITT Research Institute ($87 million), Brigham and Women's Hospital ($82 million), South Carolina Research Authority ($71 million), and Scripps Clinic & Research Foundation ($69 million), all affiliated R&D institutes; SRI International ($63 million) and Battelle Memorial Institute ($63 million), both independent R&D institutes; and the National Research Council, Transportation Research Board, which administers the Strategic Highway Research Program for the Department of Transportation ($76 million).

INDEPENDENT R&D INSTITUTES

In terms of their total investment in R&D, independent R&D institutes, including independent R&D laboratories, private research hospitals, and independent medical research centers, constitute one of the largest elements of the fourth sector. Research hospitals and medical research centers comprise almost half of the group. Information concerning the R&D activities of this group is relatively abundant compared with that concerning the other categories studied.

Most independent R&D institutes are quite small. Only 89 institutes listed in the Gale's Research Centers Directory (Gale Research, 1996) had a staff of more than 100 or an annual research budget of over $10 million. As Table 2.17 illustrates, more than half of the 85 large "hard science" R&D institutes are focused on research in the medical and health sciences, with most of the remaining institutes equally divided between those focusing on multidisciplinary sciences, the biological and environmental sciences, and engineering and technology research.

Data on the R&D budgets of these 85 large institutes are incomplete. The 35 institutes that provided budget data spent a total of $1.62 billion on research in 1994. Among the very largest independent institutes (measured in terms of research budgets or total staff) are: Midwest Research Institution, SRI International Inc., Southwest Research Institute, Research Triangle Institute, RAND Corporation, MITRE Corporation, Memorial Sloan-Kettering Cancer Center, Fred Hutchinson Cancer Research Institute, Fox Chase Cancer Center, Dana-Farber Cancer Institute, the Howard Hughes Medical Institute, and the World Wildlife Fund (Gale Research, 1996).

In 1992, the top 5 recipients of federal R&D funds were SRI International (with $63 million in federal funds), Battelle Memorial Institute ($63 million), Fred Hutchinson Cancer Research Institute ($56 million), Dana-Farber Cancer Research Institute ($50 million), and Research Triangle Institute ($50 million).

The technology transfer activities of independent R&D institutes vary widely. Some independent institutes, particularly those focused on health and medical science, appear to perform only basic research. The results of this research are

TABLE 2.17 Distribution of 85 Large Independent R&D Institutes by Research Focus, 1994

Research Focus	Number of Institutes	Percent of Total
Agriculture, Food and Veterinary Science	2	3
Biological and Environmental Sciences	12	14
Health and Medical Sciences	43	51
Astronomy and Space Sciences	1	1
Computers and Mathematics	1	1
Engineering and Technology	12	14
Physical and Earth Sciences	0	0
Multidisciplinary Institutes	14	16

SOURCE: Gale Research (1996).

generally disseminated via traditional paths such as publications, meetings, and sharing among colleagues.

Other independent R&D institutes, particularly the large ones, do contract work for clients and transfer the majority of the technologies they develop to them. In addition, internally developed intellectual properties may be licensed to a wider market by the institutes or the research client. Independent R&D institutes also carry out technology transfer through other mechanisms, for example, by sharing information at conferences, providing technical assistance, and employing unique R&D facilities and capabilities.

In addition to transferring their own internally generated technologies, some independent R&D institutes transfer technology developed by other organizations. For example, Research Triangle Institute has contracts with the Ballistic Missile Defense Organization (BMDO) to facilitate the transfer of technologies developed in BMDO's R&D programs to private industry.

As is the case with R&D, comprehensive sources of data that could be used to measure the technology transfer activities of these institutes are scarce. The 26 independent and affiliated R&D institutes that responded to a survey of the Association of University Technology Managers (AUTM) (1994), received roughly $74 million in royalty income in 1993 on a total of 409 licenses. This compares with over $242 million in royalties received on 3,413 licenses by 117 universities reporting to AUTM.

The six largest independent, nonprofit, applied R&D/engineering institutes in the United States are Battelle Memorial Institute, Midwest Research Institute, Research Triangle Institute, Southern Research Institute, Southwest Research Institute (SwRI), and SRI International. Originally established to provide R&D and technical assistance to industries within a defined, local, or regional geographical

TABLE 2.18 The Six Largest Independent, Nonprofit, Applied R&D Institutes in the United States

Name of Institution	Date of Incorporation	Number of Employees (FY 1994)	Source of R&D Funds (FY 1994) (Government/Industry [%])
Battelle Memorial Institute (Columbus only)	1929	2,599	78/22
Southwest Research Institute	1947	2,400	42/58
SRI International (Menlo Park only)	1946	1,900	60/40
Research Triangle Institute	1958	1,450	84/16
Midwest Research Institute	1944	1,350	73/27
Southern Research Institute	1941	477	75/25

NOTE: Data as of May 1995.

SOURCE: Southwest Research Institute, unpublished data, 1995.

area, these institutes now have clients throughout the world. Although all six perform some contract research for private companies, four rely on government contracts for more than 70 percent of their business. Only one, SwRI, receives a majority of its contract work from private-sector clients (Table 2.18).

These six institutes vary considerably in size. Each has its own peculiar multidisciplinary research focus, organizational structure, and ways of doing business. For example, SwRI, which conducts research in over 28 different fields from automation through fluid dynamics and hydraulics, has a special organizational structure that allows other independent or federal-government-owned contractor-operated labs to be integrated into SwRI as separate departments. SwRI claims no patent rights on its output. Rather, the rights are always given to the clients.[81]

Affiliated R&D Institutes

Most of the organizations in this group are affiliated with universities, research hospitals, or other medical research institutes. It is difficult to estimate the total R&D volume of these organizations, since so few of them report their budgets separately from those of their parent institutions.

As in the case of independent institutes, most affiliated institutes are small. Only 35 have a staff of more than 100 or an annual research budget of over $10 million (Gale Research, 1996). Together, these 35 institutes spent a total of $250.7 million on R&D in 1994. The research focus of these large institutes is shown in Table 2.19.

Here again, half of the large affiliated R&D institutes are focused on medical and health sciences research. Among the very large affiliated institutes are IIT Research Institute (IITRI), the H. Lee Moffit Cancer Center and Research Institute, and the St. Jude Children's Research Hospital. IITRI is a separately incorporated nonprofit research organization affiliated with the Illinois Institute of Tech-

TABLE 2.19 Distribution of 35 Large Affiliated R&D Institutes by Research Focus, 1994.

Research Focus	Number of Institutes	Percent of Total
Agriculture, food, and veterinary science	0	0
Biological and environmental sciences	4	11
Health and medical sciences	16	44
Astronomy and space sciences	2	5
Computers and mathematics	1	3
Engineering and technology	6	17
Physical and earth sciences	1	3
Multidisciplinary institutes	6	17

SOURCE: Gale Research (1996).

nology that conducts applied R&D and engineering research. Eighty-five percent of its work is sponsored by government agencies and 15 percent by industry.

CONSORTIA AND RELATED ORGANIZATIONS

For the purposes of this report, R&D consortia are defined as groupings of two or more organizations that fund or perform collaborative R&D. R &D consortia may be permanent organizations consisting of institutions that exist primarily for some other purpose but also perform R&D on behalf of their members (e.g., trade organizations) or created specifically to engage in R&D on behalf of the members (e.g., SEMATECH). Consortia may also be temporary organizations created for a specific R&D project or projects that dissolve when the project is terminated.

Consortia can be formal partnerships or less-formal groupings. The nonprofit corporation is a common type, although for-profit consortia exist as well. Consortia may perform R&D within their own facilities, coordinate R&D done in some or all members' facilities, contract to a nonmember to perform R&D, or engage in some combination of all three activities. In addition to consortia focused on industrial needs, there are a number of consortia that conduct basic research. Not surprisingly, such consortia have a large proportion of university members.

Consortia are created for many reasons, including the desire to achieve efficiencies from shared facilities and shared costs, to pool scarce talent, to increase synergy within or diversify a participant's technology portfolio, to facilitate standards setting, to market products, or to foster exchange of precompetitive R&D results. It is important to note that formal consortia have become a part of the U.S. R&D enterprise largely because they were deemed to have been successful elsewhere, especially in Japan.[82] However, Japan may have needed consortia more than the United States because of the different nature of informal technology transfer in the two countries. In the United States, labor, particularly high-tech labor, is highly mobile and much technology moves between firms by that route. In countries where labor is less mobile (Japan being the extreme example), other mechanisms may be required to foster technology flow.

The U.S. federal government has encouraged the formation of consortia in recent years through changes in law and provision of financial incentives. Prior to 1984, firms participating in collaborative R&D arrangements were exposed to the possibility of treble damages should the arrangement be judged in violation of U.S. antitrust laws. Not surprisingly, this legal climate dampened the enthusiasm of many firms for collaborative R&D efforts. In 1984, Congress passed the National Cooperative Research Act (NCRA, P.L. 98-462), which removed the threat of treble damages for consortia that registered with the DOJ. Congress extended this protection to joint production ventures in 1993 with the passage of the National Cooperative Production Amendments (P.L. 103-42).

In addition to the legislative changes, the federal government has also used financial incentives to encourage the formation of research consortia. SEMATECH (Box 6) was established with support from the Defense Advanced Research Projects Agency (DARPA, now ARPA) of approximately $100 million per year. Federal funding of SEMATECH was discontinued in 1996 by mutual agreement of the consortium and DARPA. The federal government is a financial and technical partner in a number of other consortia as well, such as the National Center for Manufacturing Sciences and the Gas Research Institute. However, most federal contributions to consortia are less than the $100 million invested in SEMATECH.

The Technology Reinvestment Project (TRP) and the Advanced Technology Program (ATP) encourage the formation of consortia among organizations submitting proposals to these programs.[84] In addition, some agencies that engage in CRADAs have begun to emphasize working with consortia, some of which have been formed expressly for that purpose. This is particularly true of the Department of Energy (DOE), which has the largest CRADA program in the government.

Between January 1, 1985 and December 31, 1995, 575 separate JRVs (joint research ventures), involving a total of 9,136 entities or "members," had been registered with the DOJ (Figure 2.17). Research on JRV findings indicate that most JRV members (86 percent) are profitmaking companies. Private nonprofit organizations including colleges and universities represented 10 percent of memberships, and government agencies and organizations constituted 4 percent of JRV members. About one-third of the members of JRVs are foreign based. Over the 10-year period since passage of the NCRA, the average number of members in a JRV has been 15.9. As of 1995, 30 percent of all registered consortia had only 2 members, 45 percent had more than 5 members, and nearly 13 percent had over 20 members. Participation in JRVs is highly concentrated. Whereas more than two-thirds of all identified JRV members (roughly 8,000 of the total 9,136 entities) have participated in only one JRV, 28 entities have participated in 21–50 JRVs, and 10 entities were involved in more than 50 JRVs each (Vonortas, 1996).

Most research performed by JRVs has been process oriented. With respect to technology focus, the largest single group of consortia was in telecommunications (22.8 percent), followed by environmental technologies (9.7 percent), advanced materials (9.2 percent), energy (8.7 percent), transportation (7.7 percent), software (6.8 percent), chemicals (6.6 percent), and 10 other technology areas with between 4.7 and 0.5 percent (Table 2.20). Few registered consortia are engaged in research in areas where intellectual property rights are well enforced, for example, biotechnology, pharmaceuticals, and medical equipment.[85] Similarly, defense-related research has received attention from only a small number of JRVs (National Science Board, 1996).

Data on the total volume of resources invested in these consortia are not available. The amount of resources devoted to collaborative R&D in the United

> **BOX 6**
> **SEMATECH**
>
> SEMATECH is a consortium of 10 U.S. semiconductor manufacturers representing roughly 75 percent of the semiconductor revenue in the United States. The group was formed in 1987 as a cooperative effort between the U.S. Department of Defense and the semiconductor industry. In 1996, the consortium became entirely privately funded. There are roughly 700 employees at SEMATECH, about 200 of whom are assigned from member companies for periods ranging from a few weeks to a few years. The SEMATECH budget, about $200 million per year since 1987, dropped to $135 million in 1996.
>
> SEMATECH programs are focused on semiconductor manufacturing technology, hence SEMATECH's name (SEmiconductor MAnufacturing TECHnology). The largest portion of its budget is spent on programs that relate to equipment improvement. Quite often these programs are carried out at equipment supplier sites or in member company sites. Teams of SEMATECH engineers as well as on-site engineers are involved in these programs.
>
> The focus at SEMATECH has been to take the "transfer" out of technology transfer. This is accomplished by having researchers who have been assigned to SEMATECH for a specific project take newly developed technology back to their own companies. The second method of technology transfer is to perform work on equipment at the site where the equipment is going to be developed and manufactured. This again tends to take the "transfer" out of technology transfer. SEMATECH also produces reports and holds meetings to provide technology-related information to its member companies. In 1994, for example, SEMATECH had over 600 meetings and entertained over 25,000 visitors.
>
> SEMATECH programs have been directed toward the development of precompetitive manufacturing technology. None of the programs at SEMATECH involve work on specific products or specific production processes. This is the competitive arena of the consortium's member companies. Rather, SEMATECH programs are oriented toward generic manufacturing technology.

States ranges from 1.7 percent to 7.3 percent of total U.S. company-financed R&D.[86] The volume of collaborative R&D appears to be increasing. It is important to note that JRVs (consortia and other R&D joint ventures registered with the Department of Justice) represent only a small fraction of total cooperative R&D ventures (Hagedoorn, 1995).

> **BOX 6—Continued**
>
> An important part of these programs is the measurement of results. Earlier in its history, SEMATECH established a return-on-investment (ROI) measurement for each program.[83] In recent years the average ROI has been above 4 for member companies; the range was from about 8 ROI to about 2 ROI. In 1994, a new measurement system, based on user satisfaction and user support, was established. The goal is to have 70 percent of SEMATECH programs receiving 50 percent or higher ratings of customer support and customer satisfaction.
>
> The key elements of the SEMATECH technology transfer approach are:
>
> - having company-assigned employees involved in program definition, operation, and evaluation. These users have the responsibility for returning that technology to their member companies;
> - developing specific metrics for each program. Such metrics are deemed essential for successful technology development. Programs are evaluated quarterly at SEMATECH;
> - stopping what does not work. Usually senior management in the member companies of SEMATECH are more inclined to stop programs. Engineers quite often believe that if the program can continue for only a short time longer, the problems can be solved and the program will be of great value. The involvement of senior management is essential to getting programs stopped;
> - focusing on programs where there is a high return. The users of technology are the best judges of that. Support and satisfaction are two measures; SEMATECH also has experimented with return on investment. There are other metrics that can be used, but these are essential to determining which programs should be continued and which should be stopped.
> - instituting processes for choosing technology development programs that are of interest to the member companies. These processes should be continually evaluated and updated.
>
> SOURCE: W. J. Spencer, Chairman and Chief Executive Officer, SEMATECH.

In general, U.S. consortia have been more successful at achieving research results than at transferring the fruits of their research back to members (and to a lesser extent from the members to the consortium). Several studies of the Microelectronics and Computer Technology Corporation (MCC) have identified technology transfer as the consortium's most serious problem (Gibson and Rogers,

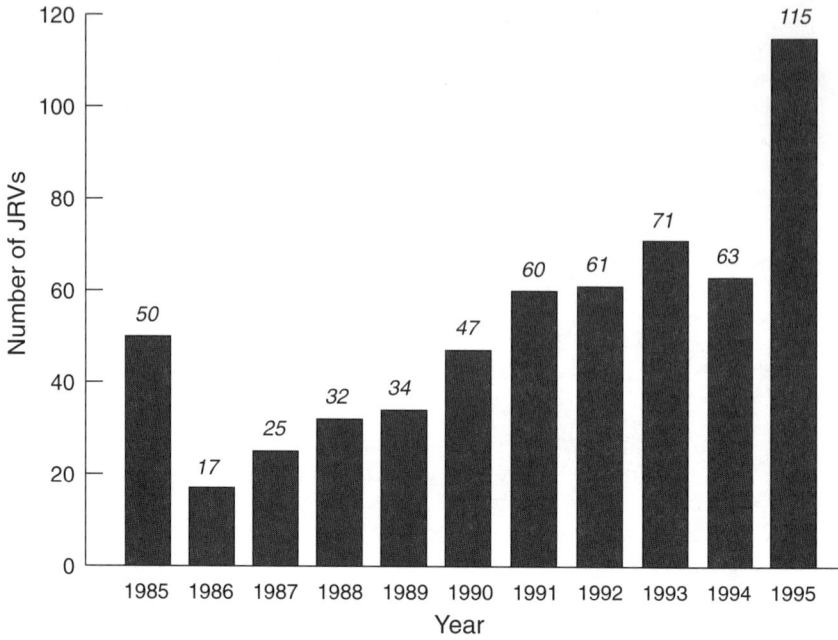

FIGURE 2.17 New joint research venture announcements. SOURCE: Vonortas (1996).

1994; Hill, 1995). Most other consortia also have problems with technology transfer, even though R&D carried out in consortia is "demand pull," that is, defined by consortia members themselves according to their perceived needs. Largely as a result of weak technology transfer links, many consortia participants have judged their membership in consortia as not worth the cost and effort and have expressed concerns over the return on their consortia investments.

A number of large consortia have come together as the Council of Consortia CEOs to solve common problems. "The Council's mission is to sustain the vitality of collaborative technology development, transfer, and application as a proven means of both maintaining and advancing North American competitiveness in key industries" (Council of Consortia CEOs, 1997). The council currently has 16 members drawn from among the largest U.S. and Canadian R&D consortia, including Bellcore, the Electric Power Research Institute (EPRI), GRI, MCC, SEMATECH, and the Semiconductor Research Corporation (SRC). The council maintains permanent and ad hoc working committees that analyze and report on issues important to consortia management, including technology transfer. The council itself meets twice a year at the CEO level to discuss these and other problems faced by consortia. (See Annex II, pp. 237–240, for an interesting consortium-based case study of successful technology transfer by the Electric Power Research Institute.)

TABLE 2.20 Primary Technical Areas of Joint Research Ventures (JRVs), 1985–1995

Technical Area	1985	1986	1987	1988	1989	1990	1991	1992	1993	1994	1995	Total JRVs	Percent
Telecommunications	8	1	6	8	10	15	17	17	23	15	11	131	22.78
Environmental	9	1	3	2	0	6	9	3	5	6	12	56	9.74
Advanced materials	3	5	3	4	5	2	2	5	6	5	13	53	9.22
Energy	5	1	2	1	4	6	9	14	7	0	1	50	8.70
Transportation	8	3	2	0	0	1	4	3	5	9	9	44	7.65
Software	1	0	1	2	4	2	3	1	4	3	18	39	6.78
Chemicals	2	2	2	2	7	5	8	4	1	3	2	38	6.61
Subassemblies and components	5	0	1	2	0	1	1	1	3	6	7	27	4.70
Manufacturing equipment	1	2	2	1	2	3	1	1	1	3	9	26	4.52
Factory automation	2	0	1	3	1	2	0	5	3	3	2	22	3.83
Photonics	1	0	1	0	0	2	1	2	3	2	9	21	3.65
Test and measurement	0	1	1	2	1	0	4	4	1	1	6	21	3.65
Computer hardware	1	0	0	1	0	0	1	1	4	1	4	13	2.26
N/A	1	1	0	1	0	0	0	0	1	2	5	11	1.91
Biotechnology	1	0	0	3	0	1	0	0	1	1	3	10	1.74
Medicals	1	0	0	0	0	1	0	0	2	3	3	10	1.74
Pharmaceuticals	1	0	0	0	0	0	0	0	1	0	1	3	0.52
Total JRVs	50	17	25	32	34	47	60	61	71	63	115	575	100.00

SOURCE: Vonortas (1996).

In sum, consortia offer attractive ways to leverage R&D in principle. In practice, many consortia have been deemed successful by participants as well as outside observers, while others have failed to live up to their initial promise. Clearly, much more remains to be learned about how to manage research consortia successfully and to transfer their scientific and technological results back to members.[87]

Organizations That Transfer or Facilitate the Transfer of Technology Created by Others

The second category of fourth-sector institutions is a diverse group of organizations sometimes referred to as technology transfer intermediaries. Although some of these organizations are large, most are small. They perform a wide variety of functions that assist the technology transfer process in some way, mostly by providing information, expertise, and/or money. Four types of intermediaries are reviewed below: organizations that provide technology transfer referrals and information; technology brokers, technology transfer consultants, law firms, and conference organizers; technology business incubators and research parks; and organizations not otherwise classified.

PROVIDERS OF TECHNOLOGY TRANSFER REFERRALS
AND INFORMATION

This group is made up of institutions and individuals that facilitate technology transfer among technology suppliers and buyers. These organizations are not necessary to every technology transfer. In fact, they are probably involved in only a small percentage of new technology transfer interactions, particularly where transfer between private firms is concerned. In most cases of technology transfer, the parties have met and gotten to know one another through a variety of mechanisms (e.g., personal relationships among technical employees and managers, membership in common organizations). However, organizations (particularly small organizations) that need technology or have technology to offer may benefit from the presence of a referral or information provider. Furthermore, as both government and industry scale back their R&D efforts, more and more companies are attempting to leverage their R&D resources by seeking external sources of technology. Technology transfer referral organizations will become increasingly important to that effort.

Most referral organizations deal in information about availability of and need for technologies, and they often create or publish technology databases to assist in this process. They do not generally become involved in technology transfer beyond facilitating the initial contact, and they do not generally broker technologies. (The role of technology brokers is treated separately below.) The activities of referral organizations range from publication of technology newsletters focused on specific technology areas to the creation and publication of technology databases provided to clients on a self-search or assisted basis. Some of these organizations are government funded, established primarily to facilitate access to federal technologies; others, in the private sector, promote technologies and expertise from universities, private firms, and federal laboratories. Government-funded information/referral organizations are discussed at pages 140–141.

Private-Sector Organizations

Databases containing information about university- and industry-based technologies and expertise tend to be relatively small, diverse, and fragmented. University technology managers have discussed the possibility of constructing a central repository of available university technologies, but no such project has ever gotten off the ground. However, a number of technology transfer referral organizations (mostly in the private sector) are independently constructing such databases.

NERAC, Inc., a private organization with functions similar to those of the NTTC, has been operating since the mid-1980s. NERAC, the former NASA New England Research Applications Center, is an independent, nonprofit organization supported by subscription fees paid by firms interested in accessing external technologies. NERAC's information specialists use the center's extensive databases to help client firms find sources of technology. NERAC began primarily as an interface to technology developed by federal laboratories. More recently, it has broadened its technology scanning activities to include universities, international sources, as well as private databases and other commercial-sector sources (NERAC Inc., 1997).

The Community of Science (COS, formerly Best, North America) maintains a large database (50,000 entries) of university researchers and their specialties. COS has also created two other databases, one focused on licensable university technologies (4,500 entries) and the other on university research facilities (1,700 entries). COS provides its paying clients (primarily researchers in universities and industry) with rapid access to the scientific expertise of other researchers. While not focused specifically on technology transfer, the COS databases nonetheless facilitate contact that can lead to transfers.

Knowledge Express Data Systems (KEDS) is a database producer and on-line service provider for the technology transfer community. KEDS produces databases of technologies available for licensing from universities, firms, and some federal laboratories and agencies. KEDS works directly with the technology transfer offices of universities and federal laboratories to acquire information on available technologies and capabilities. In addition, the organization uses a number of databases developed by the NTTC. KEDS databases also contain information on the needs and capabilities of high-tech companies, as well as reference material and technology transfer news.

There are also a large number of small firms that provide technology databases and expertise that focus on particular industrial sectors or technology areas. These organizations generally serve as consultants to their clients (usually for a noncontingent fee), helping to locate technologies and sometimes facilitating deals between technology suppliers and users. Seventy-two percent of the 120 organizations listed in "Consultants & Brokers in Technology Transfer," published by the Licensing Executives Society (LES) (1993), engage in finding technologies, while 83 percent engage in locating potential licensees for available technologies. Most of these organizations indicate that they work in only a few technology areas and industry sectors. These brokers and consultants represent only a small fraction of the total number of such organizations and individuals. Nevertheless, the LES sample indicates that a large percentage of brokers and consultants are engaged in locating technologies and providing referrals.[88]

TECHNOLOGY BROKERS, TECHNOLOGY TRANSFER CONSULTANTS, LAW FIRMS, AND CONFERENCE ORGANIZERS

This category consists of organizations and individuals that perform some of the many services of a technology transfer intermediary. The value they add to the technology transfer process is thought to be substantial, although there is no way to measure this impact except indirectly by tracking the revenue these intermediaries bring in. Even these figures are elusive, however.

The category includes nonprofit and for-profit organizations, sometimes performing similar services. There are relatively few large organizations, primarily brokers and patent law firms, and a large number of small organizations and individuals. Overall, data on the activities of this category, with the exception of the larger organizations, are scarce.

Technology Brokers

Technology brokers are organizations or individuals that market or assist in marketing technologies developed by others, primarily through licensing and/or formation of new companies. Brokers are compensated through contingent or success fees, usually a portion of the royalties and/or equity in a new firm. Brokers typically bear some or all of the cost of bringing the new technology to market, including but not limited to the costs of patenting, patent defense, marketing, and portfolio management. There are a few relatively large broker organizations and a much larger (but unknown) number of smaller brokers, many of whom are individuals.

The largest and one of the oldest broker organizations is Research Corporation Technologies (RCT). RCT was spun out of its parent organization, Research Corporation (RC), in 1987. Because of its tax paying for-profit status, RCT has more flexibility than RC to engage in start-up ventures. Research Corporation itself was founded in 1912 to commercialize inventions by university scientists, using the income generated (after payment of inventors' royalties) to fund research grants at universities. Likewise, RCT eventually intends to use its surplus revenues to enter the grant-giving arena. RCT has relationships with a number of universities that allow it to market some or all of their inventions. In doing so, RCT bears all the costs of preparing an invention for commercialization, including patenting, and it pays a portion of the royalty revenue (usually 60 percent) to the originating university. RCT had revenues of almost $60 million in 1994, of which $38 million was distributed to institutions and inventors. That year, RCT had relationships with 146 universities and appraised 671 invention disclosures from these institutions. RCT has a staff of just under 30, half of whom are technology transfer professionals.

British Technology Group, USA (BTG USA) is the U.S. subsidiary of the former British government corporation created to market British university technologies. BTG was privatized in 1992 and now operates as a for-profit firm traded on the London Stock Exchange. It markets a significant proportion of all U.K. university technologies as well as technologies from the U.K. private sector. The U.S. subsidiary was established in 1990. Typically, BTG USA takes possession of intellectual property by assignment or exclusive license and bears all the costs of acquiring and/or maintaining the patent, marketing the technology, and managing any revenues. In return, it receives a share of royalties from licenses or equity in start-up companies that it launches. Generally, income is divided equally between BTG USA and the technology source organization. BTG's focus is private-sector technologies and early-stage university technologies. In 1994, BTG worldwide had revenues of over $46 million, 33 percent of which came from U.S. sources, and it owned over 9,000 patents and 470 licenses relating to 1,300 separate technologies. BTG deals with a wide range of technologies, including those related to pharmaceuticals, electronics and telecommunications, dentistry, aerospace, chemicals and plastics, and automotive and medical engineering.

Competitive Technologies Inc. (CTI) is the result of the merger of several similar, smaller organizations. CTI was created and founded by Lehigh University and now exists as a private for-profit firm traded on the American Stock Exchange. The majority of CTI's revenues are drawn from royalties and from gains on equities CTI holds in start-up companies. CTI has three areas of activity: (a) joint ventures with universities, which function much like university technology transfer offices in granting licenses but can also take equity and make investments in start-up firms; (b) intercorporate licensing of private-sector technologies in which CTI serves as a licensing agent without taking title or bearing all the costs of patent protection; and (c) participation in state-sponsored venture or seed-capital organizations, in which CTI has both a management and an investor's role.

In addition to the three organizations outlined above, there are a large number of smaller firms, as well as many individuals, engaged in brokerage services. These smaller brokers are less likely to make investments in the technologies and less likely to undertake major expenses for patent protection. However, their reward structure (e.g., receipt of a percentage of royalties) would still be based on the success of the transactions they broker. Anecdotal evidence suggests that small brokers tend to occupy tightly focused technology or market niches.

Technology Transfer Consultants

In addition to brokers, consultants are also important to the technology transfer activities of their clients. As with brokers, there are a few large firms and many more small firms and individuals. Unlike brokers, who charge contingency or success fees, consultants are compensated on an hourly or flat-fee basis for performing services that help companies, universities, or federal laboratories license technologies or spin them off into new firms. They perform some of the services provided by brokers, but do so without a financial commitment. In exchange, they receive a guaranteed fee, but they cannot participate financially in major technology transfer successes. The American Consultants' League has over 40,000 members (most in fields other than technology transfer, of course) and estimates that the total of all consultants in the United States may exceed 10 times that number.

Consultants bring considerable expertise about one or more parts of the technology transfer process to their clients, expertise that would be difficult for all but the largest and most active clients to develop in-house. However, there are few, if any, consultants who are charged with independently executing the entire technology transfer process, as brokers do. After all, few technology generators are willing to give an outside consultant complete responsibility for bringing a technology to market on a fixed-fee basis with no incentive for success or penalty for failure.

Nonetheless, consultants perform almost every individual function in the technology transfer process, particularly where the formal transfer of intellectual property is involved. Consultants help ferret out marketable technologies within their clients' laboratories. They perform technology evaluations to estimate relative values in technology portfolios and assist in patenting decisions. They do market assessments and surveys, carry out marketing function, help locate potential licensees, and assist in negotiating and executing licenses and intellectual property transfer agreements.

Law Firms

Although a number of commercial firms, universities, and federal laboratories have legal expertise in the technology transfer area, there is still a very active commercial market for such legal services. A large number of law firms and individual attorneys provide services related to technology transfer, including patenting, licensing, and other traditional business-related legal advice. There is, however, a much smaller yet growing cadre of law firms that offer a broader range of new technology-related value-added services.[89]

A number of U.S. and European law firms have recently formed the TechLaw Group, a nonprofit network of major law firms that provide services for technology-oriented clients. TechLaw conducts educational programs, joint research projects, study sessions, and exchanges of information and materials, and serves a liaison function with private and government groups involved in promoting technology. TechLaw member firms provide a wide range of technology services to clients.

Many technology-oriented law firms have resident technical expertise on their legal and support staff. It is not uncommon to find attorneys in these firms who also have advanced degrees in the sciences. In addition, the firms often employ consultants to provide specialized expertise. For the most part, law firms are compensated by fixed fees for their technology-related services, but other alternatives, such as outcomes-based fees (e.g., through equity positions) are being considered in this relatively new arena of the legal profession.

Technology Transfer Conference Organizers

Conferences introduce suppliers and buyers of technology and help initiate the process of technology transfer. A few specialized technology transfer organizations sponsor conferences at which representatives from universities and federal laboratories gather to display their wares—technology capabilities and licensable inventions—to prospective licensees/sponsors, generally commercial firms.

Technology Transfer Conferences (TTC), a nonprofit firm located in Nashville, Tennessee, is a major organizer of such conferences. TTC sponsors six conferences per year in the United States, Canada, and abroad and has hosted 125 such meetings in the last 15 years. TTC invites universities and federal laboratories as well as some small firms to display their technologies to potential buyers from national companies. These national companies tend to be larger firms, but smaller companies are becoming part of TTC's clientele as well. The technologies showcased in TTC conferences include those in the life, physical, material, and environmental sciences. Companies that attend are generally interested in applied technologies, but TTC reports that more and more firms are investigating sources of basic research and looking to develop contacts in specific technology areas for future use.

Another technology transfer conference sponsor is the International Society of Productivity Enhancement (ISPE). ISPE is a nonprofit, membership organization founded in 1984 to accelerate the international exchange of ideas and scientific knowledge. ISPE sponsors two technology transfer conferences per year, each of which attracts between 125 and 150 participants. As with the TTC conferences, ISPE events attract representatives from institutions with technologies to sell as well as potential buyers of technologies.

TECHNOLOGY BUSINESS INCUBATORS

The National Business Incubator Association (NBIA) defines business incubators as "assistance programs targeted to start-up and fledgling firms. They offer access to business and technical assistance provided through in-house expertise and a network of community resources; shared office, research or manufacturing space; basic business support such as telephone answering and clerical services; and common office equipment including copy and fax machines. Business incubators support emerging businesses during their early, most vulnerable stages. They promote new firm growth, technology transfer, neighborhood revitalization, and economic development and diversification" (National Business Incubator Association, 1997).

While almost all incubators have one or more high-tech firms, not all business incubators are technology incubators. This latter term is usually applied to incubators that are primarily focused on commercializing new technologies through entrepreneurial ventures. According to NBIA, most technology incubators are associated with universities.[90] Other technology incubators are associated with federal laboratories, high-tech firms, or some combination of these institutions.[91] The NBIA estimates that of the approximately 550 incubators operating as of early 1997, 90 to 100 were true technology business incubators.

In response to a 1991 NBIA survey (National Business Incubator Association, 1992), incubators ranked "economic development" and "diversification of the local economy" as their first and second most important objectives, respectively. They ranked the "commercialization of research" and the "transfer of . . . technical capabilities to local businesses" as their third and fourth most important objectives. Furthermore, over 27 percent of all incubator clients were engaged in "technology products" or "research and development" in 1991. The largest groups of clients were "service firms" and "light manufacturers." [92]

According to recent work by Tornatzky et al. (1996), all technology incubators have ties to external sources of technology, since they rarely have expertise available in house. They provide this vital service through several types of arrangements. Many technology incubators have arrangements with a nearby university; some are even sponsored by or integrated into a university. The key service provided by incubators to their clients is access to faculty, graduate students, and, to a lesser extent, facilities. Some incubators have similar relationships with federal laboratories or high-tech firms. In a fewer number of cases, technology business incubators are integrated into the technology commercialization function of a parent R&D organization.

Some technology business incubators do more than serve as first homes for new high-tech businesses, they actively search out potential clients. Some incubator programs have developed aggressive efforts to locate potentially commercializable technologies and budding entrepreneurs within the research programs of nearby laboratories. This activity takes many forms, from consciousness raising, to establishing networks to identify research with commercial potential, to active searches in which "ferrets" knock on laboratory doors to access the commercial potential of ongoing R&D.

A number of federal laboratories have relationships with incubators, but

NASA recently became the first federal agency to directly enter the incubator business. NASA and the IC2 Institute in Austin, Texas, have entered into a 3-year experimental joint project designed to facilitate the commercialization of NASA-developed technologies. They have established two business incubators near the Johnson Space Flight Center in Houston and the Ames Research Center in Silicon Valley, California. (A third incubator has recently been established at the Stennis Space Center in collaboration with the state of Mississippi.) These incubators (NASA calls them Technology Commercialization Centers, or TCCs) focus on the technologies from the adjacent centers that can be commercialized within 1 to 2 years. The Johnson and Ames TCCs house start-up companies and assist them by drawing upon a regional network of entrepreneurs, business and technical experts, capital and market know-how, as well as the talent and technology pool of NASA. These two TCCs currently house a total of 30 companies. NASA currently pays the incubators' operating expenses, but expects them to become independent within a few years.

Research Parks

A research park is a real estate development designed to serve the needs of research-oriented companies. Most research parks generally provide space and facilities as part of their services. They are often located near technology sources in universities or other high-technology institutions. Furthermore, because they cluster growing high-technology firms together, they provide a significant opportunity for spontaneous technical interaction and technology transfer.

The Association of University-Related Research Parks defines a research park as a property-based venture that has:

- existing or planned land and buildings specifically designed for private and public research and development facilities, high-technology and science-based companies and support services;
- a contractual and/or operational relationship with a university or other institution of higher education;
- a role in promoting research and development by the university in partnership with industry, assisting in the growth of new ventures, and promoting economic development; and,
- a role in aiding the transfer of technology and business skills between the university and industry tenants. (Association of University-Related Research Parks, 1997)

As of 1980, university-related research parks accounted for a minority of all U.S. research parks. Of the 27 parks established by academic institutions between 1951 and 1980, 16 had failed, 5 were judged marginally successful, and only six were classified as unqualified successes by the U.S. General Accounting Office (1983). The 1980s witnessed a resurgence in the establishment of university-related research parks. By 1989, 115 university-related parks housed an estimated 2,100 companies and 173,000 workers (Matkin, 1990). As of 1995, there were 136 U.S.-based university-related research parks housing 4,765 companies and employing over 253,000 people (Association of University Related Research Parks, 1995).

TECHNOLOGY TRANSFER ORGANIZATIONS AND MECHANISMS NOT ELSEWHERE CLASSIFIED

Technical/Professional Associations

A number of technical and professional societies conduct activities designed to stimulate cooperative research and/or facilitate transfer technology. Relative to the entire technology transfer enterprise, the impact of the societies is small, but some of these programs are well established and fill an important niche in specific fields.

The American Society of Mechanical Engineers (ASME), for example, has a Committee for Research and Technology Development (CRTD) that has operated for 86 years. ASME does not fund R&D itself, rather it serves as a catalyst to facilitate research activities that involve multiple performers and funding sources. At the moment, ASME manages $15 million worth of contract research. The committee approves a research problem for action, raises funding, and locates scientists and/or engineers to carry out the research. Historically, about half of the funding for CRTD projects comes from industry (including industry associations) and half from government. The performers of CRTD-sponsored R&D are universities, federal labs, industry, nonprofits, or some combination of these. Most of the research sponsored by ASME is "paper studies," in which results are transferred through publication or presentation at meetings. A few research projects involve lab, or "metal-bending," work. Transfer of these results occurs via the participants themselves (who tend to be the interested parties) and through publication.

The American Society of Heating, Refrigerating and Air-Conditioning Engineers maintains a separate research arm that was founded almost a century ago. With a research budget of $2.6 million in 1994, the society sponsors research projects at universities and private firms in areas of interest to its members (Gale Research, Inc., 1995). Projects include evaluation of distribution losses in hot-water systems, filtration of indoor allergens and biological toxins, and computer algorithms for moisture loss and latent-heat loads in bulk storage of fruits and vegetables.

The Civil Engineering Research Foundation (CERF), an independent, nonprofit foundation created by the American Society of Civil Engineers (ASCE), began operation in 1989. CERF's mission is to unite diverse groups within the civil engineering community in industry-led R&D programs by serving as the "facilitator, coordinator, and integrator" of civil engineering research. Although CERF has primarily a coordinating role in civil engineering research, it adds some of its own funds to these efforts. Since 1989, the foundation has contributed $11 million (representing money donated by ASCE members) to engineering research programs that it sponsors. CERF uses a variety of means to carry out its objectives, including cooperative research programs, consortia, technology evaluation centers, surveys, and prototype demonstrations. CERF organized and now administers the National Council for Civil Engineering Research, consisting of over 60 civil-engineering-related research organizations, which fosters cooperation to advance the interests of the civil engineering profession through research. In 1993, CERF led the construction-materials trade associations in launching a $2 billion to $4 billion research program with the ambitious title, "High-Performance Construction Materials and Systems: An Essential Program for America and its Infrastructure." The program goal is to improve U.S. competitiveness and revitalize the nation's aging infrastructure by exploiting advanced construction materials.

Engineering, Design, and Architectural Firms

Engineering, design, and architectural firms play a major role in transferring technology, both within the United States and internationally. (See, for example, Freeman, 1968.) Many of the larger firms were established originally as engineering design departments of major manufacturing firms and were later spun off as independent companies offering services to a wide range of clients. Over time, these firms have become inventors and systems integrators in their own right, assuming roles as developers of new technologies that they then market along with their more routine design services.

Engineering design firms invest relatively little in separately identified R&D, especially in the United States—a circumstance for which they have often been criticized. However, they nevertheless conceptualize new technologies, and, working in conjunction with manufacturers, reduce these to practice while retaining title to the patents and know-how that result. For example, the M.W. Kellogg Corp. for a number of years was the source of most of the new process technology for producing synthetic anhydrous ammonia from natural gas. Kellogg designed and built plants for numerous U.S. and international firms, often on a "turn-key" basis.

Many engineering, design, and architectural firms specialize in process and production technologies, as well as facilities design and construction. They are less likely to develop or own proprietary product technology. Thus, since many producers (e.g., manufacturers, public utilities) only infrequently build new facilities and cannot afford in-house process development and design staffs, these firms fill an important niche in the marketplace.

In addition to developing and transferring their own technologies, engineering, design, and architectural firms are important conduits for diffusing new technologies developed by others. They do this by specifying the use of those new technologies within the designs they sell to their customers. In this way, they act as gatekeepers for new technology, encouraging customers to invest in processes that are most efficient and effective and least likely to pose undue risks of functional or financial disappointment.

Some engineering, design, and architectural firms are quite large, employing hundreds of people and annually engaging in projects whose total costs are in the billions-of-dollars range. Bechtel, Fluor, Kellogg, A.T. Kearney, and Stone and Webster are in or near this class. Others are much smaller, with highly specialized expertise in certain narrow but essential fields of technology and may subcontract for design work with larger firms. Some small firms may have one or more staff who have very broad experience in a sector and can offer a one-stop source of expertise to smaller client firms that need broad-based technical help to solve a problem or to expand capacity or product line. There are also a number of single-person firms and individuals who operate as consultants in this area. There are no data that distinguish those smaller engineering, design, and architectural firms from the many thousands of other types of consultants.

Venture Capital Firms

The classic function of venture capital (VC) firms in the technology transfer process is to invest in the growth of new start-up or spin-off technology companies. A great many do little more than that. Most VC firms perform their own analyses (e.g., of technologies, markets) before making a commitment to invest, and most are prepared to take remedial actions with their companies (e.g., recruiting new management) to protect their investment. However, some VC firms play a more active role in company development.

The most important element of any new firm is the quality of its staff, particularly its management. Many VC companies assist their client firms with the identification and recruitment of key management team members. Another method for strengthening new company management is to create networks with other, more experienced, firms, particularly suppliers, customers and neighbors, who can provide informal guidance to managers of the start-up. VC firms may also assist their client companies with other traditional business services, such as finance and accounting, organization, and office space.

A number of VC entities limit their investment to specific technology niches. This permits them to acquire technological know-how and to assist their client firms in the technology arena as well as in the financial and management areas. This assistance can take the form of expert market analysis as well as location of sources of complementary technological expertise for the new firm's technical staff. Technology savvy VC firms can also measure and monitor the technological progress of their clients better than investment firms with a more general portfolio.

In 1996, there were more than 600 venture capital funds in the United States.

That year, these funds collectively invested roughly roughly $10 billion in approximately 2,000 companies in all stages of growth from start-up to turnaround, including more than 280 initial public offerings. Also in 1996, 44 percent of venture capital went to information technology companies and another 31 percent to health-care-related enterprises, and another 25 percent went to non-technology companies (Horsley, 1997; VentureOne, 1997). It is estimated that only about one-quarter of VC-funded U.S. start-up companies succeed. Nevertheless, the average return on investment is 20 percent for the VC industry as a whole.[93]

The Internet

The Internet was initially developed for the express purpose of enabling rapid communications and the transfer of large amounts of data and visual representations for the purposes of R&D. It has performed this function admirably for over a decade, and it is slowly acquiring a similar enabling role in technology transfer.

Most U.S. R&D-performing institutions have a presence (usually a home page) on the World Wide Web, and most of those without home pages reported that they were in the process of building one. The quality and utility of these home pages vary widely. As users figure out what works and what does not work on the Internet, and as problems of access to intellectual property and payment for Internet-based services are resolved, more and more institutions (particularly high-tech concerns) will establish not only a presence on the Web, but also will provide sophisticated access to real information and other things of value.

This access will undoubtedly include some interactive functions, for example, introducing potential providers and purchasers of technology that are key to technology transfer. Outsourcing and outplacing of technologies will be facilitated by on-line databases using sophisticated search engines operated directly by searchers or with the help of human intermediaries. Additional technological advances in communications and processing will permit users to exchange 3-D images, video, and sound over networks, significantly enhancing the quality of presentations and available information.

SUMMARY: ORGANIZATIONS THAT TRANSFER TECHNOLOGY
CREATED BY OTHERS

As the preceding discussion makes evident, the number of privately held, non-R&D-performing U.S. organizations involved in technology transfer to industry is huge, and the spectrum of technology transfer services they provide is extensive. The collective performance of these diverse players, however, has been criticized severely in recent decades. Indeed, the poor performance of many U.S. companies in more traditional manufacturing industries, particularly small and medium-sized enterprises (SMEs), suggests that there are significant gaps in the scope and/or quality of technology transfer services provided to SMEs by public-and private-sector organizations.[94] The perception that this vast collection of privately held organizations is not meeting the technology-transfer/industrial-modernization needs of U.S. SMEs was a driving force behind the establishment and expansion of public-sector technology-transfer/industrial-extension initiatives by state governments and federal agencies during the 1980s and early 1990s.

Conclusion

Privately held, nonacademic organizations form the smallest of the four sectors of the U.S. technology transfer enterprise in terms of R&D performed or quantitatively measurable technology transfer (patents and royalties). This sector is also the least well documented and measured. However, its importance to the nation's innovation system should not be underestimated. As discussed above, these organizations vary widely in size, function, and contribution to the U.S. R&D and technology transfer enterprise.

The highly heterogeneous population of U.S. privately held independent and affiliated R&D institutes fills some important gaps in the U.S. R&D enterprise, addressing unique R&D and evaluation needs of certain industries or subsectors (particularly in biomedical fields) that universities and federal laboratories do not address. Nevertheless, it would be inaccurate to characterize most of these institutes as industry-oriented.

The vast majority of independent and affiliated research institutes receive most of their funding from federal mission agencies or private foundations and are focused primarily on basic research. More than half of these institutions are concentrated in the health and medical fields, and these institutes collectively account for a significant share of all health and medical R&D performed in the United States. While the R&D activities of many institutes in this group constitute a critical link in U.S. drug testing and evaluation, and directly benefit health- and medical-related industries in many other ways, the research agendas of these institutes are not driven or shaped to any significant extent by the day-to-day R&D needs of industry.

Even within the relatively small population of independent and affiliated engineering R&D institutes, many of which were originally established to serve the needs of regional industries, there are today relatively few institutes whose R&D activities are substantially geared to the applied R&D needs of private industry. Five of the seven largest independent engineering R&D institutes perform the vast majority of their R&D in service to federal agency missions, not the R&D needs of private companies. In short, these institutes do not fill perceived gaps in basic or applied R&D that are directly relevant to needs of more traditional, technologically mature manufacturing industries, gaps that many have identified as a significant weakness of the U.S. R&D enterprise (National Academy of Engineering, 1993).

Industry-led research consortia, both publicly and privately funded, have assumed greater significance in the U.S. R&D enterprise in the past decade, partially filling some of the aforementioned R&D gaps in selected high-tech and technologically mature industries. However, coverage in terms of industries and technology areas (and the share of firms within a given industry or technology field) remains very limited, as does the claim of these consortia on public and private R&D resources overall. Furthermore, in part because of the diversity and highly autonomous nature of U.S. industrial R&D consortia, there has been remarkably little knowledge transfer concerning organizational and operational practices among the rapidly growing population of U.S. consortia.

The U.S. population of privately held, non-R&D performing organizations involved in technology transfer to industry is large, diverse, and highly autonomous, and the range of technology transfer services provided is extensive, though uneven among industrial sectors. Indeed, in the absence of significant public-sector involvement, this diverse group of technology transfer intermediaries, together with private vendors of hardware and software and large industrial firms/ customers, have long constituted the primary sources of new technology, technical assistance, and advice for U.S. small and medium-sized enterprises (SMEs) in most manufacturing industries.

During the past decade, however, the slow pace with which many U.S. companies, particularly SMEs in more traditional manufacturing industries, have adopted new production technology suggests that significant gaps exist in the scope and quality of industrial-modernization services provided by this vast amalgam of private companies and private technology transfer intermediaries (National Academy of Engineering, 1993; National Research Council, 1993a).

Admittedly, in recent years, several industry-led initiatives, some with limited public funding, have begun to address some of the innovation and technology diffusion challenges that face SMEs as well as larger firms in a number of technologically mature U.S. industries. For example, through increased industry self-organization and support from federal agencies and university-based researchers, segments of the U.S. textile and apparel industry have successfully applied modern information technology to achieve a major revitalization of their entire design, supply, and marketing chain—largely in response to the new "lean" retailing strategies of major retail distributors, strategies also enabled by advances in information technology (Abernathy et al., 1995).[95] Similarly, through a combination of firm-specific and industrywide initiatives, often in partnership with federal agencies or academic researchers, the U.S. automotive industry has successfully met many of the manufacturing challenges that hit the industry in the late 1970s and early 1980s.[96] Other successful or promising industry-led efforts to meet the manufacturing and other technology diffusion needs of SMEs include the National Center for Manufacturing Sciences and SEMATECH's work with semiconductor equipment and material manufacturers.[97]

Particularly promising in the judgment of the U.S. delegation are the recent technology road mapping exercises of the Semiconductor Industry Association and a coalition of organizations from the U.S. chemicals industry that inventory the two industries' sources of technology and forecast technological needs throughout the industries' respective value-added chains (American Chemical Society et al., 1997; Rea et al., 1996). These road mapping efforts show potential for advancing both the development and diffusion of new technology in industries where technological advance is more evolutionary than revolutionary. From the perspective of firms involved as well some outside observers, the semiconductor industry technology road map effort has been successful at focusing the attention and resources of the industry and the federal government on a shared conception of technological challenges and opportunities. The more recently developed technology road map for the chemical industry was launched, in part, by a request from the White House Office of Science and Technology Policy.[98]

In addition to these industry-specific initiatives, state and federal governments have attempted to strengthen the existing but relatively weak network of private and public service providers with more comprehensive industrial-modernization and technical-extension programs. (See Part II, pp. 76–79 and Annex II, pp. 201–213.)[99]

Nevertheless, while the experience and promise of these and other private- and public-sector (and joint public-private) initiatives are encouraging, it is important to recognize that the reach of these efforts, in terms of companies, industries, and technology areas (and the share of firms within a given industry or technology field) remains very limited, as does their claim on public and private R&D resources overall (National Academy of Engineering, 1993; Shapira, 1997).

ANNEX
II

Case Studies in Technology Transfer

BIOTECHNOLOGY

Simon Glynn and Arthur E. Humphrey

Biotechnology is literally a new technology, enabled by rapid expansion of our understanding of cell biology, especially of DNA, and the development of techniques that use this new understanding to physically change the genetic content of cells. The United States dominates in the biomedical sciences and is the source of the vast majority of basic information in biotechnology. The United States has also dominated early efforts to realize the potential of biotechnology. This paper is intended to review the technology flows that have enabled this success.

Defining the Scope of Biotechnology

THE TECHNOLOGIES

Biotechnology is defined by technologies, not outputs. These technologies, especially the sequencing and decoding of genes on a large scale, have transformed our understanding of the function of DNA in cells. These advances also enable researchers to manipulate genetic information in cells. For example, using recombinant DNA technologies, the human gene that codes for insulin (a protein) can be isolated and then inserted in a bacterium. The bacterium can be made to synthesize human insulin, which may then be used to treat diabetes. Genetically engineered cells can produce not only human hormones such as insulin or growth hormone, but also blood products like clotting factors, vaccines, and new antibiotics.

These new technologies can also be used to create a class of proteins called monoclonal antibodies that are especially useful in diagnostics. These proteins are not created using recombinant DNA techniques, but by fusing a tumor cell to a white blood cell and then cloning this new cell. The resulting cells produce antibodies that are chemically identical. Monoclonal antibodies are used widely in research to identify the presence of specific types of molecules and to detect the presence of disease.

HUMAN THERAPEUTICS AND DIAGNOSTICS

Data on biotechnology revenues are inconsistent, but total annual revenues to U.S. companies from products developed using biotechnology appear to be about $10 billion and are projected to increase 15 to 20 percent each year over the next few years. Human therapeutics and diagnostics represent over 90 percent of these revenues (U.S. Department of Commerce, 1993). Table A-1 shows the number of drugs currently in development that use biotechnology techniques. In 1994, there were only 19 biotechnology-based drugs approved for use in the United States. (See Table A-2.) These drugs as a group rely on human hormones that were either understood or thought to be therapeutically useful in the treatment of diseases such as diabetes, anemia, and multiple sclerosis. These drugs have about $9 billion in annual global sales, or less than 5 percent of total global sales for pharmaceuticals (Merrill Lynch, 1996). As of February 1992, 640 diagnostic kits using monoclonal antibodies, DNA probes, and recombinant DNA

TABLE A-1 Biotechnology Drugs in Development, 1989–1993

	1989	1990	1991	1993
Approved medicines	9	11	14	19
Medicines or vaccines in development				
Phase I	26	38	48	41
Phase I\II	12	13	16	22
Phase II	23	32	46	53
Phase II/III	8	6	7	6
Phase III	11	15	18	33
Phase not specified	5	3	2	4
Application at FDA for review	10	19	21	11
TOTAL medicines or vaccines in development	95	126	158	170

NOTE: Total medicines or vaccines in development reflects medicines in development for more than one indication.

SOURCE: Pharmaceutical Manufacturers Association (1993).

TABLE A-2 Biotechnology Medicines or Vaccines Approved for Use by the Food and Drug Administration as of 1993

Product	Indication(s)	Company	Year Approved
Beta interferon	Multiple sclerosis	Chiron	1993
DNAse	Cystic fibrosis	Genentech	1993
Factor VIII	Hemophilia	Genentech, Genetics Institute	1993
IL-2	Renal cell cancer	Chiron	1992
Indium-111-labeled antibody	Cancer imaging	Cytogen	1992
Aglucerase	Gaucher's disease	Genzyme	1991
G-CSF	Adjunct to chemotherapy	Amgen	1991
GM-CSF	Bone marrow transplant	Immunex	1991
Hyaluronic acid	Ophthalmic surgery	Genzyme	1991
CMV immune globulin	Prevention of rejection in organ transplants	MedImmune	1990
Gamma interferon	Chronic granulomatous disease	Genentech	1990
PEG-adenosine deaminase	Immune deficiency	Enzon	1990
t-PA	Myocardial infarction, pulmonary embolism	Genentech	1990
Erythropoietin	Anemia associated with renal failure, AIDS, cancer	Amgen	1989
Hepatitis B antigens	Diagnosis	Biogen	1987
Alpha interferon	Cancer, genital warts, hepatitis	Biogen, Genentech	1986
Hepatitis B vaccine	Prevention	Biogen, Chiron	1986
Human growth hormone	Deficiency	Genentech	1985
Human insulin	Type I diabetes	Genentech	1982

SOURCE: Read and Lee (1994).

techniques had been approved by the U.S. Food and Drug Administration (FDA), including screening tests for the AIDS and hepatitis C viruses (U.S. Department of Commerce, 1993).

The current generation of biotechnology drugs relies on major advances in biotechnology to identify and decode genes. This large-scale sequencing of genes is being done globally and is coordinated through gene databases on the Internet. Sequencing of the entire human genome may be completed by 2005 (Washington Post, 1996). Two examples of protein drugs based on these techniques are Amgen's obesity drug Leptin and a protease inhibitor for AIDS that has had a dramatic clearing effect on the HIV virus (Merrill Lynch, 1996).

NONMEDICAL USES OF BIOTECHNOLOGY

Nonmedical uses of biotechnology are also apparent. Using biotechnology techniques, researchers hope to transfer into plants specific beneficial traits (e.g., resistance to pesticides, tolerance of hostile environmental conditions such as salinity or toxic metals, or higher nutritional content) (National Research Council, 1987). Bioprocess technologies are also expected to help in diverse sectors of the economy. In the petroleum industry, for example, bioprocessing has potential to degrade wastes or toxic substances (National Research Council, 1992c).

Revenues from these nonmedical uses of biotechnology (agriculture, specialties, environmental) are less than 10 percent of total revenues in biotechnology (U.S. Department of Commerce, 1993). There are several reasons for this. First, the use of biotechnology in areas other than human therapeutics and diagnostics presents unique research and technical barriers not addressed by biomedical research. Second, the use of biotechnology is constrained by economics. Drugs developed using early biotechnology techniques have tended to be exceedingly expensive. But new opportunities will require technologies to synthesize and purify the biological products at sharply lower cost and higher capacity (National Research Council, 1992c). Finally, commercial development in biotechnology in the United States (so far) is directed by the size of the opportunity. The most immediate consequence of this is the current focus on human therapeutics and diagnostics, where the returns to investors are expected to be largest; there is relatively less focus on agricultural or industrial applications. For these reasons, many of the nonmedical uses of biotechnology are not expected to be commercially available before 2000 (Table A-3).

New Biotechnology Companies

Two different types of firms are pursuing the commercial potential of biotechnology: new biotechnology firms (NBFs), started specifically to exploit opportunities using biotechnology techniques; and large companies in pharmaceuticals, chemicals, and other sectors for which biotechnology has important implications.

The biotechnology sector included 1,272 biotechnology companies in 1993, of which 235 were public (Read and Lee, 1994). More than 100 of these companies were started in the last 2 years, and 70 percent are less than 10 years old (Read and Lee, 1994). A large proportion of these NBFs, but certainly not all, are developing human therapeutics and diagnostics. Compared with the larger pharmaceutical sector, NBFs as a group are relatively small. According to a survey by Ernst and Young (1993), revenues for biotechnology companies were about $7 billion in 1992, compared with revenues of $114 billion for pharmaceutical companies. The biotechnology sector is nonetheless a very large funder of biomedical research. According to the survey, NBFs spent nearly $5.7 billion on R&D in 1992, about half the R&D expenditures of the pharmaceutical sector. As is obvious from these levels of R&D spending, the overwhelming majority of biotechnology companies are research organizations with essentially no revenues. Nearly one-third of NBFs have no approved products, and 70 percent had rev-

TABLE A-3 Selected Nonmedical Uses of Biotechnology

Animals	
Vaccines	Colibacillosis or scours (1984), pseudorables (1987), feline leukemia (1990)
Therapeutic MAbs	Canine lymphoma (1991)
Diagnostic tests	Bacterial and viral infections, pregnancy, presence of antibiotic residues
Plants	
Diagnostic tests	Diagnose plant diseases (turfgrass fungi)
Biopesticides (killed bacteria)	Kills caterpillars, beetles (1991)
Bioprocessing	
Diagnostic tests	Diagnose food and feed contaminants (salmonella, aflatoxin, listeria, campylobacter, and *Yersinia entercolitica*)
Chymosin or renin	Enzyme used in cheesemaking (1990)
Alpha amylase	Enzyme used in corn syrup and textile manufacturing (1990)
Lipase	Enzyme used in detergents (1991)
Xylanase	Enzyme used in pulp and paper industry (1992)
Luciferase	Luminescent agent used in diagnostic tests
Environment	
Diagnostic test	Detect legionella bacteria in water samples

NOTE: MAbs = monoclonal antibodies.
SOURCE: U.S. Department of Commerce (1993).

enues of less than $5 million in 1992. Moreover, with very few exceptions, development efforts in the majority of these biotechnology companies are several years from approval.

Almost all of these small companies will run out of money before their ideas are transferred to clinical practice. This problem becomes critical as the amount of R&D required to move sophisticated medical technologies to commercialization increases. The investment in R&D is also risky. For example, failure to win FDA approval for their sepsis products cost investors in three companies—Synergen, Centocor, and Xoma—about $2.5 billion (Humphrey, 1995).

To finance their research and development efforts, the new biotechnology firms have used a variety of funding mechanisms. The most important of these have been investments from venture capital firms, through public financing, and from larger companies.

VENTURE CAPITAL

The development of the U.S. biotechnology industry has largely been financed by venture capital firms. Venture capital is available to NBFs because the opportunity to exploit new advances in biotechnology for human therapeutics and diagnostics creates liquidity in public markets (as initial public offerings). Indeed, biotechnology attracted more venture capital financing—$261 million invested in 95 companies—than any other sector of the economy in 1992, except software and services (Venture Economics, 1994).

Venture capital firms are an important reason for the success of NBFs in the United States. In this country, nearly 75 percent of NBFs started as independent firms, compared with only 5 percent of NBFs in Japan, where venture capital is essentially nonexistent (National Research Council, 1992b). This difference may be an advantage for U.S. firms, since venture capital allows NBFs to form earlier and closer to intellectual capital in universities than would otherwise be possible (Zucker et al., 1994).

PUBLIC FINANCING

Public markets have also been a valuable source of financing for the higher-quality, larger-capitalization NBFs. In the early 1980s, several start-up biotech firms (Genentech, Cetus) set Wall Street records when they first went public. These firms have also been able to return to the public markets to finance production scale-ups and clinical trials.

It is important to realize that health care reform and regulation impact the availability of venture capital and public financing, since investors focus on the anticipated returns on their investments. For example, regulations that require a certain number of clinical trials to determine the expected time to market of new drugs and therefore the cost of developing them. Health care reform efforts also play an important role by increasing the uncertainty with regard to biotechnology. Buyers of biotechnology-based drugs are now less often individual physicians than health care corporations, and third-party payers are becoming more restrictive, increasing the risk for investors and venture capital.

LINKS BETWEEN NBFs AND LARGE COMPANIES

Large pharmaceutical companies are an especially important source of funding for new biotechnology firms. Large pharmaceutical companies have been investing in NBFs at an unprecedented pace. These cash infusions are especially important for equity investors in new biotechnology companies, because they reduce the future dilution they face.

Linkages to NBFs are important to large pharmaceutical companies for several reasons. First, major pharmaceutical firms are looking increasingly for new, unique drugs for which there is no analog to treat diseases for which there currently are few or no effective drug therapies. These are diseases such as cancer, Alzheimer's, and AIDS that account for $500 billion in medical expenses each year in the United States (Merrill Lynch, 1996). Because the technology for developing these drugs is concentrated in the new biotech firms, NBFs have a comparative advantage in developing these new drugs.

Second, it is important to recognize that the distinction between pharmaceutical companies and biotechnology companies is blurring as pharmaceutical firms are increasingly using biotechnology techniques to develop new drugs. According to a study by the Boston Consulting Group (BCG) (1993), 33 percent of research projects in major pharmaceutical companies in 1993 were based on biotechnology, compared with only 2 percent in 1980. In some larger pharmaceutical companies, up to 70 percent of the research projects were based on molecular biology techniques. Equity investments have enabled larger companies to access the technology in these NBFs and to develop internal capabilities in biotechnology.

Finally, the special strengths of the large pharmaceutical companies continue to be in traditional drug discovery, manufacturing, marketing, and distribution. Large pharmaceutical firms also are experienced in the drug approval process, which is especially difficult for NBFs. Linkages to large pharmaceutical companies thus let NBFs exploit these competencies. For example, Humphrey (1993), at the inaugural meeting of the American Institute of Medical and Biological Engineering in 1992, observed that failure to integrate process design and engineering expertise into the development process for biotechnology drugs prior to phase III clinical trials resulted in many nonoptimal bioprocess designs that did not use leading-edge technology.

LINKAGES TO FOREIGN FIRMS

Technological links are also expanding between new biotechnology firms in the United States and large foreign firms. Foreign pharmaceutical companies understand that a global orientation is required to ensure long-term competitiveness and financial returns, and they recognize that the United States is the world's largest health care market. Foreign firms also seek access to advances in biotechnology developed in the United States (National Research Council, 1992b). From the perspective of NBFs, the need for cash infusions to fund R&D encourages linkages with large, cash-rich foreign firms (National Research Council, 1992b).

These linkages so far serve to transfer technology from the United States to foreign countries, although Japan's strength in enzyme related bioprocessing technologies is a potential opportunity for future technology transfer from Japan to U.S. biotechnology firms. Japan's Kirin Brewery provided U.S. biotechnology firm Amgen critical robotic bioprocess technologies for the production of Epogen and Neupogen (Box 1). But the implication may be that these transfers represent a future competitive advantage for foreign firms in the U.S. and global markets[1] (National Research Council, 1992b).

The Importance of Universities

The important advances in biotechnology—so far—have been made disproportionately by researchers in large U.S. research universities and then have diffused to the commercial sector, usually through NBFs. In this sense, U.S. universities perform an incubator role for the biotechnology sector in the United States.

> **BOX 1**
> **AMGEN**
>
> Amgen, the largest independent biotechnology company in the world, is headquartered in Thousand Oaks, Calif. The firm has research centers in Boulder, Colo., and Toronto, Canada; an international distribution center in Louisville, Kentucky; clinical research centers in Cambridge, England, and Melbourne, Australia; a fill-and-finish facility in Juncos, Puerto Rico; and a European regional headquarters in Lucerne, Switzerland. Other international operating facilities are located in Australia, Belgium, Canada, China, France, Germany, Italy, Japan, the Netherlands, Hong Kong, Portugal, Spain, and the United Kingdom.
>
> Founded in 1980 by a group of scientists and venture capitalists, Amgen was able to attract a prestigious scientific advisory board that included several members of the National Academy of Sciences. In the autumn of that year, George B. Rathmann, formerly of Abbott Laboratories, was named Amgen's chairman and chief executive officer—he was the company's first employee.
>
> Amgen commenced operation in early 1981 with a private-equity placement of approximately $19 million, involving venture capital firms and two major corporations. The company chose its Thousand Oaks location to be near such major research centers as the University of California at Los Angeles, the University of California at Santa Barbara, and the California Institute of Technology.
>
> The company raised capital through public stock offerings in 1983, 1986, and 1987. Amgen's stock is traded on NASDAQ's National Market System under the symbol AMGN.
>
> Using techniques of recombinant DNA and molecular biology to create highly specialized health care products, Amgen scientific achievements have positioned the company at the forefront of the biotechnology industry.
>
> As a result of this technology, Amgen has developed several human biopharmaceutical products. Two have been key moneymakers. Its re-

PUBLIC FUNDING OF BIOMEDICAL R&D

Health R&D now accounts for a rapidly growing share (16 percent in 1995, or $11.4 billion) of the government's total R&D investment (National Science Foundation, 1994). Health research also received the single largest share–4 percent–of federal basic research dollars in 1995, or $6.3 billion. In comparison, general science, which included funding for the National Science Foundation (NSF) and for the research portion of the now-canceled Superconducting Supercollider, accounted for only 20 percent, or $2.9 billion, of estimated federal basic

ANNEX II

> BOX 1—Continued
>
> combinant human erythropoietin, EPOGEN® (Epoetin alfa), stimulates and regulates production of red blood cells; and its recombinant granulocyte colony-stimulating factor (rG-CSF), NEUPOGEN® (Filgrastim), selectively stimulates the production of a class of infection-fighting white blood cells known as neutrophils.
>
> Amgen received its first patent for EPOGEN® on October 27, 1987, and a product license application (PLA) was filed with the FDA 2 days later. On June 1, 1989, EPOGEN® was approved by the FDA for treatment of anemia associated with chronic renal failure.
>
> Besides the development of key proprietary recombinant DNA methods for the production of EPOGEN and NEUPOGEN, two major process technologies were important to the commercialization of these products. To achieve rapid and successful commercialization of these products, AMGEN made the decision to sharply scale up its roller-bottle technology. To do this, it entered into a joint venture—Kirin Amgen—with Japan's Kirin Brewery Co., Ltd., in 1984 for the commercial development of recombinant human erythropoietin. Through this joint venture, robotic bottling technology was adapted to roller-bottle manufacturing,
>
> A second key factor in the firm's success was the in-house modification of the roller-bottle cap, allowing not only easy removal in the robotic process but also an increase in gas-mass transfer across the cap, resulting in a greater than tenfold improvement in productivity.
>
> That this joint technology transfer has been eminently successful can be seen by the financial success of the corporation. Total AMGEN revenues for the year ended December 31, 1993, were $1.4 billion, primarily from sales of EPOGEN and NEUPOGEN. Revenues in 1992 were $1.1 billion. Net income for fiscal 1993 was $383.3 million, or $2.67 per share on a primary basis. In just 13 years, starting from scratch, AMGEN has become a Fortune 400 trading corporation. Adaptation of enabling technology was important to achieving this success.
>
> SOURCES: Amgen, Inc. (various years).

research authorizations (National Science Foundation, 1994). The overwhelming majority of this biomedical funding is directed to U.S. research universities and academic medical centers.

To a considerable extent, this support has been concentrated on the emerging genetic engineering techniques in biotechnology, especially for AIDS research. In 1993, the U.S. administration, stating its intention to strengthen the FCCSET (Federal Coordinating Council for Science, Engineering, and Technology) process, included funding for six presidential initiatives in its initial 1994 budget proposal.[2] The largest of these initiatives was for biotechnology research, and more than three-quarters of this funding was controlled by the National Institutes of Health (NIH) (National Science Board, 1993).

UNIVERSITY-INDUSTRY RELATIONSHIPS

In quite a few instances, the mechanism for technology transfer in biomedical R&D has been the establishment of (usually single-product) biotech start-ups, often with individual scientists and their graduate students literally moving from academia to industry. For this reason, universities have been the locus of innovation in biotechnology. For biomedical firms, locating near U.S. research universities provides access to state-of-the-art research in fields essential to their continued success. Indeed, about half of all NBFs in the United States are grouped around three major centers of academic biotechnology research: 21 percent of NBFs are close to Stanford, University of California at Berkeley, and UC San Francisco; 18 percent are near MIT and Harvard in Boston; and 12 percent are located near the NIH campus in Bethesda, Md. (Humphrey, 1995).

The diffusion of basic information and expertise from U.S. universities to new biotechnology firms is essentially complete. Indeed, these laboratory technologies are now widely disseminated, since virtually all of the research that enables biotechnology was performed in U.S. universities and academic medical centers using public money. There are few valuable strategic positions in these techniques (although separation and purification techniques, and process control are critical, as they create an economic advantage) (Gaden, 1991).

Nonetheless, quite a number of interesting case studies seem to indicate that both the number and variety of alliances in biomedical R&D between academia and industry are increasing dramatically. In a recent study, Cohen et al. (1994) identified more than 1,050 research centers at U.S. universities, representing an aggregate budget of $4.12 billion in 1990, exactly half of all federal expenditures on academic R&D that year. Of these university-industry centers, 232, or 22 percent, conducted biotechnology research. Nearly 45 percent of expenditures were for basic research, although this actually represents less of an emphasis on applied research than academia as a whole (Cohen et al., 1995).

Data on individual participation suggest that relationships between researchers in academia and industry are even more pervasive than information on university-industry alliances indicates. For example, many NBFs have also established scientific advisory boards that include research scientists from U.S. universities and academic medical centers. Blumenthal (1992) found that 47 percent of biotechnology faculty consulted with industry, that 23 percent were involved in formal university-industry relations, and that 8 percent had received equity based on their own research.

Biotechnology companies encourage these relationships. Genentech, for example, provides several million dollars of free recombinant materials to academic researchers every year. As a condition of receiving these materials, Genentech requires that any research findings be reported to Genentech, and Genentech asserts the first right of refusal on any commercial applications developed (Personal communication from H. Niall, chief scientist, Genentech, to Simon Glynn, research associate, National Academy of Engineering, August 10, 1993).

These dynamics are important, because federal and industry funding of biomedical research are not quite the same thing. Industry and universities have increasingly diverging research agendas in biotechnology, and this is reflected in the priorities of academic researchers (Box 2). Of the individuals interviewed for the Harvard biotechnology project, 30 percent of biotechnology faculty with industrial support said that their choice of research topics had been influenced by the likelihood that results would have commercial application. This compared to only 7 percent of faculty without commercial funding who said so. The terms of funding are also different: For extramural grants from the NIH, 92 percent are for 3 years or longer; for industry-funded research in universities, the majority of grants are for 2 years or less, consistent with the shorter time horizon of applied research (Blumenthal et al., 1986b).

ECONOMIC INCENTIVES

The institutional environment in which academics live is extremely important for technology linkages. In this respect, the changes in medicine have been faster and more dramatic than in other areas. Few, if any, examples of basic research in academic medical centers attracted commercial interest (unlike physics and chemistry and even music) until the early 1970s and the acceleration of genetic research. Even then, at Stanford, there was significant culture shock (and in some cases even outright hostility) when patents and commercial interest intruded into these medical departments after the first successful recombination of DNA by Cohen and Boyer in 1973. This is in sharp contrast to the current view of biotechnology at Stanford. Observed a prominent scientist, "The problem [now] is not pushing technology out of the lab; the problem—and this is a problem—is pushing the technology too early. Technology advances too fast from academia to commercialization. I have a staff of nine, and everyone has a pet cure for cancer that they are pushing" (Personal communication from D. Botstein, Stanford University, to Simon Glynn, research associate, National Academy of Engineering, August 9, 1993).

A critical element in this culture is the use of programs to provide financial incentives to support and encourage innovation by academics.[3] At Stanford, for example, 15 percent is subtracted from total license revenues for the technology licensing office budget (this is usually excessive; the excess then goes to the Dean of Research for a research incentive fund for researchers without sponsorship). The net royalties are then divided, one-third to the inventors, one-third to their department, and one-third to the school of medicine (Personal communica-

> **BOX 2**
> **The Monsanto-Harvard Agreement**
>
> In 1982, Monsanto Corp. of St. Louis, Missouri, opened new facilities to expand its fundamental biotechnology research program. These laboratories were aimed at developing new products in animal nutrition, agriculture, and human health, as well as developing and expanding the basic understanding of new biotechnology techniques. Through its program in molecular biology, the company was seeking a window on developing biotechnology through internally conducted research and collaborative research with universities and small start-up companies. The latter included such firms as Genentech, Genex, Biogen, and Collagen.
>
> As part of the program, in 1974 Monsanto entered into a landmark agreement with Harvard University to fund purely basic research for a period of 12 years. The company cosponsored studies on the molecular basis of organ development and tumor angiogenesis. The agreement assigned publication rights to the participating Harvard researchers and commercialization rights to Monsanto.
>
> After the initial 12 years, the program was not renewed; rather, it was scaled down and agreements were designed to support the research of individual Harvard scientists whose work was of specific interest to the company. Considerable insight was gained from the research about the carcinogenic behavior of many compounds produced by Monsanto. However, greater commercial advantage was gained through the support of cooperative research with small or emerging biotechnology companies. One such cooperative venture was the joint program with Genentech to produce recombinant bovine and porcine growth hormones.
>
> SOURCE: Genetic Engineering News (1982).

tion from H. Wiesendanger, Office of Technology Licensing, Stanford University, to Simon Glynn, research associate, National Academy of Engineering, August 10, 1993). A similar policy is in effect at MIT and UC Berkeley. These licensing fees are an important alternative to government funding at the major U.S. research universities and especially in the emerging field of biotechnology, with companies supporting up to 16 percent of university research in this area (Blumenthal, 1992). The patents on DNA recombinant techniques by Boyer and Cohen are an example: The $14.6 million earned by the Cohen-Boyer patents in 1991 represented 58 percent of total income from all patents held by Stanford (Personal communication from H. Wiesendanger, Office of Technology Licensing, Stanford University, to Simon Glynn, research associate, National Academy of Engineering, August 10, 1993).

Interaction With NIH and NIH-Funded Investigators

NIH is the largest funder of biomedical research in the world. The agency funded $3.5 billion in biotechnology-related R&D in 1992, about 80 percent of all federal spending for biotechnology (National Research Council, 1992a). Relationships between researchers at NIH and university scientists receiving NIH funding are therefore an important dimension of technology transfer in biotechnology.

NIH helps industry to develop this research for commercial use in four ways: participating in cooperative R&D agreements (CRADAs), licensing patented materials, training post-doctoral students and research fellows, and publishing.

CRADAs involve NIH researchers and facilities in industry-directed research. This lets NBFs leverage these NIH resources. Several important products have resulted from these collaborations, including the AIDS drugs AZT and DDI, and the HIV antibody tests. Nearly 1,000 CRADAs have been negotiated between the U.S. Department of Health and Human Services and industry since 1987 (U.S. Department of Commerce, Office of Technology Policy, unpublished data, 1996). NIH also facilitates technology links by licensing materials developed by NIH and by training post-doctoral students and research fellows. These technologies and researchers interested in collaboration are listed in an electronic bulletin board funded by NIH.[4] NIH researchers also publish about 7,000 technical journal articles per year as well as present research at scientific meetings (National Research Council, 1992b).

THE QUESTION OF FOR-PROFIT FUNDING AND RECIPROCITY

Funding from for-profit organizations is seen as a potential problem in these relationships, especially if these relationships involve foreign competitors of U.S. firms. The 1993 agreement between the Scripps Research Institute, which receives substantial NIH funding, and Swiss-based Sandoz Pharmaceuticals is an example of this. Under the terms of the agreement, which was scheduled to go into effect in 1997, Sandoz would give Scripps $300 million over 10 years and an option to extend the contract for another 10 years. In return, Scripps, the largest private biomedical research laboratory in the United States, agreed to give all its discoveries to the Swiss firm for the next 20 years (Hilts, 1993a,b).

The Scripps-Sandoz agreement was widely attacked by NIH and Congress for giving Sandoz substantial control over the Scripps research laboratories and their findings, and for encouraging the commercial development of federally funded research by non-U.S. companies. Scripps spends about $100 million per year on research and receives $70 million of this funding from NIH. In response, NIH announced that the exclusive rights to patents from biotechnology discoveries made using these federal funds would be removed. Scripps subsequently agreed to accept a substantially reduced contribution from Sandoz and to modify the terms of the agreement (Hilts, 1993a,b).

Importance of Fellowships, Conferences, and Specialized Journals

An assessment of the factors that facilitate technology transfer must also give attention to the circumstances under which innovation happens. The discovery of DNA cloning, for example, derived from basic research. But the discovery occurred in quite exceptional circumstances—an environment conducive to scientific discovery and the exchange of information.

Inventions almost never happen in isolation. The collaboration by Cohen and Boyer that led to the discovery of DNA cloning was proposed at a U.S./Japan scientific meeting held in Honolulu, Hawaii.[5] Technology transfer in biotechnology depends to a very large extent on the ability of individual scientists or groups of scientists (as opposed to institutions) involved in research to interact with each other. Consequently, the imperative for NBFs is to create close interactions with these academic researchers.

According to Hugh Niall, chief scientist for Genentech, this requires establishing a culture in NBFs as similar to universities as possible. Critical to this is the ability of researchers to move back and forth between industry and academia, and to build their academic credentials by doing this. The immediate consequence of this is a large network of alumni—senior researchers who leave Genentech are usually retained as consultants, for example. A second consequence is that NBFs are able to recruit new scientists to continually renew the research organization. For example, Genentech employed 40 to 50 post-docs (out of 330 researchers) in 1993 in a 2-to-3 year fellowship program funded almost entirely by the company. These post-docs do curiosity-driven research, and many of them go on to careers in academia (H. Niall, personal communication, 1993).

Conferences, professional organizations, and journals are also very useful for technology transfer. Large organizations, for example the Federation of American Societies for Experimental Biology (FASEB), attract 15,000 to 20,000 people to their meetings (National Research Council, 1992b). Many of these organizations also contribute to the internationalization of biotechnology. For example, more than one-quarter of the members of the American Society for Microbiology (ASM) come from outside the United States. These foreign members are seen as active contributors. Japanese members are especially active in molecular biology and fermentation technologies, for example. Foreign authors are also significant contributors to the ASM's many scientific journals (National Research Council, 1992b).

Policy Questions

INTELLECTUAL PROPERTY RIGHTS

Perhaps the most important policy question in university-industry relationships relates to intellectual property rights. Universities are the recipients of the majority of federal funding in basic research, but they are not appropriate institutions for the development and transfer of these findings to clinical practice. To address this, the 1980 Bayh-Dole Act gave universities new patent rights for all discoveries resulting from federally sponsored research, thus recognizing a critical element in the transfer of technology: Industry must have reasonable expectations of being able to recover product development costs (which are extremely high in biotechnology) or it will not participate. Patents and licenses on intellectual property developed at universities are, consequently, an absolute prerequisite for the transfer of biotechnology to clinical practice.

The apparent effect of this initiative on universities is impressive. Before the enactment of Bayh-Dole in 1980, only about 4 percent of the more than 30,000 patents held by the federal government were ever licensed. Now, nearly 50 percent of patents are licensed (National Research Council, 1992a). Leading research universities also expanded their efforts to transfer technology to industry and to enhance their licensing activities. Indeed, in biotechnology, universities were more efficient in generating patents than private industry. Biotechnology companies in the 1980s were realizing more than four times as many patent applications per dollar invested from university research than from their own labs' investments (Blumenthal, 1992).

The problem with these patents is the requirement that patented inventions be described in enough detail that they can be reproduced without undue experimentation. Because microorganisms generally cannot be described in such detail, courts have stipulated that this requirement must usually be met by submitting a sample of the microorganism to a depository. But this gives competitors direct access to the microorganism, increasing the opportunity for patent infringement. Differences between the U.S. first-to-invent system of patents and the first-to-file system used in Japan and most other countries also create problems (National Research Council, 1992b; Olson, 1986).

If the acquisition or enforcement of a patent seems difficult, NBFs may rely instead on trade secrecy laws to protect a product or a process. There are several disadvantages to trade secrecy laws, however. First, they offer no protection against someone who independently discovers the secret. Such a discoverer may then patent the finding and prohibit the original party from using it. Second, trade secrecy prohibits scientists from publishing the results of their own research in the scientific literature. Finally, theft of a trade secret is often difficult to prove (Olson, 1986).

REGULATION

The use of recombinant DNA has been regulated since the technology's inception. In 1976, NIH issued guidelines for genetic research. These guidelines banned certain types of experiments, reflecting the perceived risks. More typically, experiments had to be performed using various levels of physical and biological containment. As experience with recombinant DNA increased, the NIH guidelines were successively revised. Today, the overwhelming majority of experiments using recombinant DNA are exempt from these guidelines (Olson, 1986).

Several limitations are apparent in the NIH guidelines. First, they apply only to institutions that receive federal funds, and the penalty for violating the guidelines cannot extend beyond canceling this funding (although several regulatory agencies do require that NIH guidelines be observed). Second, NIH guidelines focus on research, not on commercial development. The scientific review process used by NIH is inadequate to deal with the volume of commercial development (Olson, 1986).

NIH guidelines have also come in conflict with the rules of several regulatory agencies. Approval for the release of genetically engineered microorganisms into the environment, for example, involves regulatory channels outside the NIH, including those of the Food and Drug Administration (FDA), the Environmental Protection Agency (EPA), and the Department of Agriculture (National Research Council, 1992b).

There are other regulatory problems, as well. These agencies, in many instances, seek several distinct objectives. They have the responsibility to protect human health and the environment from any potential dangers posed by biotechnology. The FDA, for example, requires pharmaceutical companies to demonstrate through a variety of means, including clinical tests on humans, that a new drug is "safe and effective." In the instance of drugs developed using recombinant DNA, the FDA requires them to undergo the entire approval process irrespective of identical approved or existing substances manufactured using identical techniques. The reason for this is concern over the possibility of undetected contamination by drugs or chemicals, or the possibility of genetic instability in a recombinant organism.

Receiving approval for a new drug developed using recombinant DNA techniques is consequently a long and expensive process. Approval of a new drug application (NDA) usually takes 2 years, and the average cost for FDA approval exceeds $200 million (Humphrey, 1995; Olson, 1986). In certain instances, the process can be accelerated. For example, important new drugs for AIDS have recently been approved for use in less than 1 year (Reingold, 1995).

But the FDA is also increasingly under pressure to encourage and facilitate the expansion of commercial biotechnology in the United States. By imposing burdensome regulations, winning approval for biotechnology products will take longer. The current lead the United States enjoys in converting the results of biotechnology into commercial products may therefore be lost to biotechnology firms in other countries with less restrictive regulations.

THE ORPHAN DRUG ACT AND PRICE REGULATION

The viability of the Orphan Drug Act and the pricing of emerging biotechnology drugs are important questions for regulation. Congress passed the Orphan Drug Act in 1983 to encourage development of drugs that, although clinically useful, had no commercial appeal due to very high development costs or that were intended to treat diseases from which fewer that 200,000 people suffered

(Tregarthen, 1992). The Orphan Drug Act provides two incentives, if the drug is approved by FDA for use. First, a company that wins "orphan" designation for its product may receive tax credits for up to 50 percent of the cost of developing and marketing the drug. Second, the company receives an exclusive 7-year monopoly to market the drug for the specific orphan disease (Mossinghoff, 1992; Tregarthen, 1992). FDA has awarded nearly 500 orphan designations. As of 1993, 60 orphan drugs had been approved for use by the FDA (Mossinghoff, 1992).

Seeking FDA approval is risky and expensive. But several important drugs developed under the Orphan Drug Act—for example Taxol, used to treat ovarian cancer, and the AIDS drug AZT (Retrovir)—have demonstrated outstanding commercial potential. Both of these drugs were the result of federally funded research (Tregarthen, 1992), and Congress has expressed concern over the high prices charged by drug companies for medicines developed in this way. In 1986, Wellcome priced AZT at $10,000 for a year's supply but in response to intense public pressure has subsequently reduced the price of the drug to about $2,500 (Tregarthen, 1992; U.S. Centers for Disease Control and Prevention, 1996; Whitehead, 1992).

The Future of Biotechnology

The current view of the future of biotechnology is of continued U.S. leadership in basic research and of technology transfer out of U.S. universities and NBFs into larger pharmaceutical companies in the United States, Japan, and Europe.

Large pharmaceutical companies will continue to expand their presence in biotechnology as the new generation of biotechnology drugs now in development by NBFs enters clinical trials. Many of these pharmaceutical companies will be foreign. Swiss-based Sandoz, for example, has acquired an interest in two U.S. NBFs (Genetic Therapy and Systemix) as well as access to advanced technologies through its long-term agreement with the Scripps Research Institute. Indeed, virtually every European country, as well as Canada, Japan, and the Organization for Economic Cooperation and Development, has developed programs to exploit this "new" biotechnology (Gaden, 1991). These programs can be expected to increase international technology transfer in biotechnology, although—with perhaps a few exceptions—the pattern of technology transfer in these programs will continue to represent a net flow of technology out of the United States.

THE DEVELOPMENT AND TRANSFER OF MANUFACTURING AND PRODUCTION TECHNOLOGIES TO U.S. COMPANIES

Robert K. Carr

Introduction

Technology for manufacturing and production is difficult to discuss as a homogenous entity. Unlike software, biotechnology, and electronics, the three other sectors studied in this document, manufacturing and production technologies overlap a number of traditional academic and industrial categories. Manufacturing and production technologies can be product, firm, or industry specific. Broadly defined, they can include parts of associated technologies, such as materials and environmental cleanup, and are an important feature of industries as diverse as biotechnology, automobiles, and microchips. Furthermore, a modern definition of production technology includes not only machinery and other hardware, with associated software and computing and communications, but also a range of "soft" technologies, including just-in-time production techniques; lean, flexible, and agile production; total quality management; and a host of other new ways of doing things better. Finally, while the research base for manufacturing and production technologies is centered in engineering, it draws heavily on advances in materials, software, electronics, and other academic disciplines.

The following case study examines the infrastructure supporting the development and transfer/diffusion of production and manufacturing technologies to private industry in the United States. It is important to recognize from the outset that the vast majority of the research and technology transfer of production and manufacturing technologies takes place entirely within the private sector. The research and technology transfer resources at the disposal of private-sector firms will be examined only briefly, however. The focus here will be on the roles of nonmanufacturing research institutions and intermediaries (e.g., universities, federal laboratories, and nonprofit R&D institutions) in the development and transfer of production and manufacturing technologies through public-sector technology transfer networks and other nonmanufacturing technology transfer agents.

It may be useful to distinguish between large and/or R&D-intensive manufacturing firms on the one hand and small and medium-sized firms (SMEs) on the other. The majority of industrial R&D in the United States is performed by large firms. The top 10 R&D-intensive firms perform fully 25 percent of all industrial R&D, and all large firms (with 5,000 employees or more) perform 73 percent (National Science Foundation, 1996c). Furthermore, these large manufacturing firms have their own infrastructure for developing or acquiring manufacturing technologies and for interfacing directly with academic, government, and nonprofit R&D entities. Thus, they tend to be less involved in the organizational frameworks that exist to support the transfer of manufacturing technology from the government, nonprofit, and academic sectors.

On the other hand, the 375,000 or so SMEs, which constitute fully 98 percent all of U.S. manufacturers, perform little or no R&D. Many of these firms do not produce finished products and are not well known in the marketplace or among U.S. exporters. Nonetheless, they are critical elements in the "food chain" of the manufacturing sector, accounting for up to 60 percent of the cost of manufactured goods (National Research Council, 1993). Therefore, their efficiency and productivity have a major impact on the overall competitiveness of the manufacturing sector. For this reason, the focus in this case study will be on technology transfer programs where small and medium-sized firms are the recipients.

Developing New Manufacturing Technologies: The R&D Base

Given the diverse nature of production and manufacturing technologies, the wide range of industries in which they are used, and the very large number of manufacturing firms, it should come as no surprise that gathering precise data on manufacturing R&D poses difficulties. In academia and government, R&D tends to be classified according to academic discipline or government mission, rather than by crosscutting categories such as manufacturing. In industry, data on R&D is collected (primarily by the NSF) according to industry as defined by SIC codes, and by firm size. Nonetheless, some data for production and manufacturing R&D can be identified.

PRODUCTION AND MANUFACTURING R&D IN INDUSTRY

Industry funds and performs the lion's share of U.S. research and development, providing 57 percent of all support for U.S. R&D and accounting for 72 percent of U.S. R&D performance. According to the National Science Foundation (1996c), manufacturing firms accounted for 74 percent, or roughly $87 billion, of the R&D performed by industry in 1993. However, process-oriented R&D is not distinguished from new-product R&D in the NSF data. Furthermore, not all process and manufacturing technology R&D is carried out in the manufacturing sector; service firms that support the manufacturing sector are also active in this area.

Compared with other nations, the level of industrial R&D devoted to production processes in the United States is low. In a 1988 study, Edwin Mansfield stated that American firms "devote about two-thirds of their R&D expenditures to improved product technology and about one-third to improved process technology." He contrasted the ratio of U.S. process-to-product-oriented R&D with that of Japan, where the proportions are reversed (Mansfield, 1988). Another study (National Science Board, 1992) found that only 19 percent of U.S. industrial R&D was devoted to process innovation. These two studies indicate that U.S. industry performed between $23 billion and $40 billion in process-related R&D in 1995 (National Science Foundation, 1996c).

Recent data from an ongoing study (Whiteley et al., 1996) by the Industrial Research Institute and Lehigh University's Center for Innovation and Management Studies (CIMS) have confirmed these estimates and provided additional survey data on the division between product and process development in several industry areas (Figure A-1). The IRI/CIMS survey divides R&D into basic and applied research, product and process development, and technical services (support for existing products/processes in the field). For the 87 firms surveyed in 1994, 22.5 percent of their R&D efforts were process related and 41.8 percent were product related. Basic and applied research consumed 17.1 percent of their efforts and technical services consumed 18.5 percent. The figures for 1993 were quite similar. However, as can be seen from Figure A-1, the ratio of product-to-process-related R&D varies widely by industry group.

FEDERAL SUPPORT FOR PRODUCTION AND MANUFACTURING R&D

The federal government's role in production and manufacturing technologies includes both funding research in the university, private, and nonprofit sectors, as well as conducting process-oriented R&D within federal facilities. The available data do not permit one to precisely separate federal funding and federal performance of production and manufacturing R&D. However, the nature of funding agency missions provides a general idea. In 1993, the Federal Coordinating Council for Science, Engineering and Technology (FCCSET) prepared several S&T initiatives for the FY 1994 budget. One of these initiative areas was Advanced Manufacturing Technology, while another, Advanced Materials and Processing, was supportive of manufacturing technology. These two initiatives no longer exist in their former form, but the budget figures reported for FY 1994 provide a good sense of the level of federal activity in this area. To be sure, much of federal R&D in advanced manufacturing as well as in materials was intended to meet unique federal requirements in defense and space. Nonetheless, there have frequently been significant spin-offs from such activity.

The FY 1994 federal total for advanced manufacturing was $1.385 billion and for advanced materials was $2.061 billion (Federal Coordinating Council for Science, Engineering and Technology, 1993). The agencies with the largest budgets for advanced manufacturing programs were the Department of Defense (DOD) ($596 million, including the Technology Reinvestment Project, or TRP), the Department of Energy (DOE) ($367 million), the National Institute of Standards and Technology (NIST) ($141 million), and the National Science Foundation ($130 million). In the advanced materials initiative, DOE had the largest budget ($946 million), followed by DOD ($422 million), NSF ($328 million), the National Aeronautics and Space Administration (NASA) ($131 million), and the Department of Health and Human Services ($93 million). Almost all of the figures for NSF represent federal funding of extramural research. Most of the other agencies performed the lion's share of their advanced manufacturing and materials research in their own laboratories, with a small part sent to external performers.

The National Science and Technology Council (NSTC), the successor to FCCSET, changed the nature of the FCCSET initiatives, but there are still activities relevant to manufacturing technology. The Manufacturing Infrastructure Initiative is designed to support R&D and other activities that support the entire manufacturing sector. The initiative consists largely of ongoing programs, most of which are described separately in this section. The National Electronics Manufacturing Initiative (NEMI) was launched in response to industry interest in forming a partnership with the federal government to assess the technology needs of electronics manufacturing. NEMI is not a set of specific programs or projects,

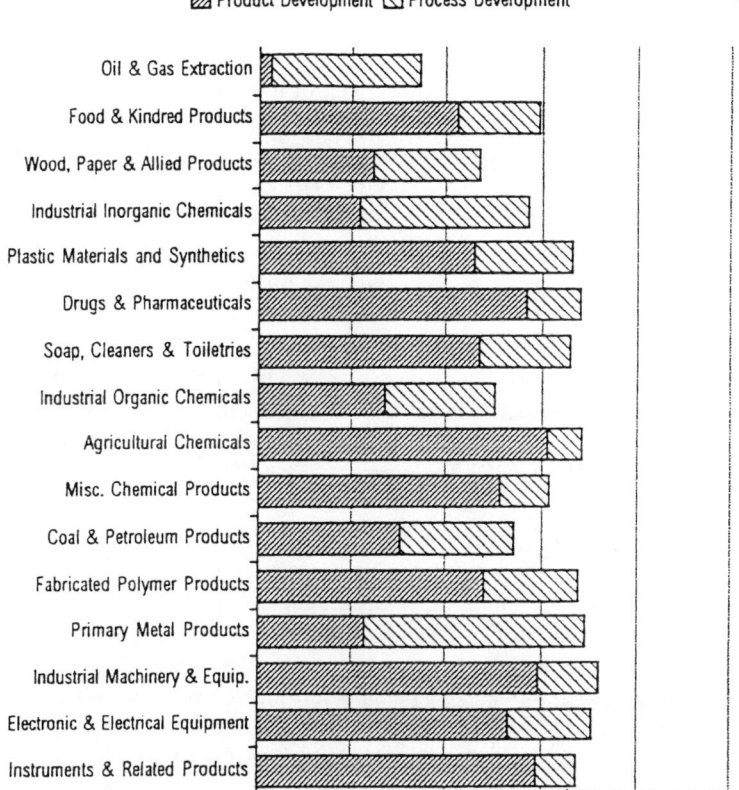

FIGURE A-1 Allocation of R&D funds for different industries: product vs. process development, fiscal year 1994. SOURCE: Whiteley et al. (1996).

but rather a way of doing strategic planning and partnering. The Materials Technology Initiative inherited by NSTC has been folded into other NSTC initiatives where its subcommittees provide support to other activities. For the most part, these initiatives include few, if any, new R&D programs among federal agencies or significant new budgets for work in these sectors. They are best thought of as frameworks for budget presentation, coordination, and reporting of manufacturing activities in diverse federal R&D programs.

Defense Manufacturing Programs

The major DOD activities in manufacturing R&D include the TRP and the Manufacturing Science & Technology program (MS&T). TRP funded a number of technology extension activities in 1993 and 1994, but has recently been converted to a military-specific program and no longer funds industrial extension programs. Although the TRP funded some projects that included the development of new production technologies, most of its activities in support of manufacturing were in the deployment arena. The MS&T program, a successor to the ManTech program, consists of manufacturing R&D activities by the services, the Defense Logistics Agency (DLA) and the Defense Advanced Research Projects Agency (DARPA). The MS&T program is managed by the Director for Defense Research and Engineering (DDR&E) and, in 1995, was funded at $329 million. MS&T "matures and validates emerging manufacturing technologies to support low-risk implementation in industry and DOD facilities." MS&T programs are carried out with all types of organizations, including DOD contractors, suppliers, hardware and software vendors, industry centers of excellence, industrial consortia, universities, and research institutes. Cost sharing is part of the MS&T program, particularly with industrial partners, who may bear a considerable portion of their own costs for an MS&T program.

Department of Energy Laboratories

Most of the large multiprogram laboratories of the Department of Energy have manufacturing programs. Two DOE laboratories stand out in this regard: Sandia National Laboratories and Oak Ridge National Laboratory. Sandia is primarily an engineering laboratory, and therefore manufacturing programs (particularly those related to weapons manufacture) have been critical to its mission for many years. Oak Ridge, particularly its Y-12 facility, has long been engaged in manufacturing research. The Oak Ridge Center for Manufacturing Technology has been formed to coordinate manufacturing technology programs at that facility. DOE laboratories tend to use CRADAs to work with industry in manufacturing as well as in other areas. Although most manufacturing R&D in DOE laboratories supports weapons manufacture, much of it (particularly in areas such as precision machining) is transferable to civilian uses. DOE laboratories also have substantial activities in the area of materials. Some of this work is unique to government (e.g., plutonium), while other aspects of it, like advanced manufacturing, is transferable.

NIST Laboratories

NIST is the only federal laboratory that has industry as its principal client. An important part of NIST's service to industry is the Manufacturing Engineering Laboratory (MEL). With a staff of 300, MEL works with U.S. manufacturers to develop and apply technology, measurements, and standards. MEL operates the National Advanced Manufacturing Testbed, the successor to the Automated Manufacturing Research Facility established over a decade ago. In addition to the MEL, NIST also operates the Materials Science and Engineering Laboratory.

Much of the work of this facility is relevant to production and manufacturing technology.

National Science Foundation

NSF has a collaborative manufacturing research effort among several NSF directorates that supports manufacturing in several ways. NSF spends about 4 percent of its budget (about $300 million in 1995) on manufacturing-related grants and programs (National Research Council, 1995b). Investigator-initiated R&D projects that support development of the fundamental science and engineering base underlying manufacturing technology are part of NSF's traditional peer-reviewed grant program. In addition, NSF funds two types of engineering research centers that support manufacturing. The centers, described below, are expected to become fully self-supporting.

Engineering Research Centers (ERCs) are engaged in cross-disciplinary research and education activities that are important for U.S. competitiveness. ERCs are located at universities and promote links between research and education. The ERC program was begun in 1985 and currently supports 23 centers involving 100 participating academic institutes and almost 600 nonacademic partners. NSF contributed $51 million to the centers' operation in 1995, while all other sources contributed $96 million. Almost 5,000 people (researchers and students) utilized center facilities in 1995. The 18 ERCs in operation at the end of 1994 were distributed according to their major technology focus as follows (National Academy of Engineering, 1995a):

Design and Manufacturing	5
Materials Processing for Manufacturing	3
Optoelectronics/Microelectronics and Telecommunications	4
Biotechnology/Bioengineering	3
Energy and Resource Recovery	2
Infrastructure	1

Industry/University Cooperative Research Centers (I/UCRCs) and State/Industry University Cooperative Research Centers (S/IUCRCs) are also located in universities and focus on fundamental research areas recommended by industrial advisory boards. S/IUCRCs are similar to I/UCRCs, but are more closely focused on state or regional economic development and are initiated by states with industrial support. I/UCRCs have been in existence since 1973, with the first S/IUCRC added in 1991. In 1995, NSF contributed $8 million to 67 cooperative research centers, while all other sources contributed $79 million. Almost 2,300 people (researchers and students) utilized I/UCRC and S/IUCRC facilities in 1995.

PRODUCTION AND MANUFACTURING R&D IN ACADEMIA: UNIVERSITY INDUSTRY RESEARCH CENTERS (UIRCs)

Although research relationships between industry and universities date back to the late 19th century, the wholesale formation of centers involving industry-university collaboration is a relatively recent phenomenon, dating from the 1970s and 1980s. As of 1991, there were over 1,000 UIRCs with a total estimated budget of $4.12 billion, of which $2.53 billion was devoted to R&D (most of the balance was for educational activities) (Cohen et al., 1994). A relatively small number of centers receive support from the NSF. Overall, government provided 46 percent of UIRC funding (federal sources accounted for 34.2 percent; state sources for 12.1 percent) while industrial participants provided 30 percent of the centers' financial backing (representing over 70 percent of industry's financial support for academia) as well as additional noncash support.

UIRCs are quite diverse in their size, organization, relationship to industrial needs, and research activities. Some have a traditional academic orientation, pursuing research for its own sake, while others (about 25 percent) are focused on the goals of industry. According to Cohen et al. (1995), "More than one-quarter of UIRCs conduct R&D relevant to the manufacturing sector exclusively, while more than two-thirds conduct R&D that is relevant to both the manufacturing and non-manufacturing sectors."

The technology focus of UIRCs includes significant emphasis on manufacturing (Cohen et al., 1995). Almost 20 percent of UIRCs carried out research in manufacturing technologies, while 30 percent of centers were involved in environmental technology and waste management and 27 percent in advanced materials, the latter two technologies being of considerable interest and concern to manufacturers. Another study (Dickens, 1995) identified 1,030 university-based engineering research units (including UIRCs) at 154 universities. A survey of the directors of these units revealed that 45 percent were working in materials, 42 percent in energy and environmental technologies, 29 percent in manufacturing, 27 percent in information and communications, 17 percent in transportation, and 13 percent in biotechnologies and life sciences.

Although UIRCs are the most visible type of university-industry research cooperation, the establishment of a UIRC is not essential for such interaction to occur. Many universities and their departments (particularly engineering) have relationships with industry, receiving support for research activities and contributing knowledge to industrial sponsors.

OTHER CENTERS OF MANUFACTURING R&D

In addition to government, universities, and industry, independent nonprofit institutions also engage in R&D related to manufacturing. They make up by far the smallest of the four groups in terms of their numbers and R&D budgets. Their manufacturing R&D activities are proportionately smaller, in part since over half of nonprofit independent R&D institutes conduct work primarily in biomedical and other areas not directly relevant to manufacturing. Nonetheless, there are some significant centers of excellence in manufacturing research in organizations such as the Battelle Memorial Institute, SRI International, and Southwest Research Institute. These types of institutions are frequently called upon to solve industrial problems relating to production and manufacturing R&D.

Transferring Manufacturing Technology to Industry

There is substantial need for and substantial barriers to the acquisition of modern technology by small firms. At the end of the 1980s, over three-fifths of the machine tools used by U.S. manufacturers were over 10 years old, and more than one-quarter were over 20 years old (Shapira, 1990). Furthermore, more recent data indicate that while larger firms are modernizing, smaller firms continue to lag behind. Table A-4 shows the percentage of large and small U.S. firms that have adopted nine key types of modern production technology. Small U.S. firms lag larger firms substantially in this regard. Both large and small U.S. firms lag their Japanese counterparts in all but one category. As well as lacking modern equipment, U.S. small manufacturers tend to have neither highly trained staff nor modern operating methods. They are often content with this arrangement because it is similar to that of their nearby competitors, and it is often still perceived as sufficient for corporate survival.

Transferring Manufacturing Technology from Federal Laboratories and Universities

Federal laboratories transfer technology through a number of mechanisms, but three are particularly relevant to manufacturing: licensing, cooperative R&D, and technical assistance. These three mechanisms are used to different degrees by different laboratories. NASA centers have tended to focus on technical assistance, while DOE laboratories have preferred cooperative R&D. Licensing is the least-used mechanism, particularly for manufacturing technologies. The DOE reported in late 1994 that manufacturing technologies accounted for 18 percent of all DOE CRADAs. Two closely related areas, advanced materials and instrumentation, and pollution minimization and remediation, accounted for 18 percent and 12 percent, respectively, of CRADAs (U.S. Department of Energy, 1994).

Many federal laboratories have active technical assistance programs for manufacturers. While some labs have been offering technical assistance to firms in their immediate vicinity for many years, more and more federal laboratories are becoming technology sources in state-run industrial extension services, and their technical assistance activities will begin to be reflected in the data collected by these state extension services.

The National Technology Transfer Center (NTTC) refers callers (mostly from the business community) to sources of technology in federal laboratories. In one

TABLE A-4 Use of New Technology in Manufacturing, Japan and the United States, 1988

Type of New Manufacturing Technology Japanese Definition (closest U.S. definition in parentheses)	Technology Users as a Percentage of Small and Large Manufacturing Enterprises[a]				Large/Small Ratio		Japan/U.S. Ratio	
	Japan		U.S.		Japan	U.S.	Small	Large
	(a) Small	(b) Large	(c) Small	(d) Large	(e)	(f)	(g)	(h)
Numerically controlled and customized numerically controlled machine tools (NC.CNC machine tools)	57.4	79.4	39.6	69.8	1.4	1.8	2.0	1.1
Machining centers (FMS cells or systems)	39.4	67.4	9.1	35.9	1.7	4.0	7.4	1.9
Computer-aided design (and computer-aided engineering)	39.1	75.2	36.3	82.6	1.9	2.3	2.1	0.9
Handling robots (pick-and-place robots)	22.6	62.2	5.5	43.3	2.8	7.8	11.2	1.4
Automatic warehouse equipment (automatic storage and retrieval)	10.9	44.9	1.9	24.4	4.1	13.1	24.1	1.8
Assembly robots (other robots)	8.3	41.4	3.9	35.0	5.0	8.9	10.6	1.2

NOTES:

[a] The comparisons between Japan and the U.S. are approximate since differences exist in technology definitions and employment size categories. Additionally, the Japanese data are enterprise-based, while the U.S. data are establishment-based. The Japanese define a small manufacturing enterprise as having less than 300 employees and a large enterprise as having 300 employees or more. The U.S. government defines small enterprises as firms with 50 to 499 employees and large enterprises as firms with 500 employees or more.

(e) = (b)/(a).
(f) = (d)/(c).
(g) = (a)/(c).
(h) = (b)/(d).

SOURCES: After Shapira et al. (1992, p. 5).
(a),(b) Ministry of International Trade and Industry (1989).
(c),(d) U.S. Department of Commerce (1989).

13-month period (from June 1, 1994, to June 30, 1995), requests for assistance in the area of manufacturing technology ranked second, accounting for 9 percent of all inquiries, while materials sciences and environmental pollution and control accounted for 16 percent and 7 percent, respectively. Of the firms making requests during that period, over 70 percent were small businesses, most with fewer than 100 employees.

Small Business Development Centers (SBDCs), operated by the Small Business Administration (SBA), provide educational and research resources to small businesses. There are over 900 SBDCs in operation, providing direct counseling to small business owners and managers. SBDCs are initiated at the state level and funded by state governments as well as the SBA. SBA funds go to a state university or economic development agency, which serves as the "lead center" in the state, with subcenters established at other educational institutions and chambers of commerce. In 1995, federal funding for the SBDC program was $73.5 million, while matching state funds amounted to $81.6 million. Individual SBDCs vary according to geographic area and in terms of clients and services offered, but many actively support small manufacturers. SBA has a collaborative working relationship with the Manufacturing Extension Partnership (MEP) at NIST. Many, if not most, SBDCs are integrated into state technology extension programs and are part of the network of service providers available to small businesses. In states where active industrial extension networks resolve manufacturers' technology problems, SBDCs tend to focus on managerial issues such as finance.

Universities transfer technology primarily through licensing, the formation of spin-off companies, faculty consulting, cooperative R&D (particularly in UIRCs), and the flow of graduates to private firms. There is little information on the technology areas of university licensing and faculty consulting, and therefore it is difficult to know what percentage of these activities is related to manufacturing. It is likely that most of the manufacturing technology flows from engineering programs through faculty consulting and graduates as well as from cooperative R&D in UIRCs. It is probable that much of this technology is at the high end, useful primarily to large or technologically sophisticated firms. As is the case with federal labs, some university centers and engineering departments have become resources for state extension networks. Community colleges are even more frequently involved in state industrial modernization and extension systems.

UIRCs are active in technology transfer to their industrial sponsors. According to the Cohen et al. (1995) study, almost two-thirds of the UIRCs indicated that transferring technology to industry was "important," even though only a small percentage of the effort of centers was specifically devoted to technology transfer. The centers reported that collaborative R&D, exchange of research personnel, delivery of prototypes or designs, and informal contacts were the most effective technology transfer mechanisms. Although not surveyed as a technology transfer mechanism, the flow of students to industry is clearly another important avenue of transfer. The 1995 study indicates that industrial sponsors of UIRCs hire a substantial number of center graduates students.

Transferring Manufacturing Technology: Industrial Extension Programs

THE AGRICULTURE MODEL

Development of the agricultural research and extension system began in 1862, when the federal government established agricultural colleges, run by the states, to offer practical instruction in agriculture. Fifteen years later, the federal government established a system of state agricultural experiment stations, again under the state auspices. Finally, the Cooperative Agricultural Extension Service (AES), a partnership among federal, state, and county governments, was created by the Smith-Lever Act in 1914. The AES has grown into a nationwide system with more than 9,600 county agents and 4,600 university researchers. Funding is spread among federal, state, and local governments.

In an era of farming by large agribusiness, some believe the extension service no longer plays a critical role. Still, when the AES first came into existence, farmers were small businessmen who ran into many of the same barriers to technological advancement as do small manufacturers today. Thus, when manufacturing extension was first considered, the Agricultural Extension Service was an obvious model. However, in spite of the parallels, there are some critical differences between the model and the realities of modern manufacturing. For one thing, farmers in a local area tend to have the same problems, while manufacturers' problems are often very different.

BOTTOM-UP APPROACHES: STATE AND LOCAL PROGRAMS

Some states have long recognized that small manufacturers, like farmers, had much to gain from technical assistance. While most state efforts in the technology area initially focused on research and development, a few states created technical assistance programs. North Carolina began such a program for manufacturers in 1955, and Georgia followed in 1960. Both of these programs are still flourishing. North Carolina's cooperative technology program is the country's largest, with an annual budget of $37 million, while Georgia's now ranks fourth at $30 million (Coburn, 1995). Pennsylvania followed suit in the mid-1960s with the Pennsylvania Technical Assistance Program (PENNTAP). By the end of the 1970s, Maryland, Massachusetts, Michigan, New York, Ohio, and Virginia had begun their own extension programs.

By 1994, 40 states had technology extension programs (Coburn, 1995) in addition to other efforts to assist manufacturers. Approximately half of the state programs were operated by educational institutions, with the balance managed by nonprofit organizations or state agencies. These programs offer different types of services, including supply of technical information, seminars and workshops, demonstrations, referral of consultants and other experts, and in-plant consultation. However, intensive field assistance (generally agreed to be the most effective technique) was provided by only a few programs (Shapira et al., 1995).

In *Modernizing Manufacturing*, Philip Shapira (1990) groups state industrial extension programs into four categories:

Technology broker programs focus on providing technical information and referrals for client firms. Typically, these programs have large numbers of requests, each of which receives modest attention from program staff (generally less than a day). In 1993, PENNTAP assisted 490 client firms (the majority of them small firms) and handled 700 requests with a staff of eight at no cost to the requesting firms.

University-based field office programs generally make engineers available to assist firms in a wide range of problem areas. By virtue of being university based, these programs can easily access engineering faculty or R&D centers for assistance. Service is normally provided for free. One of the oldest and largest such programs, the Georgia Industrial Extension Service, is operated by the Georgia Institute of Technology. Founded in 1961, it has 13 regional offices (being expanded to 17), which served over 1,000 companies and communities in 1994 using $1.55 million in state funds.

Technology centers and state-sponsored consulting services are not part of a university, although they may have links to one. Their focus is generally on technology modernization (i.e., technology assessments, upgrade recommendations, implementation, training, etc.). Assistance is often provided by private consultants subsidized or paid for by the state program. Pennsylvania's Industrial Resource Centers (IRCs) are an example. Established in 1988, the IRCs are private nonprofit corporations operated by private-sector boards. Seven of the IRCs serve the traditional manufacturing sector, and one is devoted to the support of the biotechnology industry. They provide assistance with their staff and through private consultants. Initial assistance is generally free, but in-depth assessments and services by outside consultants require some company payment.

Manufacturing networks are regional networks of firms that cooperate in technology diffusion, training, design, finance, and marketing. To a certain extent, these networks have been influenced by the successful small-firm networks in Northern Italy. In this country, the Southern Technology Council (STC) has established networks in North Carolina and Arkansas, which involve community colleges and economic development authorities along with local firms. The Arkansas Industrial Networking Project was created by the Arkansas Science and Technology Authority with a $90,000 grant from the STC. The project's goal is to improve the competitiveness of small manufacturers by facilitating cost sharing for R&D, purchasing, training, and expensive technology, and encouraging cooperation on contract bidding. The Arkansas networks, which are focused on the wood-products and metal-working industries, involve about 100 companies.

Shapira (1990) surveyed state industrial extension programs for information about the types of assistance they provided to client firms. The most frequently offered services, in descending order of popularity, were: (1) improve/solve problem with existing production technology; (2) identify vendor of new technology/software; (3) specify new production/process technology; (4) refer client to training source; (5) improve quality control/statistical process control; (6) improve existing plant/layout operations; (7) identify new markets; (8) address waste management/environmental problems; and (9) improve/debug and existing product. Service type (1) was by far the most frequently sought, while types (2), (3), and (4) were roughly equal in importance.

In addition to industrial extension programs, the states engage in a number of other activities in support of their technology base. Coburn (1995) identifies a number of different models, including the following:

University-industry technology centers (UITCs), which feature interdisciplinary research in areas relevant to industry. These centers are either university based or are operated by a nonprofit in close association with a university. They are often supported by federal agencies, as in the case of the NSF engineering centers. Kansas, New Jersey, New York, and Ohio have strong center programs.

Government-industry consortia are groups of firms that, with other R&D institutions, such as universities, focus on research in a given area. Although some states sponsor such consortia, they are more typical at the federal level.

University-industry research partnerships are similar to consortia but are project centered and have a start and end date. Arkansas, Connecticut, Kansas, and Maryland sponsor such programs.

Equipment and facility access programs provide state firms access to expensive and sophisticated equipment and facilities and associated staff expertise. Colorado, Maryland, North Carolina, and Pennsylvania, among other states, have such programs.

Technology financing programs provide capital to firms and to specific projects under a broad spectrum of arrangements including grants, low-cost loans, guarantees to third-party lenders, and investment in exchange for equity.

Start-up assistance includes state-supported incubators for new businesses as well as research parks, where high-tech companies can obtain both research space and (usually) access to a nearby source of technology such as a research university.

TOP-DOWN APPROACHES: FEDERAL MANUFACTURING TECHNOLOGY PROGRAMS

From the 1960s to the 1980s, industrial modernization programs were found only at the state and local level. Federal manufacturing programs provided only limited and uncoordinated support for these efforts and were focused primarily on basic research and the defense sector (Shapira et al., 1995). In 1988, Congress enacted a mandate for the federal government (through NIST) to assist state industrial development efforts, and, in 1992, the Clinton administration expanded substantially the manufacturing technology efforts that NIST had undertaken up to that date.

Manufacturing Extension Partnership

The 1988 Trade and Competitiveness Act established at NIST a new MEP program. By 1992, seven Manufacturing Technology Centers (MTCs) had been established under MEP, initially with the goal of transferring advanced technology from NIST's Advanced Manufacturing Research Facility and other federal labs. However, it quickly became obvious that what most manufacturing firms needed was proven, off-the-shelf technologies. Therefore, the MTCs shifted their emphasis to helping small and medium-sized firms adopt less-advanced technologies, including "soft" technologies such as training, management, and networking.

As of October 1995, 42 states and Puerto Rico had established or were planning to establish manufacturing extension centers, and over 60 individual centers are currently affiliated with MEP. These centers employ 2,500 agents in over 250 field offices. MEP also operates the State Technology Extension Program (STEP), which provides grants to states to plan and begin manufacturing extension services, although by the end of 1995, only a few states were still without industrial extension programs. To assist states to evaluate and improve their centers, MEP is developing a uniform system of program evaluation.

Since the program's inception, the resources devoted to MEP have increased dramatically. Direct appropriations to the MEP program can be seen in Figure A-2. However, the *total* funding for extension activities, including MEP and TRP as well as state, local, and private matching funds was over $250 million in 1994, a threefold increase in industrialization funding from just 2 years earlier (Shapira et al., 1995). MEP support for individual manufacturing technology centers is supposed to end after 5 years, presumably after the centers have become self-sufficient. However, recent evaluations have called into question whether the centers can continue to operate for a longer (perhaps indefinite) period without federal funding.

While each center tailors its services to meet the needs dictated by its location and manufacturing clients, some services are common to most extension centers. These include:

- assessment of technology and business needs
- definition of needed changes, and
- implementation of improvements.

Many centers also assist companies with quality programs, employee training, workplace organization, business systems, marketing, and financial issues.

MEP has created links with a number of affiliated organizations that can provide assistance to manufacturing extension programs or directly to small businesses. For example, MEP and EPA have launched a program to assist smaller manufacturers solve environmental problems before they become regulatory concerns. MEP also conducts research to better understand the barriers to modernization faced by small manufacturers and to discover additional ways to overcome these barriers.

The Technology Reinvestment Project

The Technology Reinvestment Project (TRP), created by the 1992 Defense Authorization Act, made a substantial down payment on the U.S. industrial modernization system. TRP had three areas of focus: technology development, technology deployment and diffusion, and manufacturing education and training. In its first 2 years, TRP funded the majority of the manufacturing extension centers that make up the MEP. Since 1994, TRP has ceased providing grants for technology deployment (i.e., extension) activities, but during FY 1993 and FY 1994, TRP awarded $223 million to 95 separate projects. This funding was in addition to MEP direct appropriations (Figure A-2). Required matching funds totaled somewhat more than that figure, meaning that nearly half a billion dollars worth of new technology extension activities were funded in a 2-year period as the result of TRP grants (U.S. Department of Defense, 1995).

TRP also funded projects to increase manufacturers' access to federal technologies, particularly those in federal laboratories. TRP no longer exists, and its remaining activities and reduced funding were recently transferred to the Dual-Use Technology Office of the Office of the Secretary of Defense. The new program has been redirected toward more military-specific projects.

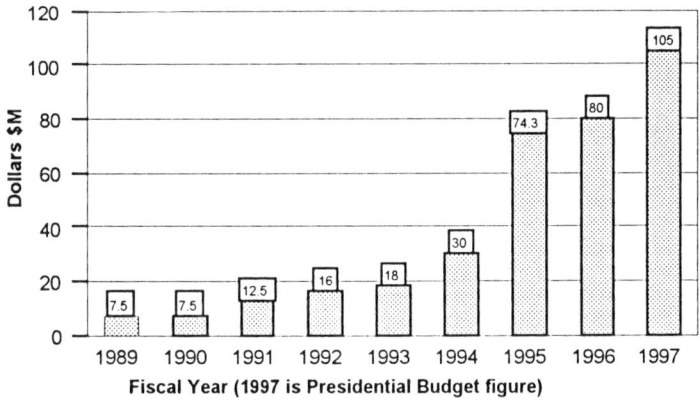

FIGURE A-2 MEP appropriations, including 1995 recision and 1996 continuing resolution. NOTE: 1997 figure is from President's budget. SOURCE: Unpublished data from National Institute of Standards and Technology.

Federally Sponsored Consortia

Federally sponsored consortia include groups such as SEMATECH, a consortium of microchip manufacturers that has received funding for about half its operating costs from DARPA, as well as the National Center for Manufacturing Sciences (NCMS), which receives funding from the Defense Department's MS&T program and elsewhere. Both of these consortia are working to develop new production technologies in different industry sectors. Although federal laboratories play an active role in the research of these consortia, the primary federal input is money. Technologies developed by these consortia are transferred to their member firms. Although the members are generally active in planning consortia R&D activities and are presumably interested in the results, effectively transferring the results of those activities has proved to be a difficult management issue. Most federally funded consortia are registered under the National Cooperative Research Act of 1984 (P.L. 98-462), although they represent a small percentage of all the ventures so registered.

The National Cooperative Research Act

The National Cooperative Research Act of 1984 grants special antitrust treatment to joint research and development ventures and consortia that conduct research, analysis, experimentation, or testing. Industrial participants can protect themselves from possible treble damages imposed under antitrust laws by registering with the Department of Justice. In 1993, the act was amended (P.L. 103-42) to add the same antitrust protection to industrial participants in joint production ventures. This legislation has facilitated the formation of several R&D consortia, including the federally funded consortia described above, that are engaged in research related to manufacturing. The NCMS is one example of a consortium whose formation was made possible by the act.

TRANSFER OF MANUFACTURING TECHNOLOGY WITHIN THE PRIVATE SECTOR

Several mechanisms exist for the transfer of production and manufacturing technologies within the private sector. The most commonly used are:

- Vendors and suppliers of manufacturing technology, which are often an excellent source of information and assistance for their clients. Obviously, their interests are not always identical to those of their clients, since they are primarily interested in selling their equipment. Many smaller manufacturers do not know how to evaluate proposals from vendors and have difficulty making an informed choice among many competing suppliers. However, if a choice can be made, suppliers are often the source of considerable assistance in plant reorganization, training, and other changes that will let customers take full advantage of new equipment, even though their support tends to diminish with time.

- Professional and trade associations, which are a potential source of technical information for some firms. Professional associations (particularly engineering associations) have considerable information about advanced manufacturing technologies. For example, the Society of Manufacturing Engineers produces and markets a newspaper, a refereed academic journal, two trade magazines, seven special-interest technical newsletters, hundreds of video-based training programs and reference books, and thousands of technical papers. The society also features a manufacturing-oriented library and an on-line electronic information service. However, as noted above, many manufacturing engineers do not belong to professional and trade associations.
- Consultants and service firms, which are a major source of new process technology for U.S. manufacturers. Such firms tend to be hired by larger manufacturing companies, and thus their approach to problem solving will tend to reflect solutions appropriate to larger manufacturers. Firms at the smaller end of the spectrum are often not able to afford such consultants, and those that can are often not able to implement the consultant's recommendations without continuing assistance. As noted above, some state-sponsored manufacturing extension programs fund or subsidize private consultants to deliver services. The state of New York spends several million dollars annually on private consultants for small manufacturers (National Research Council, 1993).
- Supplier development programs, which have been used by a number of large manufacturers to improve the quality and efficiency of their suppliers. Since as much as 80 percent of the cost of products such as airplanes, automobiles, and computers may be purchased from outside suppliers, large manufacturers have a strong interest in the performance of the supplier community. Studies indicate that close relationships between suppliers and their customers may induce suppliers to adopt more modern technology (National Research Council, 1993). Although supplier development programs can be an excellent source of assistance in adopting new technologies, they generally do not reach below the first tier of the supplier chain. The nature of these customer-led programs varies widely, from general reviews of supplier progress and advice to comprehensive customer-mandated programs to help suppliers meet mandated quality and other standards.
- The American Supplier Institute (ASI) was formed in 1981 to provide training to suppliers to the automotive industry and is now chartered as a nonprofit educational institute in Michigan. Its board of directors includes representatives of major automotive companies and American universities. ASI focuses on quality and management improvement programs, using the ideas of W. Edwards Deming, Genichi Taguchi, and others. It does not provide technical assistance, per se, for manufacturers, but the

quality and management programs generally lead to improvements in all areas of firm activity.

Analysis and Trends

The organization of manufacturing R&D and extension services in the United States reflects the decentralized nature of America's governmental structure and of the nation itself. The diverse state and federal programs designed to assist industry have frequently been criticized as uncoordinated and confusing to access, although states are beginning to establish single points of contact for industrial modernization programs. Such actions may improve coordination from the point of view of the company. However, on the federal level, programs are authorized, funded, and operated in different areas of the government, and coordination is likely to continue on an informal level at best.

Although a model for public-sector manufacturing extension programs existed for nearly a century in the agricultural sector, no pressing need was felt to extend it to manufacturing until global competition began to challenge U.S. economic performance. Until that time, federal policymakers were not generally concerned by the state of the U.S. manufacturing base, and public-sector industrial modernization programs of the time were efforts of individual states to protect their firms against competition from other states. Even now, one might ask why government should not leave the task of modernizing small firms entirely to the private sector and the marketplace.

However, as shown in Table A-4, the market has not brought the same rate of utilization of new production technologies to SMEs as it has to large companies. The slow pace of small-firm modernization has an impact on the economy far beyond the community of small manufacturing firms. Lack of modern equipment, techniques, and management practices reduces the productivity and quality of small manufacturers and, since these firms account for such a large portion of the value of final products, of the entire manufacturing sector, a critical part of the U.S. economy. Why is the rate of modernization among SMEs slower than among large firms? Primarily it is because there are structural barriers to small-firm modernization that large firms do not face.

At the firm level, these barriers include lack of financing; lack of awareness of available proven technologies; fear of change; insufficient time to study and implement changes; lack of skill and training of technical personnel; paradigm shift in new equipment (numerically controlled versus mechanical); inability to perform comprehensive cost analysis; prior bad experience with new technologies; and inability to select the correct product and vendor (Shapira, 1990). These typical characteristics of small firms are not shared by larger manufacturing companies.

Furthermore, the short-term nature of most supplier relationships in the United States and the absence of effective supplier networks is another barrier to modernization faced by small firms, particularly in comparison with similar firms in Germany and Japan (Shapira, 1995). Trade associations are sometimes a source of modernization assistance, but in the United States, associations generally focus less on technical information and assistance and more on influencing government policies. Therefore, in the absence of public-sector industrial modernization activities, equipment vendors are often the sole source of modernization information for small firms. However, the vendor's interest (closing the sale) is often quite different from that of their clients. Small firms lack the wherewithal of large firms to recover from an investment mistake, and, particularly for firms that have already made a bad investment, suspicion of the impartiality of vendors and dealers is a critical barrier. Many clients of industrial extension programs, when surveyed, have said that impartial advice about modernization options and available equipment is one of the most valuable services provided by those programs.

Do industrial extension services work? Are they a cost-effective way of boosting the competence and competitiveness of small manufacturers? Do they provide what their clients need? There is a large body of anecdotal evidence suggesting the answer to all three questions is yes. The fact that states have begun and continued such modernization services in the face of their own financial problems supports the notion that such programs are successful. However, the amount of systematically collected data and analysis currently is inadequate to prove this. NSF has devoted considerable time and resources to the process of evaluating the university-based centers it supports (including ERCs and I/UCRCs), and a number of centers have been defunded following unsatisfactory evaluations. For its part, the MEP has an active program to develop evaluation methodologies that can be used by state and local extension programs to evaluate their efforts. It includes activities to define, measure, analyze, and report on short- and long-term impacts of MEP centers on the operations of client firms.

In 1995, the General Accounting Office (GAO) issued a report based on a survey of clients of 57 extension programs in 34 states that had received at least 40 hours of assistance in 1993. The 551 completed questionnaires covered the most common types of MEP center services. Most manufacturers responding (73 percent) believed that MPE assistance had affected positively their overall business performance. A minority (15 percent) said the assistance had not affected their business performance, and the rest said it was too early to tell or they had no way to estimate. The majority of respondents said that the impacts of MEP center assistance had a positive effect on their use of technology (63 percent); improved quality of their product (61 percent); and improved productivity of workers (56 percent). About half of the respondents indicated that MEP assistance had a positive impact on their customers' satisfaction, their profits, and their ability to meet production schedules. In a related survey of small firms that did not use the MEP, most (82 percent) told the GAO they were unaware of MEP services, while another 10 percent were aware of but did not need MEP services.

Although federal technology programs (particularly the Advanced Technology Program and the TRP) have come under fire since the election of the Republican majority in the 104th Congress, the MEP itself has relatively few critics.

The MEP has succeeded in developing strong bipartisan support in states and localities, which have in turn been effective in persuading their federal legislators that the MEP should continue as an active program. While the growth of MEP funding has slowed in the past year, the program has nonetheless enjoyed modest increases in the face of cuts nearly everywhere else, and it has grown substantially since its inception.

In spite of the apparent success of U.S. industrial extension programs, both in the field and in the political arena, the goal of government-funded manufacturing extension programs can and will probably remain modest compared with the absolute numbers of firms in the manufacturing business. Seven years ago, Shapira (1990) asserted that to have a significant effect, state and federal programs should move far beyond assisting a few thousand firms per year (the rate at that time). In fact, MEP-affiliated centers are now providing some level of service to about 15,000 new manufacturing firms per year. However, even at this rate, another 25 years would be required to reach all the small manufacturing firms in the United States. The National Academy of Engineering (1993) suggests that the goal should be to "catalyze the development of a dense national network of public and private providers of industrial modernization services that is capable of meeting the diverse technical, managerial, training, and related needs of 20-25 percent of the nation's small and medium-sized manufacturing companies by the year 2000."

What rate of industrial extension activity is sufficient to have a significant impact on the technological status of small U.S. manufacturers? As implied in the previous paragraph, it is probably not necessary to reach every small and medium-sized firm. Many firms do not need or do not want any assistance from government. Many others are far from the economic mainstream, in niche or localized markets; their competitive position is not threatened by the world economy; or their technological limitations detract little from the competitiveness of the United States. Still other firms belong to private-sector supplier networks or have found private assistance, such as consulting, on their own and are receiving the critical assistance they need. It would be difficult to gauge the sizes of these groups, but together they probably make up a significant part of the U.S. manufacturing sector. The key is to identify the most important sectors, geographical areas, and firms that can benefit most from a manufacturing extension program.

MICROELECTRONICS

Simon Glynn and William J. Spencer

Microelectronics are a vital enabling technology, one that is critical to the U.S. economy. Sales of microelectronics represent nearly 11 percent of U.S. GDP, and the microelectronics industry is one of the fastest growing sectors of the U.S. economy, increasing at a CAGR of 9.3 percent between 1987 and 1994

(Council on Competitiveness, 1996). Most of the innovations in microelectronics were first developed in the United States. The United States has nonetheless faced considerable challenges in microelectronics compared with the software or biotechnology industry, especially from Japan and the newly industrialized economies (NIEs) of Asia. This case study discusses the reasons for the United States' success in early innovation and current technology transfer in microelectronics intended to counter competitive challenges.

Defining Microelectronics

This paper focuses on technology transfer in two areas: semiconductors and flat panel displays (FPDs). These two technologies are critical competencies in microelectronics. Semiconductor content in personal computers, for example, increased from $750 to $1,500 between 1989 and 1993. Semiconductor content is also approaching 50 percent of the total cost of new weapons systems for the DOD. The world market for semiconductors increased 32 percent in 1994, to $102 billion (Council on Competitiveness, 1996).

The United States has faced considerable global competition in semiconductors. The U.S. global semiconductor market share in 1992 was 43.7 percent, nearly equal to Japan's 43.4 percent. This lead has widened in recent years, to 46 percent for U.S. semiconductor companies versus 41 percent for Japan in 1995 (Council on Competitiveness, 1996).

Flat panel displays are also key enablers of electronic systems and represent an increasing share of the total cost of these systems. For example, FPDs represent 25 to 30 percent of the cost of PCs and over 50 percent of the cost of personal digital assistants (PDAs). The world market for FPDs was $11.5 billion in 1995 and is expected to approach $22 billion by 2000 (Council on Competitiveness, 1996).

Japanese firms control 95 percent of the world's FPD market (Council on Competitiveness, 1996). The U.S. FPD industry is very small and fragmented by comparison. Many large U.S. firms exited the FPD market in the 1980s. In Japan, on the other hand, several large firms invested aggressively in today's dominant technology of active-matrix liquid crystal displays (AM-LCDs). In the United States, there are some dozen small and medium-sized firms pursuing a variety of FPD technologies. A few U.S. companies have successfully positioned themselves as materials or equipment suppliers, but there is almost no U.S. presence in the multibillion dollar LCD market (Council on Competitiveness, 1996; Saccocio, 1996).

Research and Development

SEMICONDUCTORS

The U.S. semiconductor industry spent $3.7 billion on R&D in 1994, or 13 percent of revenues. R&D spending in the underlying semiconductor equipment and materials industry is also estimated at about 12 to 15 percent of revenues (Council on Competitiveness, 1996).

Progress in semiconductors also depends—critically—on the experience of implementing advanced technologies in new semiconductor fabrication. This feedback influences the next set of technical goals in semiconductor R&D. Consequently, advanced R&D in semiconductors also requires continuous investment in new fabrication (Borrus, 1988).

But the capital-intensive nature and ever-increasing complexity of semiconductor manufacturing make large investments in R&D quite difficult. For example, the cost of building a new world-class 16MB DRAM chip fabrication facility, or "fab," is now about $1 billion (Council on Competitiveness, 1996). As many as 125 new fabrication facilities are now planned or under construction around the world (San Jose Mercury News, 1996).

These costs create enormous pressure on levels of R&D spending. "Fabless" semiconductor R&D companies emerged in the mid-1980s in response to these high capital costs. By going fab-less, these companies can increase their R&D spending to focus on design and testing. Several companies also use smaller, specialized "mini-fabs." But fab-less companies are increasingly vulnerable as global semiconductor fabrication capacity tightens.

This dynamic is complicated by declining profit margins as each generation of semiconductor technology becomes commoditized. This, in turn, reflects the rapid improvement in price/performance ratio of semiconductors. ("Moore's Law" predicts that semiconductor performance will double every 18 months, without any increase in price.)

In response to these dynamics, leading U.S. semiconductor manufacturers, including Intel, Motorola, and Texas Instruments, are participating in consortia and other cooperative mechanisms to leverage R&D, especially in generic (precompetitive) technologies (Rosenbloom and Spencer, 1996). A generic technology (e.g., superlattice or heterostructures) typically has broad applications. Cooperation to leverage R&D in these new technologies reduces the financial risk for individual competitors. Cooperation also eliminates duplication of R&D (Rosenbloom and Spencer, 1996). This approach to generic R&D largely mimics the Japanese approach to semiconductor R&D (Borrus, 1988).

The U.S. and Japanese semiconductor industries are relying increasingly on the equipment industries to help support the R&D and capital costs for new devices. Today, semiconductor manufacturing equipment is typically developed through cooperation between individual semiconductor component firms and their equipment suppliers. As a result, several different technologies are used in materials and equipment, and no single industry-wide specification exists. This is a significant problem for the industry (Council on Competitiveness, 1996).

FLAT PANEL DISPLAYS

R&D spending for flat panels is more difficult to estimate than for semiconductors. There are currently no U.S. companies that compete in AM-LCDs and other high-volume FPD markets. The U.S. FPD sector consists entirely of a group of small companies pursuing "technology-push" strategies. A large portion of the funding for these small companies is derived from federal funding and contracts, especially from the Advanced Research Projects Agency (ARPA) (Saccocio, 1996).

FPDs are perhaps the highest-potential, strategically significant competency in microelectronics. Analysts predict the integration of semiconductors directly onto FPDs, for example. In the next decade, new applications will drive the demand for FPDs, including PDAs, virtual reality, and portable computing.

As yet, there is no agreement on what FPD technologies will dominate these applications. Future FPD market growth is almost entirely dependent on technical advances in FPDs. For example, FPDs with good contrast and resolution that meet the cost and power requirements for all applications do not yet exist. Consequently, different FPD technologies are being developed to address the needs of different applications. U.S. FPD companies lead in developing many of these technologies (Saccocio, 1996).

Early Innovation in Semiconductors[6]

As recently as the 1970s, the United States dominated semiconductor technology and the manufacture of semiconductors. To a considerable extent, this dominance reflected the success of the U.S. economy is exploiting the earliest innovations in semiconductors. This success was shaped by several factors. First, AT&T/Bell Laboratories was extremely important for early innovation in semiconductors. Bell Laboratories received nearly 350 patents in semiconductors, or more than one-quarter of all semiconductor patents between the time the transistor was invented at Bell Labs in 1948 to the time the integrated circuit (IC) was developed at Texas Instruments and Fairchild in 1958. Bell Labs' rapid dissemination of these results helped develop the semiconductor industry. For example, as early as 1952, AT&T provided licenses to 35 companies under its transistor patents, even as AT&T started to fabricate germanium transistor semiconductors for internal use. Technical symposia were also held by Bell Labs to transfer technology and recent R&D developments—including silicon oxide diffusion and oxide masking, which enabled large-scale semiconductor fabrication—to these licensees.

Second, the transfer of individuals also encouraged early innovation in semiconductors. For example, Shockley, one of the inventors of the transistor at Bell Labs, left to form start-up Shockley Transistor Corporation at Stanford in 1955; researchers recruited by Shockley then left to form Fairchild Semiconductor in 1957. Fairchild itself became a source of new spin-outs, including Intel and AMD. Texas Instruments as well as Motorola recruited Bell Labs researchers.

Third, early innovation in semiconductors was directly encouraged by defense policy—especially military and aerospace demand for the new technology.

This policy had several aspects. Defense R&D programs in semiconductors in the 1950s and 1960s served as important technology transfer mechanisms linking the vast number of DOD semiconductor programs in the commercial sector. Defense funding of academic research at U.S. universities also contributed directly to early innovation. Start-ups in semiconductors tended to concentrate around these academic programs—notably in Boston (MIT, Harvard), and San Francisco (Berkeley, Stanford). In this way, learning in these federally sponsored R&D programs was transferred to commercial use.

DOD and NASA programs also created demand for advanced semiconductor technologies. For example, the two agencies were responsible for nearly 50 percent of revenues from transistor sales in 1960. This early demand was met at very high unit costs. As innovations in processes reduced unit costs, transistor technology extended into commercial uses. For example, from 1963 to 1965, DOD and NASA funded 14 programs that called for the use of ICs, notably the Minuteman II missile and Apollo spacecraft guidance systems. In 1963, these programs represented 94 percent of the market for ICs, at a unit price of $31. In 1965, DOD and NASA procurement represented only 72 percent of demand for ICs as commercial use expanded, and unit prices for ICs dropped to less than $9.

Fourth, the development of computers, and especially IBM's development of transistorized computers, was critical to the successful U.S. exploitation of innovation in semiconductors. IBM's enormously successful System 360 was the first computer not based on discrete semiconductor design, but on integrated circuits as well as magnetic tape drives and flexible software architectures that were all developed under government funding and adapted almost immediately for commercial use. IBM and its competitors created enormous demand for new semiconductor technologies, driving up profits and encouraging innovation.

Technology Flows

Technology flows in the U.S. microelectronics sector depend to an unusual extent on formal relationships between companies, equipment suppliers, universities, and the federal government.

CONSORTIA

Consortia play an especially important role in semiconductors. Collaboration between U.S. semiconductor manufacturers and equipment suppliers has helped the United States compete against the Japanese vertically integrated keiretsu (relationships between Japanese semiconductor manufacturers and original equipment manufacturers). The most important of these consortia are SEMATECH (for Semiconductor Manufacturing Technology) and the Semiconductor Research Corporation (SRC). Consortia in FPDs, notably the Microelectronics and Computer Technology Corporation, have been largely ineffective.

SEMATECH

SEMATECH is a consortium of U.S. semiconductor manufacturers, government, and academia. It was formed in 1987 as a cooperative effort between DOD and the semiconductor industry in response to the perceived targeting of the U.S. semiconductor industry by international competitors. Its purpose is to sponsor and conduct research in semiconductor manufacturing technology. SEMATECH's members include Advanced Micro Devices, AT&T, DEC, Hewlett Packard, IBM, Intel, Motorola, National Semiconductor, and Texas Instruments.

SEMATECH's focus is on developing and diffusing precompetitive manufacturing technologies and processes. (See, for example, Box 1.) Early SEMATECH programs succeeded so well that U.S. semiconductor manufacturing capabilities and equipment can now be met domestically, except in the critical field of photolithography. SEMATECH is now developing the next generation of semiconductor manufacturing technologies needed to create 300-millimeter semiconductor wafers. SEMATECH has invited domestic and foreign firms that have fabrication facilities in the United States to participate in the so-called 300 Millimeter Initiative. Results will then be transferred to consortium members. SEMATECH expects four or five U.S. companies and five or six companies from Europe and Asia to participate (Council on Competitiveness, 1996; Spencer, 1996).

SEMATECH annually received $100 million in federal funds during the Reagan and Bush administrations, which it matched with an equivalent amount for a total yearly budget of $200 million. Federal funding was reduced by about $10 million in each of 1994 and 1995 and is expected to be phased out entirely by 1997. The 300 Millimeter Initiative will be funded entirely by consortium members (Council on Competitiveness, 1996).

Semiconductor Research Corporation (SRC)

SRC was established in 1982 by the Semiconductor Industry Association (SIA) to plan and execute a program of R&D at U.S. universities in areas of interest to the U.S. semiconductor industry. SRC participants include industry and government agencies.

The SRC research program spends about $37 million annually and supports more than half of all silicon-related generic research in U.S. universities. The effect has been to dramatically increase research into silicon-based technology. In 1982, only 20 to 30 graduate students were pursuing silicon-based projects. Now, SRC funds over 350 faculty and 900 graduate students at more than 60 universities. Coordination with SRC member companies has helped faculty focus on the areas of highest commercial potential (Council on Competitiveness, 1996).

Microelectronics and Computer Technology Corporation (MCC)

MCC was created in 1982 by 10 major U.S. computer and semiconductor manufacturers with the goal of maintaining the U.S. lead in computer technologies. Members now include large companies (3M, AMD, Andersen Consulting, AT&T, Cadence, Ceridian, DEC, Eastman Kodak, GE, Harris, HP, Honeywell, Lockheed Martin, Motorola, National Semiconductor, Nortel, Rockwell, Westinghouse), associates (various companies, government agencies, and academic institutions), small business associates, and university affiliates. Unlike SEMATECH, MCC was privately created.

MCC has widely been seen as a failure compared with SEMATECH. More recently, MCC has abandoned the goal of maintaining the U.S. lead in computer technologies and has focused primarily on two areas: a high-volume electronics division that develops packaging, interconnect, and display technology; and an enterprise-integration division dedicated to building a global data highway and networking and database technologies. These initiatives have (so far) had no impact on the U.S. FPD industry.

GOVERNMENT-INDUSTRY RELATIONSHIPS

SEMATECH is not the only example of government-industry relationships in U.S. microelectronics. Indeed, the federal government has always played an active role in this sector, as noted earlier for semiconductors. Federal spending for FDP R&D over the last 5 years was about $650 million. The DOD has been the largest funder of R&D in the areas of microelectronics seen to be critical for national security. For example, DOD spending represents nearly 90 percent of federal funding for FPD technology. Key FPD initiatives funded by this spending include ARPA's High Definition Systems and Head Mounted Display programs. In 1994, ARPA awarded a 3-year, $21.4 million grant to Xerox for continued development of its AM-LCD technology (Council on Competitiveness, 1996; Saccocio, 1996).

DOD funds have also gone beyond FPD research and development. In 1994, the DOD announced a 5-year, $500 million program to support future domestic FPD manufacturing. In 1993, ARPA and Optical Image Systems (OIS) announced they would build a $100 million LCD plant in the United States to provide displays for military and commercial use. ARPA funding represents about half of the costs for the fabrication facility. Also in 1993, ARPA awarded a multiyear infrastructure development grant to the USDC, including $12 million in the first year (Council on Competitiveness, 1996; Saccocio, 1996).

Other federal agencies are also involved in government-industry relationships in microelectronics. For example, NIST has been funding research in semiconductor measurement technology. Many segments of the microelectronics industry are also collaborating with DOE laboratories through CRADAs. For example, Sandia National Laboratories has worked with SEMATECH starting in 1993 under a CRADA that covers research and development in critical areas of contamination-free manufacturing (Council on Competitiveness, 1996).

BOX 1
Cost of Ownership Technology Transfer

The Cost of Ownership (CoO) concept can be traced to the nuclear power industry in the 1960s. It was then that the notion of total system cost was conceived as a way to estimate operating expenses over the life of a nuclear power plant. Unfortunately, the implementation was not successful, in part due to the number of variables and the lack of computing power needed to manage the large number of calculations.

CoO, in the form known today, originated at Intel Corp., where it was used to examine the total cost of acquiring, maintaining, and operating purchased equipment. Dean Toombs, an Intel assignee, introduced the concept to SEMATECH. At SEMATECH, the method was first used to evaluate different lithography technologies. Lithography is used to image the semiconductor patterns onto a light-sensitive emulsion to produce the patterns that define integrated-circuit performance. The smaller the line widths, the more critical the imaging process. Each layer in the manufacturing of the semiconductor involves lithographic imaging. Today's semiconductors involve multiple layers and have requirements for a high degree of alignment with very fine resolution.

At the time, the two competing lithography technologies were projection imaging and near-contact imaging (stepper). The projection imaging system was relatively low-cost when compared with the stepper process. The apparent throughput rate was also higher with the projection process. However, the engineers, the operators, and the manufacturing personnel knew that the projection process had a lower yield and required more maintenance. There was a need for a tool that could be employed to permit an accurate evaluation of the process and provide a verifiable method of analyzing the equipment. Decisions to buy equipment are often based on purchase price and the cost of installation. These costs do not consider the effect of equipment reliability, production utilization, or product yield. Over the life of a system, these factors may have a greater impact on CoO than the initial purchase and installation costs. CoO was applied to this project to provide an accurate analysis of the life-cycle costs.

SEMATECH incorporated and expanded CoO as part of quality and equipment improvement programs. This led to the development of a method for estimating in some detail the total life-cycle cost of owning and operating equipment for a single semiconductor process step. This work was transferred from Toombs to Ross Carnes, a Motorola assignee who continued to refine the method. Joann Trego, a SEMATECH director, was responsible for training users in the correct application of the software.

SEMATECH implemented this methodology in a spreadsheet program and distributed it to member and supplier companies. CoO measures the life-cycle costs of equipment improvement or purchase for both sup-

BOX 1—Continued

pliers and users. It became possible for suppliers to measure themselves against their competition. Users could employ CoO to evaluate various supplier equipment in order to determine the best selection for their facility. CoO quickly became widely used by SEMATECH member companies in their purchase decisions. As with any product with significant market impact, copies began to appear. While there were a number of them, the quality varied and the values could be manipulated by the user. The original CoO locked critical elements of the program and prevented values from being changed. In 1991, Daren Dance acquired responsibility for the CoO effort. He guided it through numerous minor revisions and one major revision. The last version that SEMATECH produced was release B. The last release incorporates over 150 parameters that cover aspects of the equipment from tool usage through consumables and maintenance parts and support.

The SEMATECH software became a de facto industry standard, but the support diverted resources from other SEMATECH activities and the quality of competing software was suspect; so, SEMATECH guided an effort, with Semiconductor Equipment and Materials International (SEMI), to develop standard definitions and equations for CoO. SEMATECH, through Dance, worked to develop a consensus on the definitions and formulation for a generic CoO standard. The resulting SEMI E35 guidelines were accepted by worldwide balloting.

This helped the situation but did not solve the problem. Due to the popularity of the program, SEMATECH had over 1,200 copies of CoO in use and was spending a significant amount to keep the software updated and member companies trained. There was a need for a commercial supplier to be given the responsibility for maintaining the software and providing customer support.

Dance led the SEMATECH effort to find a commercial supplier. A statement of work was developed and an open bidding process commenced. This was completed with the selection of a commercial supplier, Wright, Williams, and Kelly (WWK), which incorporated the SEMATECH code into its existing interface and marketed it as TWO COOL.

The story does not end with the transfer of support responsibility. WWK developed a marketing strategy that was based on providing companies with site licenses that bundled the support costs. SEMATECH member companies, which had been receiving the software and support as part of their return on investment from their dues, now had to agree to pay an acquisition cost of less than $10,000 in order to receive the latest software and the associated support. The transition was not easy. It took WWK almost 6 months to make its first sale; after 1 year, the firm had five of the SEMATECH member companies on board.

SOURCE: W. J. Spencer, Chairman and Chief Executive Officer, SEMATECH.

UNIVERSITY-INDUSTRY RELATIONSHIPS

Industry consortia in semiconductors continue to create relationships with leading U.S. research universities. As noted earlier, SRC is an important link between the semiconductor industry and U.S. universities. SRC also administers SEMATECH's Center of Excellence program. This program directs 5 percent, or $9 million, of SEMATECH's $180 million budget to R&D in specific areas of interest to semiconductor manufacturers. All but one of these centers is located at U.S. universities. However, there are concerns at the universities that SEMATECH support for academic research may disappear when federal funding is phased out in 1997 (Council on Competitiveness, 1996).

In FPDs, ARPA has started several centers of excellence, including the Display Phosphor Center of Excellence at the Georgia Institute of Technology and the National Center of Advanced Information Components Manufacturing located at Sandia National Laboratory. Other FPD research programs at U.S. universities include the NSF Science and Technology Center on Advanced Liquid Crystalline Optical Materials at Kent State University and research programs at Temple University and the University of Michigan (Saccocio, 1996).

Interestingly, most of the important innovations in microelectronics—unlike biotechnology—have derived from industry R&D, not academic research. University programs are nonetheless exceedingly important in microelectronics, because the speed and extent of R&D in both semiconductors and FPDs depend on a relatively small number of students trained in these technologies. Demand for experienced scientists and engineers is forcing U.S. semiconductor manufacturers to compete for the people necessary to perform desired levels of R&D. U.S. FPD manufacturers are also constrained by relatively few talented scientists and engineers educated in FPD technologies in U.S. research universities.

TECHNOLOGY MAPPING IN SEMICONDUCTORS

The Semiconductor Industry Association introduced the concept of technology mapping in the 1980s to guide investment in R&D. This map has evolved into the National Semiconductor Technology Roadmap, which maps technology goals for semiconductors over the next 10 to 20 years. Technology mapping has helped to coordinate research across industry, universities, and federal laboratories, although the actual fit between the technology maps and the marketplace is quite weak (Council on Competitiveness, 1996).

SMALL FPD COMPANIES AS A SOURCE OF TECHNOLOGY

Recently, several large U.S. customers of electronic displays have formed alliances with small U.S. FPD companies to commercialize new FPD technologies. For example, Motorola has invested $20 million in a joint venture with In

Focus Systems to develop a new LCD technology. Rockwell International has formed a strategic relationship with Kopin to develop AM-LCD panels based on Kopin's Smart Slide technology (itself a result of development with Standish, another small FPD company, and the Sarnoff Research Center). Kopin has also signed a broad product-development effort with Philips North America Corp. Standish is also working with Xerox to manufacture a very-high-resolution AM-LCD (Saccocio, 1996).

International Technology Transfer

International alliances are increasingly important in microelectronics. The technical challenges of developing the next generation of semiconductors and the enormous costs of fabrication facilities are driving U.S. semiconductor manufacturers to form technology transfer agreements with foreign competitors. Several notable examples are described below.

- Texas Instruments and Japan's Hitachi formed an alliance in 1988 to share technology related to the production of 16MB DRAMs. The alliance has subsequently evolved into an arrangement to develop 64MB and 256MB DRAMs and to establish a joint-manufacturing arrangement (Council on Competitiveness, 1996).
- IBM, Japan's Toshiba, and Germany's Siemens have established two joint development agreements (one for 256MB DRAMs in 1992 and another for second-generation 64MB DRAMS in 1994). IBM and Toshiba also collaborate in other areas (Council on Competitiveness, 1996).
- In 1992, Intel and Japan's Sharp joined forces to develop, manufacture, and sell 8/18MB flash memory devices. Each company shares the R&D, and each expects to expand their flash memory market presence (Council on Competitiveness, 1996).
- Major alliances in semiconductors are also forming between Japanese companies and the newly industrialized economies in Asia. For example, Japan's Hitachi and South Korea's Goldstar are collaborating on the production of 16MB and 256MB DRAMs (Council on Competitiveness, 1996).

These alliances are also forming in the FPD market. Numerous large U.S. defense contractors have devoted considerable effort to develop displays for the DOD. Hughes, for example, successfully produced a large color FPD using LCD technology for the command and control units on Navy ships. Declining defense spending is now forcing these companies to explore commercial uses for this technology. For example, Hughes is reconsidering the commercial FPD market for its liquid crystal light-valve technology developed for the U.S. Navy. In 1992, Hughes formed a joint venture with Japan's JVC to develop, produce, and market liquid crystal light-valve projectors (Saccocio, 1996).

Outlook for the Future

Demand for microelectronics is expected to increase at a compound annual growth rate (CAGR) of 6 to 8 percent through the 1990s. Moore's Law is also expected to continue to drive the price-performance ratio in microelectronics, demanding radical advances in semiconductor design and integration of microelectronics, especially semiconductors and FPDs. In semiconductors, the direction of technology advance is relatively clear (even mapped). Unless leapfrog technologies are developed, the extensive technology flows forced by technological challenges and very high fabrication costs make it unlikely that the competitive advantage from these advances will accrue exclusively to any one company or national economy.

In FPDs, the future is less clear. Japan controls virtually all of the FPD market. U.S. FPD manufacturers may continue to lead in developing new FPD technologies, but the direction of technology development in FPDs is not at all clear. Finally, developing the infrastructure, manufacturing, and applications needed to exploit these new FPD technologies appears exceedingly difficult without Japanese involvement.

SOFTWARE

Simon Glynn

The United States has excelled in software and computers because of an exceptional ability to develop ideas. A unique convergence of institutions and relationships has created an environment that encourages not only new ideas, but also their development and application. This paper is intended to introduce the various relationships between the public and private sectors that have enabled this.

Overview of Software Research and Development

Software is now critical to many commercial and defense technologies. Indeed, any product or service enabled by computers depends on "embedded" software. One informal estimate is that perhaps 70 percent of Hewlett Packard's development engineering is concerned with software engineering (Mowery, 1996).[7] For this reason, it is difficult to communicate the scale of computer and information technologies research and development in the U.S. economy.

Spending by the federal agencies for basic and applied research in mathematics and computer science exceeded $900 million in 1991. Of this spending, nearly half was for basic research. Indeed, the rate of increase in funding for basic research in the 1980s has been faster for mathematics and computer science than for any other field—although in absolute terms the base is only 4 percent of federal spending for basic research in 1991 (National Science Board, 1993).

Most of this federal spending has been for academic research. Total expenditures for university research in mathematics and computer science were nearly $775 million in 1991 (National Science Board, 1993). Of this, federal spending represented more than $500 million, or about 70 percent of all spending for academic research in mathematics and computer science in 1991. The difference, $240 million, or 30 percent of academic spending in mathematics and computer science, came from nonfederal sources, including industry, state and local programs, and academic institutions (National Science Board, 1993).

This apparent division of intellectual effort between universities and industry is illuminated by data on Ph.D.'s employed in computer research and development. Industry employs a higher percent of Ph.D.'s in computer science (58 percent) than it does Ph.D.'s from any other science field. Nearly 90 percent of these Ph.D.'s are employed in applied research and development. In academia, on the other hand, almost 60 percent of Ph.D.'s are employed in basic computer science research (National Science Board, 1993). In this sense, academia continues to be the locus of basic scientific and technological learning.

Two observations deserve special attention with respect to this learning in software. First, the development of the software sector in the United States has been shaped by the very large contribution of federal funding to R&D in computers and software. The United States enjoys a "first-mover" advantage in software, because it is very difficult to displace a successful first-mover in software and because demand for new computer and software technologies developed first in the United States (Mowery, 1996). These first-mover advantages were created not only by commercial activity, but also by federal funding for research and development and the early development of computer science in U.S. universities (Steinmueller, 1996).

Second, the development of the software sector has involved the transfer of learning and technology beyond institutional boundaries. Innovation in computers and in software has depended on the opportunity for individuals to move among academia, the federal labs, and technology-intensive companies, for example IBM. In this respect, new, technologically innovative software companies—and the environment that encourages them—represent an increasingly important way for individuals to transfer technology.

Size and Scope of Software Sectors

Spending for software may be divided into two types: prepackaged software (SIC code 7372) and customized software and services (including computer programming services, SIC 7371; and computer systems integration, SIC 7373). Data on these sectors are presented in Tables A-5 and A-6.

Global revenues for customized software and services by U.S. companies in 1993 were estimated to be $38.7 billion and are expected to exceed $40 billion in 1994. More than 40 percent of these revenues are from markets outside the United

TABLE A-5 Revenue Trends and Forecasts, Customized Software and Services (dollars in billions), 1991–1997

	1991	1992[a]	1993[b]	1994[c]	CAGR[c] 1991–1994	CAGR[c] 1994–1997
Systems integration	16.2	17.7	19.3	20.9	9%	8%
Programming services	15.6	17.6	19.4	21.2	11%	9%
TOTAL	31.8	35.3	38.7	42.1	10%	9%

[a]Revised.
[b]Estimated.
[c]Forecast.

NOTE: Totals and percent changes are based on unrounded revenue data. CAGR = compound annual growth rate.

SOURCE: U.S. Department of Commerce (1994).

States, where experience in the technologically advanced U.S. market provides a competitive advantage. Several large U.S. players compete in these markets, including Electronic Data Systems (1992 revenues of $7 billion) and consulting firms such as Andersen Consulting and SHL-Systemhouse (Ferne and Quintas, 1991; U.S. Department of Commerce, 1994). More recently, large defense contractors, including Boeing and McDonnell Douglas, have entered these markets because of their expertise in developing large systems (National Research Council, 1992b). Larger computer makers such as IBM are also focusing on systems integration services as their customers migrate from high-end hardware to a complex environment of mainframe and distributed desktop systems.

Global spending for prepackaged software (including operating systems) was estimated to be $71.9 billion in 1993 (U.S. Department of Commerce, 1994). The United States is by far the largest geographic market for prepackaged software, representing 45 percent of spending ($32 billion). Japan is second, representing 9.6 percent ($7 billion). The individual markets of western Europe as a group invested $25.7 billion, or 36 percent of global spending in 1993 (U.S. Department of Commerce, 1994). U.S. software companies dominate this sector. According to International Data Corporation (IDC), revenues to U.S. companies were nearly $50 billion in 1992, or more than 70 percent of global spending (U.S. Department of Commerce, 1994). Accurate estimates of revenues for PC-based applications software are not widely available; however, they may be estimated from 6-month data published by the Software Publishers Association (U.S. Department of Commerce, 1994).[8] Using these estimates, spending for PC-based applications software in the United States and Canada totaled more than $6.6 billion in 1993.

It is important to understand that internal development is not included in

TABLE A-6 Global Spending for Prepackaged Software Markets, 1991–1997 (dollars in millions)

	1991	1992	1993[a]	CAGR[a] 1991–1993	CAGR[b] 1993–1997
United States	25,330	28,460	32,040	13%	13%
Western Europe[c]	21,091	23,850	25,699	11%	10%
Japan	5,270	5,967	6,938	15%	19%
Canada	1,078	1,188	1,374	13%	10%
Australia	941	980	1,094	8%	13%
Latin America[d]	1,054	1,242	1,471	18%	18%
Asia[e]	584	780	974	29%	21%
Other	1,674	1,846	2,094	12%	15%
WORLD	57,022	64,313	71,864	12%	13%

[a]Estimated.
[b]Forecast.
[c]Includes Austria, Belgium, Denmark, Finland, France, Germany, Italy, Netherlands, Norway, Spain, Sweden, Switzerland, and the United Kingdom.
[d]Includes Argentina, Brazil, Chile, Mexico, Venezuela.
[e]Includes China, Hong Kong, India, Malaysia, Singapore, South Korea, Taiwan, and Thailand.

NOTE: CAGR = compound annual growth rate.

SOURCE: U.S. Department of Commerce (1994).

these estimates, and the majority of software continues to be developed internally. Spending on internally developed software code exceeded $200 billion in 1990, compared with nearly $75 billion in spending for commercially available software (National Research Council, 1990). The preponderance of these efforts is for incremental improvement to existing software, not the development of new systems (Organization for Economic Cooperation and Development, 1985).

Factors Shaping Academic Computer and Software Research

Research performed in the leading U.S. research universities has effectively defined software as an academic and engineering discipline. The creation of a new academic discipline that is exceedingly instrument-dependent has been shaped by large public-sector investments.

Software as a concept did not exist before the development by John von Neumann of the conceptual architecture for computers in 1945. Indeed, even in 1959, there were virtually no formal programs in computer science. Yet, computer science is now an academic discipline of substantial intellectual depth: In 1989, U.S. universities produced 531 Ph.D.'s in computer science and 3,860 doctoral-level researchers. There were 5,239 people with doctorates teaching computer science in 1989 (National Research Council, 1992a).

TABLE A-7 Federal Funding for Computer Science and Engineering Research and All Science and Engineering Research, Fiscal Year 1991

Agency	Computer Science Research ($ millions)	Percentage of Total	All Science and Engineering Research ($ millions)
Defense	418.7	62	3,805
National Science Foundation	122.7	18	1,847
National Aeronautics and Space Administration	52.2	8	3,463
Energy	33.3	5	2,963
Commerce	18.4	3	444
Interior	11.4	2	549
Environmental Protection Agency	8.3	1	343
Transportation	6.1	0.9	146
Agency for International Development	3.6	0.5	290
Treasury	1.7	0.3	22
Health and Human Services	1.5	0.2	8,201
Agriculture	1.5	0.2	1,177
Education	0.9	0.1	157
Housing and Urban Development	0.2	0.03	11
Federal Communications Commission	0.1	0.01	2
Other Agencies[a]	—	—	631
TOTAL	680.6	—	24,051

[a]Other agencies that support some type of basic or applied research but not in computer science.

NOTE: All funding in fiscal year 1991 dollars.

SOURCE: National Research Council (1992a).

Funding for academic computer science is from NSF and DOD's ARPA, as well as NASA and DOE. These four agencies accounted for 92 percent of funding for computer science research in 1991, including basic and applied research (Table A-7) (National Research Council, 1992a).

NSF is the second-largest funder of computer science research, spending $122.7 million in 1991. Almost all of this funding went to universities. Indeed, NSF is largest funder of individuals in academic research in computer science (as opposed to departments or universities). These funds tend to emphasize basic research (National Research Council, 1992a). That said, NSF-funded research has contributed enormously to software development. The development of the BASIC and PASCAL programming languages was funded by NSF, for example, as well as software engineering and early object-oriented languages like CLU. NSF also funds a large computing infrastructure. The most important components of this infrastructure are the four NSF supercomputing centers and the NSFNET, which supports the Internet.

Academic computer science is also funded by ARPA. Among federal agencies, the DOD continues to be the largest funder of computer science research, spending $418.7 million in 1991; typically, about one-third of this is for academic computer science (National Research Council, 1992a). In this context, ARPA has had an extraordinary influence in defining the research agenda for academic computer science. In contrast to NSF support, which is mainly in the form of grants to individuals, ARPA support for academic programs tends to be concentrated among the leading U.S. research universities: Carnegie Mellon, MIT, Stanford, and UC Berkeley (Mowery and Langlois, 1996). The objective of ARPA funding has been to develop a basic research infrastructure in computer science that may be exploited by defense agencies. This infrastructure-building goal incorporates support for education as well as research: In 1990 one-quarter of faculty in the 40 leading U.S. departments of computer science had received their computer science Ph.D. from one of the three major universities supported by ARPA (Carnegie Mellon, MIT, and Stanford) (Mowery and Langlois, 1996).

The High Performance Computing and Communications (HPCC) program is currently a large component of this funding for academic computer science. Started in 1992, HPCC is coordinated across all federal agencies, receiving significant funding from NSF and ARPA. HPCC is defined in the context of specific applications of computing—a series of "grand challenges" in science and engineering, for example modeling global climate change and weather—that can only be solved using powerful computing. The HPCC program areas are high-performance computing systems; advanced software technology and algorithms; networking; and human resources and basic research. Funding for the individual programs is included in Table A-8. If fully funded over the proposed 5 years, the HPCC program will represent about a $2 billion investment, not including baseline spending for 1991 (National Research Council, 1992a).

Spin-On Effects from Defense-Related Research and Development

As well as funding basic research in U.S. universities, many of the initiatives funded by ARPA to develop new technologies have had surprising "spin-on" effects for commercial use. What is surprising is that these research projects, selected to complement the defense community, have had an enormous impact on the commercial sector. Examples of this research include timesharing, parallel processing, computer-enabled graphics modeling, and artificial intelligence.

Perhaps the best example of this is the Internet. The original concept for the Internet may be traced to an ARPA research project on internetworking in the early 1970s. The concept used by the Internet, of distributed computing and communication by a technology called packet-switching, was proposed in the 1960s and developed using ARPA funding. The concept was deployed as ARPANET, a secure communications network for military and university computers. Protocols were also developed to enable different networks to connect,

TABLE A-8 Agency Budgets by HPCC Program Components, FY 1994

Agency	HPCS	NREN	ASTA	IITA	BRHR	TOTAL
ARPA	151.8	60.8	58.7	—	71.7	343.0
NSF	34.2	57.6	140.0	36.0	73.2	341.0
DOE	10.9	16.8	75.1	—	21.0	123.8
NASA	20.1	13.2	74.2	12.0	3.5	123.0
NIH	6.5	6.1	26.2	24.0	8.3	71.1
NSA	22.7	11.2	7.6	—	0.2	41.7
NIST	0.3	1.2	0.6	24.0	—	26.1
NOAA	—	1.6	10.5	—	0.3	12.4
EPA	—	0.7	9.6	—	1.6	11.9
ED	—	2.0	—	—	—	2.0
TOTAL	246.5	171.2	402.5	96.0	179.8	1,096.0

KEY: HPCS = High-Performance Computing Systems; NREN = National Research and Educational Network; ASTA = Advanced Software Technology and Algorithms; IITA = Information Infrastructure Technology and Applications; BRHR = Basic Research and Human Resources.

SOURCE: Office of Science and Technology Policy (1994).

based on a now widely used protocol, TCP/IP. By the early 1980s, the success of ARPANET caused NSF to fund its own NSFNET using the same technologies. As demand for advanced computing power has accelerated, other local networks have quickly developed, linked by the NSFNET and connected to other sites and networks around the world. These technologies, as well as the protocols and standards, are collectively referred to as the Internet (National Academy of Engineering, 1995b).

These ARPA initiatives continue to demonstrate marked spin-on effects. In networking, to extend this example, current ARPA initiatives to develop new technologies are concentrated in the HPCC. One of these initiatives, the National Research and Educational Network (NREN), is expected to advance networking technology in two phases. The first phase of NREN is to increase the communication speed of NSFNET from 1.5 million bits per second to 45 million bits per second. The second phase involves research and development on "gigabit testbeds" to develop networking technology that will enable computer networks that can communicate at speeds of 1 billion bits per second (one gigabit). Most of this R&D is expected to be done in close collaboration with larger telecommunications and computer companies to encourage the transfer of these technologies to commercial high-speed data communications networks (Office of Science and Technology Policy, 1994; U.S. Congress, Office of Technology Assessment, 1993).

These surprising effects of the ARPA research initiatives illuminate a related point: Research and development are relatively "closer" in computer science than in other disciplines. U.S. universities in this sense have provided important channels for the dissemination and diffusion of these innovations in software between academia and the defense and civilian research efforts in software. Digital Equipment Corporation (DEC) is an example of the importance of these technology flows. DEC's founder, Ken Olsen, developed many of his ground breaking ideas for the minicomputer while working as a research assistant at MIT on Project Whirlwind, a DOD-funded project that was the precursor of a massive programming effort to develop the Semi-Automatic Ground Environment (SAGE) air defense system (Lampe and Rosegrant, 1992).

Indeed, some believe that a lack of interchange between military and civilian researchers and engineers weakened British efforts in computers. The Colossus machine built at Bletchley Park during World War II for code breaking, for example, was never further developed, and some aspects of it are still classified by the British government (Grindley, 1996). The very different situation in the United States enhanced the competitiveness of the U.S. computer and software efforts (Mowery and Langlois, 1996).

Mechanisms to Encourage Technology Transfer in Academic Computer Science

Other formal mechanisms have also been important in the transfer of technology from the military to the commercial sphere. For example, the federal government influenced the development of early automatic programming techniques through its support for information dissemination. The Office of Naval Research (ONR) organized seminars on automatic programming in 1951, 1954, and 1956. These conferences circulated ideas within a developing community of practitioners who did not yet have journals or other formal channels of communication. The ONR also established the Institute for Numerical Analysis at UCLA, which made important contributions to the overall field of computer science (Mowery and Langlois, 1996).

Yet another formal mechanism for technology transfer is the Software Engineering Institute (SEI) at Carnegie Mellon University. SEI was started by ARPA in 1984. In contrast to the applications-focus of many ARPA initiatives, SEI is intended to encourage the development and dissemination of generic tools and techniques for use in software engineering for defense applications. (For information on the SEI program, see their Internet home page at http://sei.cmu.org.)

Several professional societies have also influenced the development of computer science, especially the Association for Computing Machinery (ACM) and the IEEE Computer Society. The publications and conferences of the ACM and IEEE Computer Society are the major channels for dissemination of research and conceptual advances in computer science. The ACM has also shaped the development of the undergraduate curriculum in computer science (National Research Council, 1992a).

The Importance of Defense-System Acquisitions

The federal government is also a prodigious consumer of information technology. DOD programs to develop new, very complex computer systems had a tremendous influence on the development of U.S. software and computer competencies. Perhaps the most conspicuous example of this is the development of the SAGE air defense system in the 1950s, which far exceeded previous programming efforts. SAGE was developed from the Whirlwind project at MIT to coordinate the control of radar installations into a national air-defense system. Development of SAGE was directed by a division of the RAND Corporation, the System Development Corporation (Mowery and Langlois, 1996).

By 1955, RAND already employed 25 people, perhaps 10 percent of the programmers in the United States. By 1960, SDC had spun out of RAND and hired more than 800 programmers for developing SAGE. By 1963, SDC and SAGE had seeded the emerging software and computer industry with more than 6,000 individuals from the SAGE development effort (Mowery and Langlois, 1996). Indeed, by 1967, the Air Force had started divestiture proceedings to spin out SDC from the federally funded research and development centers, as competition in software made this status unnecessary.

SAGE's legacy also includes IBM's development of transistorized computers. In 1955, IBM delivered the XD-1 (patterned after Whirlwind, which also inspired Digital Equipment Corporation and the minicomputer) to serve as the "brain" of SAGE. Critical to its performance was a new memory architecture, called magnetic core memory, that later would appear in IBM's enormously successful System 360 computer. Also key were magnetic tape drives and flexible software architectures, all developed under government funding and adapted almost immediately for commercial use. IBM also recruited Emanuel Piore, head of the Office of Naval Research, as chief scientist, and increased research spending to 35 percent in the 1950s, and to 50 percent by the 1960s and 1970s. By the 1960s, IBM's computer R&D budget was bigger than the federal government's. Even as late as 1960, defense spending represented 35 percent of IBM's research budget (Ferguson and Morris, 1993).

More recently, the Software Productivity Consortium (SPC) has performed an analogous (if less dramatic) role in technology transfer to improve U.S. software and computer competencies. SPC was established by its member companies in 1985, uniting more than half of U.S. aerospace and defense firms in a for-profit consortium. The goal is to develop processes and methods that improve the design and implementation of complex software systems. This includes developing prototypes and technical reports, but not commercial products (Software Productivity Consortium, 1996). Until a few years ago, this goal was pursued with something of an ivory tower mentality by SPC, without involving consortium members and usually resulting in products that were off the mark. More recently, the SPC approach has emphasized intensive collaboration with members to develop a technical program more closely aligned to members' needs (Robert K. Carr, consultant in technology transfer, measurement and evaluation, and international technology, unpublished notes, 1993). The resulting program is seen as a useful resource for U.S. organizations to leverage investments in software development and to evaluate methods and processes.

Software Depends Critically on Innovation in Computer Technologies

Opportunities in software also depend on innovation in computer technologies. In this sense, the development of networked personal computers and workstations marks the transition to a profoundly different environment for software development. IBM is currently the world's largest supplier of software, despite its current difficulties; IBM's revenues from software (including operating systems) in 1992 were $11.1 billion. Several trends have affected IBM's thinking about the software side of their business (and by extension, the thinking of other large computer makers). First, software has developed as an opportunity that is quite distinct from computers. Operating systems and enterprise-scale applications software have become very expensive and complex to develop. For example, IBM's OS/2 operating system is estimated to have required at least 5 years and 400 programmers and cost as much as $1 billion. In addition, in recent years, independent software companies have pushed advances in several areas of operating systems and applications software.

Second, as computer hardware is increasingly commoditized, differentiation is less on the physical performance of the electronics than on the performance of the systems software and the collection of applications software and services available to users. For example, the success of Apple's Macintosh computer (whose development was mainly in sophisticated operating systems software) depended on the commercial availability of software designed and marketed by start-ups and smaller software companies. Consequently, computer makers have learned to encourage independent software companies to develop applications based on their architectures.

These dynamics contribute to a first-mover advantage for U.S. software companies. In contrast to customized software and systems integration, the personal computer and, more recently, networked computing, are radically changing the demand for software by creating very high-volume markets. Indeed, by 1984, the installed base of PCs was 23 million machines, compared with less than 200,000 for large- and medium-sized systems (Steinmueller, 1996). These high-volume opportunities easily absorb the fixed costs of software development. Also, standardization of personal computer architectures in the United States has enabled software companies to create software and operating systems that can be incorporated by different computer makers. This is in marked contrast to Japan, for example, where 6 of the top 10 software companies are tied through industrial groups to different computer makers (the top four are NEC, IBM Japan, Hitachi, and Fujitsu), each using a proprietary operating system. In the United States, by comparison, none of the top 10 software companies is tied to a computer maker (Friedland, 1993).

These trends reflect the divergence within the overall U.S. software industry between commercial and military applications. The DOD focus on systems development and on embedded systems doubtless limits the spin-on effects of these technologies for commercial use. The concept of software engineering continues to be relevant to the creation of large-scale, complex defense software systems, especially for embedded applications. But many of these systems are irrelevant to the commercial sector.

Small Companies Exploit These New Opportunities

In sharp contrast to the computer makers, most of the new independent software firms are relatively small, entrepreneurial companies. In Utah's "software valley," for example, three-quarters of the more than 1,120 technology-intensive companies have fewer than 25 employees, and 50 percent have revenues of $200,000 or less (The Economist, 1994). Several characteristics shape the opportunities for these new software companies. First, the initial capital requirements to start in software are extremely low (with the exception, of course, of intellectual capital). This is likely to be as true in the future as it has been in the past. These extremely low barriers to entry, especially in the decentralized, software-intensive low end of the hardware spectrum, limit the amount of risk a software entrepreneur must accept.

Second, these emerging software companies often exploit technologies or markets deemed too small or too risky for established players. Thus, new markets and narrow, niche markets that sometimes lead to considerably larger markets let new software companies develop the revenue stream, product, and core competencies of valuable new businesses. On the other hand, once such companies are established in a market niche, they in turn become vulnerable to new players with a better idea. Since the development cycle of sophisticated software is lengthy and requires highly focused skills, reacting to a competitive threat is usually not an easy task. As a result, software companies tend to be divided into three groups. The first group, quite rare, consists of the few that become large and develop the internal resources to have long-term staying power and to stay on the advancing technology curve. The overwhelming majority of start-ups in software are in the second group, which develops niche-market products, with company revenues in the $5 million to $15 million range. The life cycle of these companies is also quite short. Typically, they will either fail when their product life cycle has run its course or be acquired by or merged with other players to reach sustaining capabilities. The third group includes those new software companies that, for a variety of reasons, are not successful and fail.

Economic and Technological Risk is Encouraged

The dynamics of these opportunities for new companies in software are very appealing to venture capital. In 1992, software and related services attracted more venture capital financing than any other sector of the economy, including biotechnology. Some 214 software and services companies received 22 percent, or $562 million, of venture capital invested in 1992. As new software companies demonstrate the viability of new technologies or markets, the risk is less and these opportunities then become valuable to larger companies, creating liquidity by acquisition. Compared with other sectors, the valuations are also relatively high, encouraging the formation of new companies. For example, Microsoft's bid to acquire Intuit for $1.5 billion represented a breathtaking 40 percent premium over Intel's (then current) market value.

The Importance of Large, Technology-Intensive Companies

These innovative new software companies tend to be distributed according to a specific geography. That is, technology-intensive communities, for example Boston or San Francisco, that have reached critical mass in software tend to be self-perpetuating. This is because relationships with other, larger technology companies are very important to small software companies. First, the software industry is marked by a large number of spin-off companies or by entrepreneurs leaving larger software companies (or hardware companies) to create their own companies. These new software companies tend to concentrate in areas that include larger, technology-based companies where such spin-offs are common. DEC, for example, has spawned numerous spin-offs. The spin-offs are less well documented in software than in other high-tech fields but occur equally frequent (if not more so). Lotus Development, for example, spun off at least three new firms during its first 3 years, including Iris Associates (which developed the very successful Notes program using venture capital from Lotus).

As well as providing a source of entrepreneurs, large, technology-based companies also provide a critical base of new technology. There is a broad consensus that concepts are best transferred by the individuals who understand the new technology. To this end, small start-up firms have been responsible in software for an overwhelmingly large share of new commercial applications, often exploiting research and ideas developed elsewhere—usually in universities or in large, technology-based companies. The laboratories of IBM and AT&T Bell Labs especially, and also Xerox PARC, have developed software technologies that have been successfully commercialized by new software companies.

Unresolved Policy Questions

Several unresolved policy questions shape opportunities for companies in software. Concern about the domination of IBM's extraordinarily successful

System/360 architecture led the U.S. Justice Department to assert that this success represented an illegal monopolization. In response, IBM decided to "unbundle" the pricing of its systems instead of including the software in the pricing of the computer, essentially creating the opportunity for Microsoft and other independent software companies to sell competing software. Recently, similar questions have been raised in connection to the extraordinary successes of Microsoft Corp. in personal computer software.

Uncertain intellectual property rights (IPRs) are a second problem. Existing mechanisms for securing IPRs assume that something is either an expression of ideas (in which case, the expression of these ideas may be protected by copyright law, but not the ideas themselves), or a patentable process (in which case it may be protected by patent law). But software is both an expression of ideas, as lines of code, and the process that the algorithm describes—and that process is valuable. For this reason, IPRs are an imperfect mechanism (at best) for protecting innovations in software.

Intellectual property rights may disadvantage start-ups and smaller software companies. Patents, especially, present special problems in software. Many software companies are using patents to compensate for recent legal decisions denying them the copyright protection they feel they need. But patents are costly to obtain and difficult to enforce and defend. Large companies are, consequently, more likely to be able to threaten litigation and to defend against litigation. There is also ambiguity about what is and is not patentable. These problems have consequences for innovation, because small companies and start-ups are disadvantaged by the costs and uncertainties of litigation. Also, because larger companies and universities are usually the sources of the technology for spin-offs and smaller companies in software, stronger IPRs for software may actually impede innovation as patent portfolios grow but their value remains ill-defined.

Remarks on the Future

The United States will continue to lead in developing new technologies and markets in software. Most of this innovation will be centered in small software companies. (For large companies, nurturing creativity and innovation has often proved difficult, and the risk-reward equation dictating product development is typically very demanding [Hooper, 1993].) As a result, successful software start-ups will continue to spin out of larger companies led by entrepreneurs with a riskier agenda.

These dynamics create an advantage for the United States in software. But even as new technologies present opportunity for new entrants, many of these smaller companies may not have the resources to adopt these innovations and remain competitive. As software programs (including prepackaged software) have become larger and more complex, software developers have started to run into problems of quality and reliability, referred to in the literature as the "software bottleneck." Major delays in product releases, including 1-2-3 (Lotus), dBase IV (Ashton-Tate), OS/2 (IBM), and Windows NT (Microsoft) are examples of this trend (Brandt, 1991).

Attempts to address these problems include efforts to replace the current approach to software development with a more rigorous one, using code re-use and object-oriented designs (Brandt, 1991; Ferguson and Morris, 1993). Technology transfer in these new software technologies (if real) may present an opportunity for other countries to compete with the United States in software. Japan and Europe, especially, have put a premium on developing process innovations in software design and automation, although they have not yet realized any commercial advantage from these initiatives.

ELECTRIC POWER RESEARCH INSTITUTE: THE BOILER TUBE FAILURE REDUCTION PROGRAM

Jim Oggerino

Background

The Electric Power Research Institute (EPRI) has been the centralized R&D arm of the U.S. electricity industry since 1972. Its members include over 600 utilities that together provide about 70 percent of the electricity generated in the United States. EPRI manages research projects performed by contractors, including universities and large and small companies. Typically, there are over 1,000 projects in process at any given time, supported with an annual budget in excess of $400 million. Over its 23-year history, EPRI has developed many technology transfer methods and processes. This case study focuses on boiler tube failures (BTFs) in fossil-fueled power plants and the technology transfer process used to ensure that solutions to the BTF problem reach the electric-power industry.

The Issue

Roughly 40 percent of the energy consumed in the United States today is in the form of electricity, and in the next 50 years that value could grow to 60 percent. Given the extent of public reliance on electric power, any technical problem or flaw that affects the availability of the boilers that make steam to drive steam-turbine generators is serious. Historically, BTFs represented the largest single source of lost generation in fossil-fueled power plants in the nation. According to the North American Electric Reliability Council Generating Availability Data System (NERC-GADS), coal-fired units 200 megawatts (MW) or larger experienced more than 15,000 boiler tube failures during the 6-year period from 1983 to 1989. These failures represented a loss of 81 million megawatt hours (MWh) per year. At an estimated replacement power cost of $10 per MWh, this represented an annual loss of $810 million (McNaughton and Dooley, 1995).

Thus, the objective of EPRI's BTF-reduction project was to improve boiler availability at more than 800 EPRI member generating facilities, comprising about 2,000 generating units. At the time, there were 22 known mechanisms of BTF. Solutions for some of those mechanisms were available in other countries, but most had not yet been adopted in the United States. The challenge was to perform R&D on the as-yet-unsolved failure mechanisms, and to transfer knowledge about BTF solutions to EPRI members. Previous experience indicated that one could not effect technology transfer only by granting users access to the technical information. Face-to-face assistance, organizational and operational changes, and management commitment were required to succeed with a technology transfer challenge of this magnitude.

EPRI contracted with General Physics Corp. (GPC) to carry out the research on the unresolved failure mechanisms and to assist with the technology transfer program. Barry Dooley, the EPRI project manager, had come to EPRI from Ontario Hydro (a Canadian utility with a large research department), where he had done considerable work in this area. Dr. Dooley is considered a world-class expert which was, and is, an important technology transfer factor. The contract with General Physics Corp. consisted of cost-plus remuneration and was typical for EPRI at the time. In addition to EPRI and General Physics, the utilities that provided their power plants for technology demonstration were part of the solution teams that were formed.

At the time EPRI contracted with GPC, the provisions for intellectual property were that all rights were to be retained by EPRI. However, in a large number of cases, the GPC investigators were given exclusive, or sometimes nonexclusive, licenses to sell the resulting research products in any market except the U.S. utility market. Utilities receive EPRI results free through their membership. Depending on the circumstances, technology licensing for EPRI members may be cost free or require the payment of a considerable fee.

EPRI is funded by its member utilities to perform collaborative R&D and be involved in the transfer of research results to members. EPRI generally funds its research projects at the laboratory investigation level. However, at the full-scale demonstration phase, it is often necessary to use a member's generating station. It takes the highest level of trust on the part of a member utility to use an operating plant as a test bed, because any plant unavailability can result in costs of hundreds of thousands of dollars per day.

One technology transfer mechanism used to obtain sponsors for demonstration or shared R&D projects is a one-page document called a "host utility." This document is distributed through EPRI's Technical Interest Profile (TIP) system, which member utility staff join by submitting a TIP interest sheet. The interest sheet has roughly 100 technical areas to choose from. Most TIP users check three or four items and receive routine mailings on the topics of their choice. Mailing lists in each technical area contain between 2,000 and 7,000 names.

To initiate the BTF demonstration program, EPRI distributed a host utility document throughout the industry. Based on the responses it received, EPRI selected 10 utilities, representing about 40,000 MW of capacity, to begin the program. About a year after the project started, EPRI held workshops and seminars to announce interim progress to the rest of the industry. This resulted in many additional utilities volunteering to become part of the program, and so EPRI issued another host utility invitation. Two years after the first set of 10, 6 additional utilities (another 20,000 MW of capacity) were added to the project.

Prior to inviting utilities to participate, EPRI established criteria for participation. Each utility had to assign a BTF program coordinator for the project; issue a corporate BTF program mandate or philosophy statement; include in the program all fossil-generating units operated by the utility; and form cross cutting BTF program teams for which training attendance was mandatory.

After EPRI selected the first 10 utilities, it convened a meeting of BTF coordinators. All were asked to obtain senior management signatures on the corporate mandates prior to beginning activities at their utilities. EPRI then held training sessions at each utility for each BTF team, consisting of staff from the utility's engineering, operations, maintenance, and management units. The senior manager who signed the mandate had to attend the course for at least one hour. These meetings were held at home offices or at various power stations, with the selection left up to the sponsoring utility. The training material included descriptions of what actions and organizational and operational changes were required to address each of the 22 failure mechanisms. Six-month follow-up meetings were held by the EPRI team to determine if corrections or additional changes were required. It should be noted that GPC carried out essentially all of the training sessions for EPRI. GPC played a major role in the technology transfer process, freeing the EPRI project manager to focus on the R&D portion of the project.

From the outset in 1985, the target for the project was to transfer technology to achieve an average equivalent availability loss (EAL) of 1.45 percent from BTFs. This represents a nearly 60-percent reduction from the national EAL of 3.4 percent in 1985. Figure A-3 shows the EAL reductions achieved from 1985 through 1991. By 1987, the first group of 10 utilities had reduced their EAL from 2.5 percent to 1.8 percent. By 1991, that same group had further reduced their EAL to 1.5 percent. The 10 utilities predict savings of at least $41 million annually for the next 10 years. The second group, which had started 2 years later and at a much higher EAL (3.4 percent) had reduced their average to 2.0 percent.

This project demonstrated to the electric-power industry that successful technology transfer does not consist solely of being exposed to research results or technical fixes. Management commitment to support tech transfer programs and, in most cases, organizational rearrangements, also are necessary. In addition, operational changes and training programs are often required.

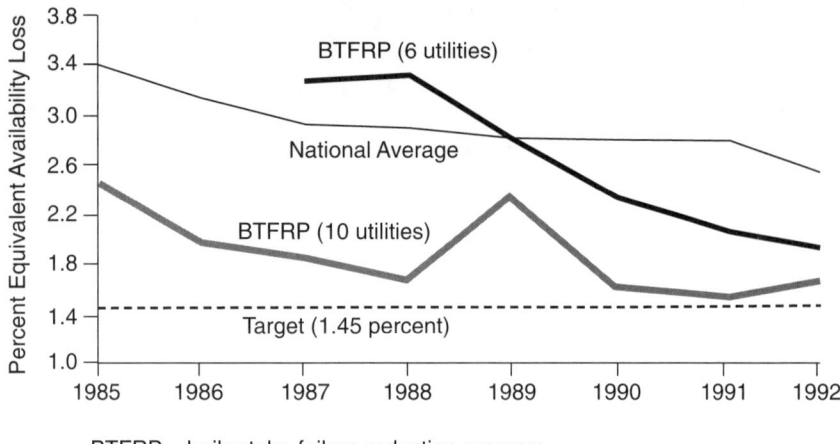

BTFRP = boiler tube failure reduction program

FIGURE A-3 Equivalent availability loss due to boiler tube failure, 1985–1992. SOURCE: McNaughton and Dooley (1995).

The demonstration program and subsequent adoption of BTF solutions by EPRI member utilities led to annual savings in the hundreds of millions of dollars. The technology transfer process was so successful it is being used for two other major EPRI programs: Plant Life Extension and Cycle Chemistry.

Perhaps more important than resolving the 22 BTF mechanisms (since expanded to more then 30), the demonstration project has resulted in utility management recognizing more fully the need to support internal product champions with funding and organizational clout. Thus, three elements—research results, a demonstration host site, and senior management support—were all required to achieve success.

PART III
TECHNOLOGY TRANSFER IN GERMANY

EXECUTIVE SUMMARY

Compared with other large industrialized countries, Germany supports a high level of research and development (R&D) activity in relation to its gross domestic product (GDP). Indeed, as a percentage of GDP, German investment in R&D is comparable to that of the United States.

A major distinguishing characteristic of the German R&D system is the existence of a broad variety of public and semipublic research institutions that complement and bridge the R&D activities of industry and universities. The most important of these institutions are the Fraunhofer Society (*Fraunhofer-Gesellschaft* [FhG]), the Max Planck Society (*Max-Planck-Gesellschaft* [MPG]), the Helmholtz Centers (formerly called *Großforschungseinrichtungen* [GFEs]), and the Federation of Industrial Research Associations (AiF). These institutions have different missions, different research focuses, and vary significantly with respect to the scope and conduct of their technology transfer activities. The participation of universities and other noncommercial research institutions in technology transfer to industry also varies greatly with respect to the four focal areas of this study: production technology (manufacturing), microelectronics, information technology, and biotechnology.

German universities' major channels of technology transfer to industry are collaborative and contract research, consultancy, informal contacts, conferences, and the provision of qualified personnel. Scientific publications are intensively used, but prove to be a less effective channel of technology transfer. In contrast to U.S. universities, the temporary transfer of personnel is rarely practiced in Germany. Another effective transfer channel for German universities is external institutions such as technology centers and particularly An-Institutes (Institute an der Universität; literally, institutes at the university). In the last 15 years, university efforts to further technology transfer have increased considerably and have reached a generally satisfying level. This assessment applies also to the transfer activities in the four focal areas. In software and especially in biotechnology, the volume of industry contacts is suboptimal, which partly reflects the lack of a sufficient number of competitive German enterprises in these areas. A new legal and institutional framework will be necessary to improve exploitation of patents at universities.

The FhG is the principal German noncommercial organization conducting industry-oriented applied research. Unlike other German public research institutions, which rely almost exclusively on institutional funding to support their research, the FhG's budget includes only 20 to 30 percent public institutional funding. Moreover, the amount of funding is linked directly to the FhG's success doing contract research for public and particularly private clients. Therefore, the FhG's research orientation is largely demand driven. The close relationship between the FhG and German universities is institutionalized through the appointment of FhG directors as regular university professors. Thus, the FhG is a real bridging institution between academic and industrial research. The future success of the FhG model depends decisively on an appropriate balance between, on the one hand, institutional funds and public-sponsored projects to build up an adequate level of research competence, and, on the other hand, private contracts to maintain the orientation toward industrial needs and to perform effective technology transfer. Except in biotechnology research, the FhG has a strong presence in the four focal areas, with special strengths in microelectronics and production technology.

Complementary to German universities, the MPG is the major institution performing outstanding basic and long-term applied research. MPG's main areas of focus are physics, biology, and chemistry. Many Max Planck institutes perform research in areas of strategic interest to industry. The most important channel of knowledge transfer is the exchange of scientific personnel. However, collaborative research with industry plays a modest but increasing role. Up to now, the intensity of contacts with industry has depended primarily on the willingness and interest of individual MPG scientists. With declining public funding, the usefulness and achievements of the MPG have to be proved, and the society has to approach technology transfer more actively.

Helmholtz Centers conduct primarily research on long-term problems entailing considerable economic risks in areas of public welfare and in fields requiring large investments. Besides the classic instrument of scientific publications, the major mechanisms of technology transfer are the participation of industry in advisory boards and committees, and collaborative research uniting industry and the centers on large projects or programs. The centers are funded primarily with public money, but industry and the federal government are striving to increase the share of industrially relevant research these centers conduct. This can be achieved by reducing institutional funds in favor of project support and broader participation of industry in the centers' research planning procedures. It is not clear to what extent these different measures suggested will be implemented. In any case, the centers will go through a process of considerable structural change within the next few years.

The institutes of the Blue List and the departmental research institutes carry out numerous research activities, mostly in basic research or applied research directed at the needs of state and federal government departments. Only a few institutes in this group have close relations with industry and perform technology transfer.

Cooperative research within the framework of industrial research associations has proved to be an effective instrument for performing projects that exceed the capacity of individual small and medium-sized enterprises (SMEs). The associations and their umbrella organization, AiF, have implemented an intensive collective evaluation procedure that guarantees effective selection of appropriate research projects. Those companies that are directly involved in the definition and supervision of projects benefit most from the results of cooperative research. Technology transfer to other member companies is limited, although many measures are undertaken to promote such activity. A major restriction of cooperative industrial research is its limitation to precompetitive problems, since several companies in the same industry have to cooperate. In some less-research-intensive industries, the share of cooperative research compared with their total R&D activities is quite large. However, in research-intensive industries like chemicals, electrical engineering, and aeronautics, the role of cooperative research is negligible.

Up to now, the volume of research activities of the European Union (EU) has been relatively limited compared with the activities of individual countries. However, the importance of EU funding is growing. Special initiatives are focusing on strategic areas in an effort to enhance European competitiveness. In areas such as biotechnology and information technology, the impact of EU-funded research is considerable. A characteristic of EU policy is a top-down approach to setting a research agenda through the use of so-called framework programs. International collaboration between industrial enterprises and research institutions is required, thus technology transfer is facilitated. In contrast, the so-called EUREKA initiative is independent of the framework programs of the European Commission. It has no framework concept, follows a bottom-up approach to setting research priorities, and pursues market-oriented research. Major areas of activity are biotechnology and, in particular, information technology. The EUREKA project JESSI (Joint European Submicron Silicon Initiative) contributes considerably to the international competitiveness of European industry in the fields of microelectronics, high-definition television, digital audio broadcasting, and other communication and information technologies.

Regarding the four focal areas, industrial R&D activities in Germany are focused on various fields of mechanical engineering. Industry gives less R&D attention to microelectronics, information technology, and biotechnology. In this respect, the German profile is almost the opposite of the American one. Only in the area of chemistry do the R&D activities of the two countries have similar structures and show positive specialization indexes.

Technology transfer to SMEs is realized through various channels. One important channel is through R&D cooperation with other companies (primarily customers of SMEs), consulting engineers, universities, and other research institutions. About half of all German SMEs that perform R&D use this channel. There is, however, considerable potential for increasing R&D collaboration between SMEs and other research institutions, particularly universities. In addition, SMEs profit from a dense network of non-R&D-performing institutions, the result of a high level of industrial self-organization. The Chambers of Industry and Commerce, industrial associations, and other institutions effectively support the diffusion of technology and know-how, particularly in technologically mature industries, through innovation-oriented consultancy and by organizing knowledge exchange among firms through journals, meetings, and informal networks.

Structural change of an economy is easier to accomplish with the help of small, new technology-based firms (NTBFs). Such firms display distinctive features connected to flexibility, innovation, and competitive advantages. There is a widely felt, obvious lack of NTBFs in Germany. NTBFs need assistance in the form of equity and managerial know-how, since most founders of such firms possess only technological skills. The failure of NTBFs to raise capital has not been reversed with the advent of new instruments like equity stock or venture capital. The supply of public funding has proved necessary. Although the risk-averse mindset of most founders of German companies cannot be overcome instantly, various steps (e.g., the formation of a transnational European stock exchange for NTBFs) are being taken to improve the current unfavorable conditions.

To summarize, highly elaborate institutionalized forms of technology transfer have been developed in Germany; examples are the An-Institutes, the AiF, and the FhG. Even universities—at least their technical faculties—show a high level of active involvement. Although the MPG and the Helmholtz Centers are developing a more active role, there is still much room for progress. A special characteristic of the German technology transfer system is that public and semipublic research institutions—including universities—have a high level of direct contract research and research cooperation with industry. This is one of the main reasons for the long-demonstrated comparative strength of the German innovation system in supporting incremental innovation and rapid technology diffusion in many industries. In the next few years, there will be an increasing need for industry to cooperate with nonindustry research institutions. At the same time, more research institutions will offer their services to industrial clients, and the competition in the research market will grow. Generally, the trend toward greater competition is a positive one. However, unfair competition among institutions due to their different levels of public base funding remains a problem. This imbalance may, under changing circumstances, disturb mechanisms that previously functioned well and require appropriate changes in national research policies.

INTRODUCTION

Part III of this binational study describes the technology transfer system in Germany. It first describes the R&D landscape in Germany in order to provide an overview of the institutional context of technology transfer. Selected R&D-performing institutions that are major players in technology transfer are then analyzed in more detail. Finally, the report examines technology transfer in four areas of technology selected by the binational panel for deeper investigation: production technology (manufacturing), microelectronics, information technology, and biotechnology. The performance of German R&D institutions in these focal areas is addressed insofar as related data and information are available.

The authors tried to present the relevant information as concisely as possible, drawing largely on the existing literature. In some sections, however, the description is more detailed because the related sources need important enlargement or additional explanation. (See, for example, the "Universities" section, below.)

THE GERMAN R&D ENTERPRISE

General Structures

In 1994, German institutions spent about DM78 billion (or 2.3 percent of GDP) on R&D. This was equivalent to $35.9 billion (in purchasing power parity), or about 21 percent of total U.S. R&D spending. The German ratio of R&D spending to GDP, though slightly lower than the American ratio (2.5 in 1994), is higher than most other large industrialized countries.[1] In recent years, the public-sector/private-sector composition of German and American R&D spending has converged. Between 1989 and 1994, the share of publicly financed R&D in Germany increased from 34 percent to 37 percent. Over the same period, the share of publicly financed R&D in the United States decreased from 46 percent to 39 percent. In Germany in 1994, about 8.5 percent of the public R&D budget was spent for defense purposes; in the United States, that figure was 55 percent. Due to the primarily civilian orientation of R&D in Germany, the share of publicly financed R&D performed by industry (13 percent) is relatively low compared with the share of such research performed by industry in the United States (31 percent) (Organization for Economic Cooperation and Development, 1995).

These general indicators give only a rough sense of the German R&D system. Particular institutional structures will be described here in more detail, following Meyer-Krahmer (1990) and Schmoch et al. (1996b). In Germany, the organization of R&D activities is shaped largely by the country's federal system of government, in which public-sector responsibilities are more evenly divided between the central government and the states (*Länder*) than is the case in the United States. German states are principally responsible for the educational sector and consequently finance the vast majority of university budgets, including more than 75 percent of academic research. The financial flows from the state-level ministries to universities are depicted by a boldface arrow in the organizational chart in Figure 3.1. Roughly 90 percent of these funds are allocated by universities for base, or general-purpose, institutional support of research. Only 10 percent of university research supported by the state is linked to specific projects.

In addition to universities, other research institutions are partially supported by state-level ministries (see Figure 3.1). These include the institutes of the Max Planck Society, the Helmholtz Centers, the institutes of the Fraunhofer Society, and "other institutions" (the Blue List institutes and independent institutions established by the states, including the An-Institutes). All these institutions are described in more detail in "Technology Transfer from Universities," below.

TECHNOLOGY TRANSFER IN GERMANY

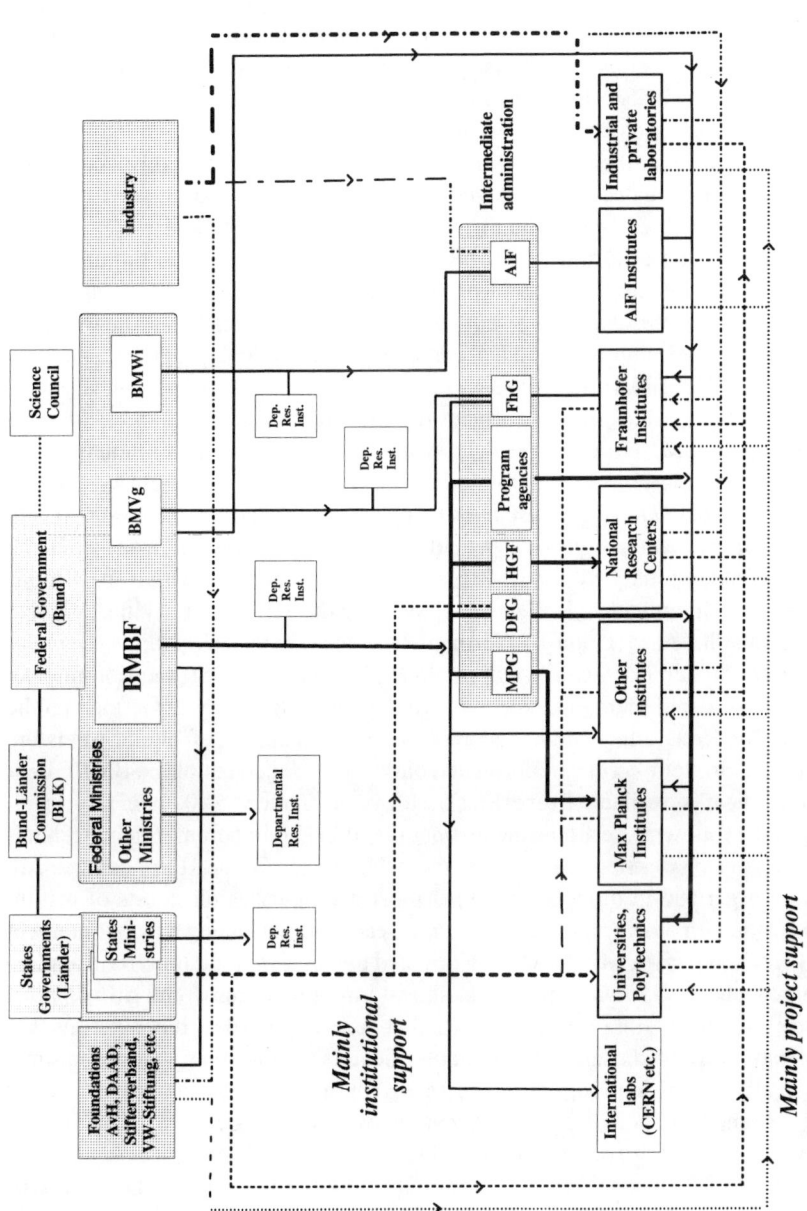

FIGURE 3.1   Organization chart of the German R&D system.   SOURCE: Schmoch et al. (1996b).

Another set of research institutions, called "departmental research institutes," are connected directly to specific state-level and federal ministries. These institutes often carry out general activities in addition to R&D activities in the area of the related ministry. Federal ministries with important ties to departmental research institutes include the Ministry of Health, the Ministry of Agriculture, the Ministry of Transport (included in "Other Ministries" in Figure 3.1), and the Ministry of Defense (BMVg). Compared with their counterparts in other countries, German departmental research institutes account for a relatively small share of total publicly funded R&D. Nevertheless, they often play an important role in the R&D landscape; some of them are leaders in special R&D sectors.

In the federal government, the most important source of R&D funding is the Ministry for Education, Science, Research, and Technology (*Bundesministerium für Bildung, Wissenschaft, Forschung und Technologie* [BMBF]).[2] The BMBF is responsible chiefly for R&D budgets and long-term research programs on a general level. The BMBF delegates more specific decisions to program agencies (*Projektträger*), which manage project-related activities for nearly all fields supported by the BMBF (see Figure 3.1). A further intermediate institution between the BMBF and R&D-performing institutes is the German Research Association (*Deutsche Forschungsgemeinschaft* [DFG]) that is responsible for supporting mainly basic research projects, especially at universities. It is noteworthy that the DFG is funded jointly by the BMBF and state governments. Other institutions that perform intermediate R&D management are the MPG, the FhG, the AiF, and the Helmholtz Association of German Research Centers (*Helmholtz-Gemeinschaft Deutscher Forschungszentren, HGF*). This variety of decision-making institutions would seem to indicate a high degree of flexibility in German public funding of R&D. In reality, however, the great majority of public funds are earmarked for long-term commitments; only about 10 percent of the BMBF budget each year is available for new tasks (Meyer-Krahmer, 1990, 1996).

In the following sections, the distinct function and important role in technology transfer of German universities, the MPG, the FhG, the HGF, and the AiF will be described in more detail. What most distinguishes these sets of institutions from each other is the focus of their research activity along the continuum of R&D activities. The MPG is chiefly oriented toward basic and long-term applied research; the FhG, toward mid- and short-term applied research; AiF supports cooperative industrial research projects that generally have a precompetitive but application-oriented character; Helmholtz Centers conduct their activities primarily in areas requiring long-term investments or entailing considerable economic risks. Some Helmholtz Centers concentrate mostly on basic research, while others work in fields of strategic industrial relevance.

Figure 3.2 depicts the general structure of the German R&D enterprise. Along the horizontal axis, institutions are classified according to their main sources of funding, whether public or private. Most of the private-sector institutions are industrial research laboratories; the number and research volume of in-

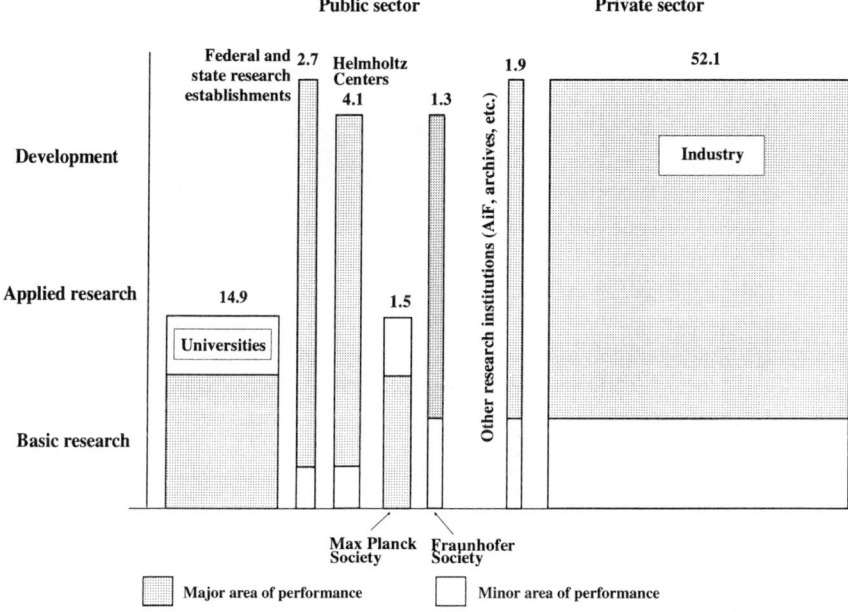

FIGURE 3.2 Main R&D-performing institutions in Germany, expenditures in billion 1995 DM. SOURCES: Reger and Kuhlmann (1995); Schmoch et al. (1996b).

dependent private research institutes are quite low. There is an intermediate class of institutions, most notably the FhG and the institutes of the AiF, which receives funding from both government and industry. The vertical axis displays the type of R&D conducted: basic research, applied research, and (experimental) development. The shading indicates major areas of performance of the different institutions. Thus, as extreme examples, the MPG concentrates on long-term basic research, whereas research in industry is mostly short term and application oriented, with time horizons on the order of 3 to 5 years. The sizes of the bars indicate the annual budgets of the respective institutions.[3]

Of course, the R&D activities of the different research organizations are not as clear-cut as Figure 3.2 makes it seem. Thus, it is not surprising that industrial laboratories perform some basic research (about 6 percent of their total internal R&D; cf. SV-Wissenschaftsstatistik, 1994) and public research institutes and universities perform some applied R&D. Nevertheless, it is important to know in which major areas the different institutions are working in order to understand relevant distributions of capital and manpower.

The greater orientation of institutions on the left side of Figure 3.2 toward basic research reflects their commitment to supporting the research needs of non-economic (in the broad sense), societal goals. The public sector tends to support earlier stages of the innovation cycle, whereas industry concentrates on later phases. Technology transfer from the left to the right side is important for the efficiency of the total system.

All in all, Germany has only a small number of completely private, profit-oriented research institutes. Instead, the intermediate position between universities and industry is occupied by nonprofit institutions, namely the FhG and AiF, both of which operate with some public funding; these institutions do application-oriented research. The MPG and the Helmholtz Centers largely supplement the activities of universities in the areas of basic and long-term research. Among the public or semipublic institutions, the sector consisting of departmental research institutes is small compared to the Helmholtz Centers, the MPG, and the FhG. The ratio of R&D expenditures in universities, other research institutions, and industry is 1:0.7:3.5. As can be seen, the institutional sector lying between universities and industry is quite large.

Industrial R&D Structures

ORIENTATION OF INDUSTRIAL R&D

To better understand the nature and dynamics of technology transfer to industry by German universities and other public and semipublic research institutions, it is important to appreciate the comparative R&D and technological strengths of German industry. In this context, European patent data offer a useful window on the relative technological strengths and weaknesses of German industry.[4] A recent study by Schmoch and Kirsch (1994) compared Germany's share of patents in 30 separate technology fields with the average share for the rest of the world in each field.[5] Using an indicator of specialization, the study identified industries in which German patenting was above or below the world average. The results show a strong orientation toward fields in mechanical engineering, such as machinery, engines, handling, and transport (Figure 3.3). Indicator values for consumer goods and civil engineering are also above the world average. Fields such as organic chemistry, basic material chemistry, and polymers generally show average or positive values, whereas biotechnology and pharmaceutical research (which is linked to biotechnology) show values distinctly below average. Finally, information technology and related fields such as audiovisual technology and telecommunications show below-average values. One can conclude from these data that German industry is marked by a strong emphasis on mechanical engineering, a conclusion that is supported by international trade statistics (Bundesministerium für Bildung, Wissenschaft, Forschung und Technologie, 1997; Gehrke and Grupp, 1994; Häusler, 1989). Indeed, in Germany there are a variety of innovative SMEs conducting research related to mechanical engineering that have a distinct focus on export.

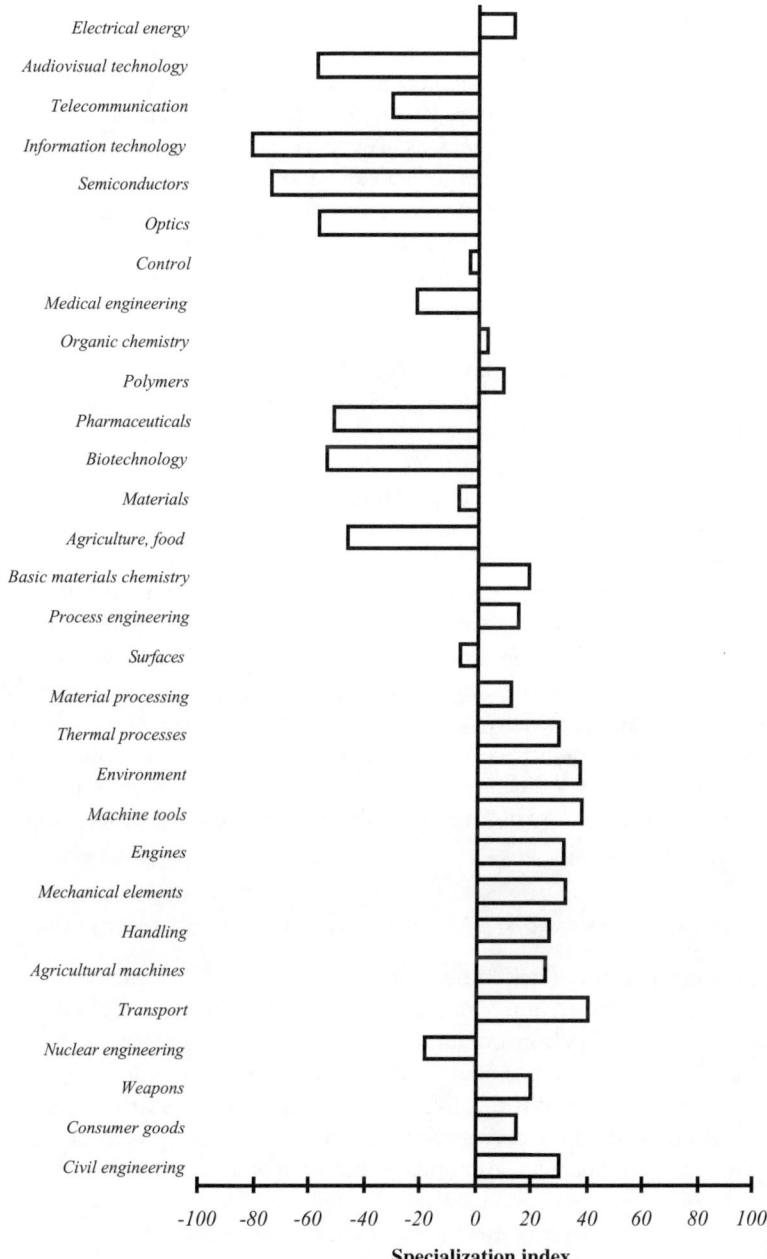

FIGURE 3.3 Specialization index of European Patent Office (EPO) patents of German origin in relation to the average distribution at the EPO for the period 1989 to 1991. SOURCE: Schmoch and Kirsch (1994).

The moderate specialization indexes for microelectronics and information technology are confirmed by foreign trade statistics: The related specialization (RCA) index in the areas of computers and semiconductor devices is distinctly below average (Gehrke and Grupp, 1994). In part, this result reflects the fact that only a few large companies—Siemens, Temic Telefunken, Bosch, and the German subsidiaries of IBM, ITT Semiconductors, Philips, and Texas Instruments—are internationally competitive in these areas of research. This means that the German public and semipublic research institutions have only a few resident industrial counterparts capable of supporting significant levels of intra- and extramural research. (For further details, see "Technology Transfer in Microelectronics," below.)

The low specialization index in biotechnology has to be interpreted in light of the very high level of U.S. activity, which largely determines other nation's average share of biotechnology R&D. Nevertheless, the moderate indicator for Germany reflects a quite hesitant start on the part of the big chemical companies. Current activities are often based on affiliations and acquisitions in the United States, whereas research in German laboratories is still at a moderate level.[6]

The patent profile of the United States differs significantly from the German one. The United States has positive index values in the fields of information technology, semiconductor devices, and biotechnology and negative ones in mechanical engineering and consumer goods (Figure 3.4). The closest correspondence to the German profile can be found in the fields of organic chemistry, polymers, and basic materials chemistry, which have above-average specialization indexes in both countries. Also in both countries, the specialization profiles are generally stable over time. In comparing the German and American technology transfer systems, these differences in the orientation of industrial R&D have to be borne in mind.

TECHNOLOGY TRANSFER TO SMALL AND MEDIUM-SIZED ENTERPRISES

Growing technological and market demands have fostered considerable growth of R&D cooperation and technology transfer between large companies and noncommercial R&D institutions in Germany. Although German SMEs face many of the same challenges that have prompted large firms to seek external sources of technology and R&D, R&D cooperation between SMEs and noncommercial R&D institutions does not appear to be as widely established as that involving large companies. Admittedly the collaborative research activities of German SMEs have not yet been studied extensively. Some analyses, however, suggest that in recent years, SMEs, especially in the manufacturing sector, are relying increasingly on technology transfer from external research institutions.

According to a joint survey by the Fraunhofer Institute for Systems and Innovation Research (FhG-ISI) and the German Institute for Economic Research (*Deutsches Institut für Wirtschaftsforschung* (Becher et al., 1989), approximately

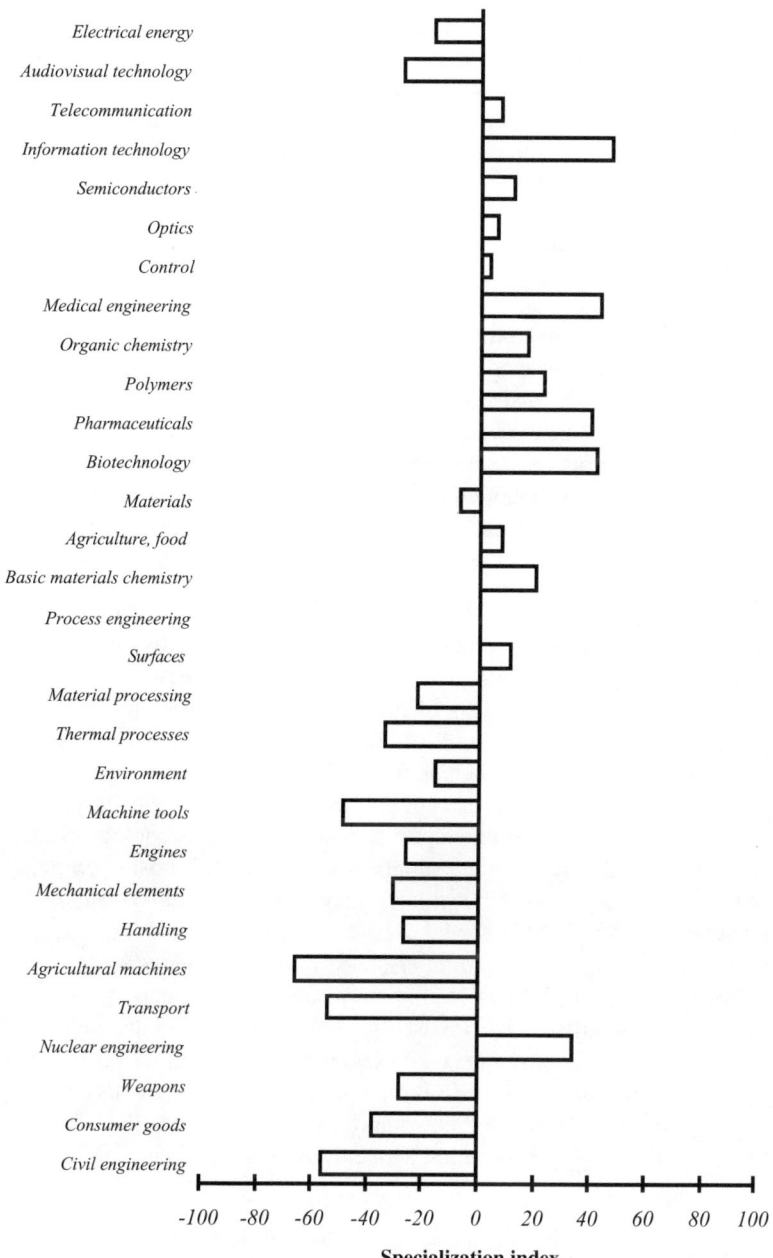

FIGURE 3.4 Specialization index of European Patent Office (EPO) patents of U.S. origin in relation to the average distribution at the EPO for the period 1989 to 1991.
SOURCE: Schmoch and Kirsch (1994).

25,000 SMEs of the former West Germany engage in R&D. A further study by FhG-ISI and Prognos (Wolff et al., 1994) estimated that in 1991 13,000 German firms were conducting cooperative R&D and about 3,000 to 5,000 R&D-performing firms offered plausible reasons why they were not engaged in collaborative R&D. Still, a significant number (7,000 to 9,000) of R&D-performing SMEs could potentially engage in collaborative research but have never done so (Wolff et al., 1994).

The FhG-ISI and Prognos study differentiated between "hard" and "soft" technology transfer, as follows:

- *R&D cooperation* (hard) consists of contract research by third parties (companies, public or industrial research facilities, universities, technical colleges, engineering offices) and joint R&D with or without a contractual basis;
- *Technology-related activities* (soft) includes informal contacts for the purpose of information exchange, performance of technoeconomic studies, joint utilization of laboratories and other testing instruments and facilities, employment of university students as trainees or interns, and the preparation of a graduation or doctoral thesis.

Whereas 50 percent of all SMEs surveyed are involved in R&D cooperation, 30 percent declared that they practice cooperation in "less active technology-related activities" (Kuhlmann and Kuntze, 1991). With respect to the importance of potential partners, there are significant differences between the two types of technology transfer. In the area of R&D cooperation, customers and consulting engineers play a vital role, whereas universities and research institutes are of medium importance, and the impact of polytechnical schools (*Fachhochschulen*) is negligible (Figure 3.5). With respect to cooperation in technology-related activities, polytechnical schools and suppliers are the SMEs' most important partners. Universities and research institutes again occupy a middle position.

Although German SMEs appear to be drawing effectively on the technology transfer abilities of customers, suppliers, and consulting engineers, some observers believe that the capabilities of university polytechnical schools and research institutes are underutilized by SMEs. In general, the knowledge generated by universities, polytechnical schools, and research institutes is sought by an SME when the firm needs to understand unfamiliar techniques, wants to make use of testing equipment, or is seeking new approaches. SMEs identified three major impediments to greater collaboration with universities, polytechnical schools, and research institutes:

- the low level of interest displayed by these research institutions in the specific research needs of SMEs (47 percent);
- the high cost to SMEs of cooperating (mentioned by 44 percent);[7] and
- the perception by SMEs that these collaborations do not lead to usable results quickly enough (42 percent) (cf. Wolff et al., 1994, p. 166).

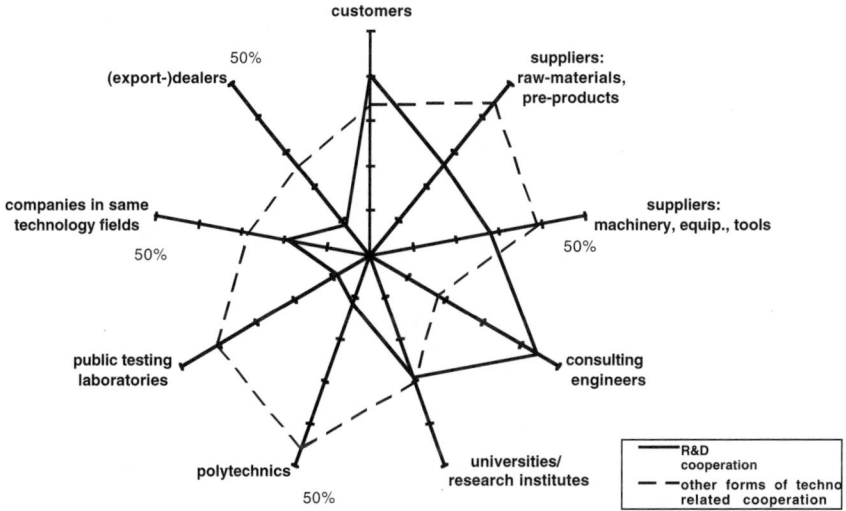

FIGURE 3.5 Partners of SMEs in R&D and technology-related activities, by percent. SOURCE: Wolff et al. (1994).

Herden (1992) asked 1,349 German SMEs about the type and frequency of their contacts with universities and other research institutes during the past 5 years. The results of this study (Table 3.1) further verify the above findings.

According to the Herden data, a mere quarter of all SMEs received technological knowledge from research institutions or universities. Among those that did, the most frequent type of technology transfer was soft contacts without R&D cooperation. In this regard, consulting on the solution to a problem was the most important channel (cited by 69.8 percent), whereas licensing, which may be viewed as another information channel, was significantly less important (mentioned by 9.1 percent). Training of qualified personnel at universities ranked second in importance (45.4 percent). The joint implementation of R&D projects (i.e., hard cooperation) ranked third (33.5 percent). Nonetheless, only about 8 percent of the 1,349 firms surveyed cooperated in joint R&D projects with academic institutions. Unfortunately, this survey did not ask firms about the temporary assignment of scientific personnel from universities to firms, therefore, the frequency of this kind of personnel transfer cannot be measured.

The level of SME cooperation with universities and research institutions is surprisingly low compared with the potential impact of the scientific knowledge that could be gained through such partnerships (Schmoch et al., 1996b). Technology transfer from universities and other research institutions to SMEs could be improved by using public subsidies to reduce the cost of collaboration. It is estimated that roughly 30 percent of the SMEs involved in R&D cooperation

TABLE 3.1 Types of Knowledge Transfer from Academia to Industry[a]

Channels for technology transfer (multiple choices possible)	Percent
Consulting on problem solution	69.8
Training of qualified personnel at universities	45.4
Joint implementation of R&D projects	33.5
Subcontracting of R&D projects	25.9
Sharing of laboratory and equipment	24.1
Information on the market potentials of new products	17.7
Directed search for R&D personnel	17.4
Directed search for recent graduates (non-R&D personnel)	14.6
Licensing	9.1
Short-term assignment of R&D personnel to universities	5.2

[a]The survey question was, "Have you directly obtained technical knowledge from research institutions and/or universities during the past five years?" Of the SMEs surveyed, 24.5 percent answered "yes," 67.1 percent answered "no," and 8.4 percent answered "not yet but planning to."

SOURCE: Herden (1992).

could benefit from public support (Kuhlmann and Kuntze, 1991). However, much more difficult to remedy is the perception among SMEs that nonindustry research institutions are not particularly interested in the research problems of SMEs. Here, technology transfer units of universities and polytechnical schools could prove their efficiency by improving mutual understanding of the different research needs and capabilities of SMEs and research institutions (Kuhlmann and Kuntze, 1991).

In general, SMEs are considered important pillars of the German innovation system (Harhoff et al., 1995). Although they contribute to new and emerging areas, their specific strength is the rapid diffusion and adaptation of existing technologies. In this regard, they can draw upon the resources of a variety of R&D-performing, transfer-oriented institutions such as Fraunhofer institutes and the research institutes of industrial research associations. (For details, see "Fraunhofer Society" and "Federation of Industrial Research Associations," below.)

Furthermore, a dense network of non-R&D-performing institutions supports technology transfer through innovation-oriented consultancy and the organization of knowledge exchange among firms. All Chambers of Industry and Commerce (*Industrie- und Handelskammern*) offer consultancy services concerning not only technological innovation and potential cooperative partners but also financial problems relating to investment and public support programs. The Chambers of Crafts (*Handwerkskammern*) offers the same services for craftsmen (Bundesministerium für Bildung, Wissenschaft, Forschung und Technologie, 1995a; Bundesministerium für Forschung und Technologie, 1993b). Both institutions are legal representatives of commercial enterprises in Germany. The Chambers of Industry and Commerce are financed completely by industry; the Chambers of Crafts receive considerable public support.

Another important institution is the Organization for Rationalization of German Industry (*Rationalisierungs-Kuratorium der Deutschen Wirtschaft*), which is jointly financed by industry, trade unions, the federal government, and the states. It supports SMEs in areas of management, organization of production, and personnel training (Bundesministerium für Bildung, Wissenschaft, Forschung und Technologie, 1995a).

The diffusion of knowledge within German industry is also fostered by about 650 industrial associations (*Industrie-Fachverbände*) representing all industry sectors (Hoppenstedt Verlag, 1995). The principal roles of industrial associations are to represent their member companies politically and to promote discussion of relevant commercial, political, or technical problems in their own journals and meetings. The 106 industrial research associations comprise an important subset of this larger group; they conduct cooperative research and are organized under the umbrella of the AiF.[8]

There are also about 400 associations for the advancement of science[9] (*Wissenschaftliche Fachgesellschaften*), representing more than 400,000 members from across the science disciplines. About half of the members are in the engineering sciences and their associations are often called Technical-Scientific Associations (*Technisch-wissenschaftliche Vereine und Gesellschaften*), most of them being organized in the German Federation of Technical-Scientific Associations (*Deutscher Verband Technisch-Wissenschaftlicher Vereine*). About 60 percent of the members of the engineering associations come from industry; the rest are from research institutions and universities. By contrast, in the natural sciences, the share of industrial members is 20 percent. The major aim of the scientific associations is to initiate discussion on recent research results and (especially in the natural and engineering sciences) to facilitate the transfer of knowledge between scientific institutions and industry (Schimank, 1988b; Wissenschaftsrat, 1992). Important instruments are the diffusion of knowledge through journals, conferences, or professional continuing education.

Finally, the states have a key position in supporting innovation by SMEs. All states support the research and technology transfer needs of SMEs, by establishing information units at universities and by providing funding for innovative projects, specific technologies, and new technology-based firms. An interesting example is the Steinbeis Foundation (*Steinbeis-Stiftung*) in the state of Baden-Württemberg, which established a network of 200 technology-transfer units at polytechnical schools. The foundation arranges contacts between companies looking for solutions to specific problems and the appropriate professors in the network. This system has proved very successful, and the Steinbeis Foundation plans to extend its network to other states.

With respect to the federal government, the most important measures for promoting technology transfer are the establishment of transfer-oriented research institutions (see "Fraunhofer Society" and "Universities, An-Institutes and Other External Institutions," below) and the cooperative research programs (*Verbund-*

forschung) of the Ministry of Education and Research (see "Universities, Statistics on General Research Structures," below). Notably, there are also 11 federal technology transfer centers, each oriented to a specific technology (e.g., biotechnology, laser technology, production, textiles). These centers were established by the Ministry for Economic Affairs.

Overall, technology diffusion in industry, particularly to SMEs, is efficiently fostered by a high level of industrial self-organization, the roots of which go back to the nineteenth century (Lundgren, 1979). These activities are backed up by and intertwined with a broad variety of public institutions and measures for supporting technology transfer.

CONDITIONS FOR NEW TECHNOLOGY-BASED FIRMS

With their adaptability and high potential for innovation, NTBFs influence the structural change of an economy. They are regarded as a stimulus for dynamic development because they

- increase the number of market competitors and therefore motivate established companies to strengthen their efforts to innovate;
- increase the demand for services that support innovation; and
- strengthen the regional suppliers of technical products, since their own manufacturing penetration is rather low (Kulicke and Wupperfeld, 1996).

NTBFs can be found in fields such as mechanical engineering, electrical engineering, electronics, process engineering, environmental technology, biotechnology, and medical technology. However, a significant number of economists and politicians deplore the lack of NTBFs in Germany in strategic fields of high technology. For example, more than 350 American biotechnology start-up companies were counted between 1971 and 1987 (Dibner 1988). The number was significantly lower in Germany. These start-ups are regarded as the basis for the outstanding position of the United States in this technological field (Kulicke, 1994).

The establishment of spin-off companies by scientists presently working for other, usually much larger industry or university research units is one of the most effective channels of technology transfer. The success of U.S. NTBFs in the computer hardware and software industry prompted the German government in the early 1980s to support establishment of NTBFs by awarding competitive grants for up to 75 percent of a start-up company's R&D costs. Although these early efforts did not meet expectations, it did demonstrate that there is a potential for NTBFs in Germany. The specific requirements of this type of start-up indicated a new strategy was needed, aiming at directly activating market forces by involving nonpublic investors (such as venture capital companies, private investors, companies, or banks) in support of NTBFs. Instead of direct subsidies to NTBFs, incentives such as refinancing, deficiency guarantees, and co-financing were offered to venture capital investors.

Because of the poor employment situation and the weak industrial base in East Germany after German unification, government support for NTBFs has been revitalized and improved since 1990 (for details, see Abendroth, 1993; Kulicke, 1993). Since 1990, more than 550 NTBFs have been established in the new federal states, but with declining rates in recent years.

Experts agree that despite these various public measures, the number of NTBFs in Germany is still rather low compared to the situation in the United States. An exact comparison of the annual rate of formation of NTBFs, however, is not possible because the available statistics and estimates are based on different definitions and demarcations. Neither country's record in this area can be regarded as optimal. The very high U.S. rate of NTBF formation is tarnished by the high failure rate of new companies, the tendency to destructive competition, and the associated costs to the U.S. economy (Florida and Smith, 1993). In Germany, the number of NTBFs is too low, but their survival rate is very high compared with other German start-ups in trade and service sectors (Kulicke, 1994).

In Germany, there are a number of barriers to the successful establishment and operation of a sufficiently large number of NTBFs (see Kulicke and Wupperfeld, 1996). Some of the most important include:

- *Limited availability of capital for firm setup, the financing of developments, and market entry.* Failure to acquire capital is the main problem for German NTBFs. Because NTBFs do not have records of market success or because their founders cannot furnish sufficient equity guarantees, banks are reluctant to loan them money. Also, many banks lack technological knowledge and for this reason hesitate to finance what they view as "risky" operations.
- *Limited managerial know-how on the part of the company founders.* Founders of technology-based firms often come from the natural sciences or engineering and have limited management, marketing, and financing experience. As a result, they fail to develop a strategic concept, which banks need to assess the risks of investment. Links (or networks) to sales partners, cooperation partners, or suppliers are often not established, a situation that cries out for a supportive management.
- *Barriers to market entry.* Because NTBF founders have limited knowledge of markets and market forces and little experience in marketing, market entry is difficult. NTBFs are further hampered by the fact that they do not possess a brand name or product image.
- *A shortage of qualified and experienced management personnel.* Most of NTBFs cannot pay high salaries; therefore, they are less attractive to potential employees, including skilled managers.
- *Unfavorable taxation.* German law does not provide preferential tax treatment of gains from venture-capital investment. However, analyses in the United States, Canada, and Great Britain show that high capital gains tax

rates are negatively correlated with investment in venture capital operations. In Germany, taxation is also a negative factor when investors sell their shares: Whereas the situation is more favorable for individuals, investor-owned corporations and other forms of "legal entity" have to pay income taxes on any profits, regardless of their form or value (see Pfirrmann et al., 1997).

The federal and state governments are working to overcome these barriers by various support programs. In West Germany, however, there is no longer a federal support program exclusively designed for NTBFs. Rather, financial support for NTBFs is now incorporated in a new program for small technology-based firms altogether. At present, most NTBFs participate in federal or state programs, and public programs are NTBFs' primary source of external financial support. Most of the support is not given as a direct subsidy, but rather as equity stock.

The promotion of NTBFs has been generally successful, and there are positive examples of successful initial public offerings (IPOs). However, there are some shortcomings of public support for start-ups, including that

- companies receive too little support. The aid generally covers 40 percent or less of their capital demand; for the remaining 60 percent, other private sources have to be found;
- public equity stock institutions offer only financial, not managerial, support;
- the programs do not offer a holistic approach; rather, they finance segments of a firm's business. Generally, such support focuses on investment in capital goods and does not cover expenditures for staff; and
- the programs aim primarily at the first stage of company set-up. The later stage of market entry, which demands considerable capital, is not covered.

Today, most of the federal states act according to a special SME policy that promotes NTBFs and SMEs through so-called SME equity stock companies (*mittelständische Beteiligungsgesellschaften* [MBGs]). MBGs receive funding from Chambers of Industry and Commerce and regional banks and work in close regional cooperation with credit institutions. Their business policy is largely determined by whether public programs offer refinancing and failure guarantees, since their own funds are rather limited (Kulicke, 1990). They also have only limited capabilities to give managerial support to their portfolio firms.

The German venture capital market is dominated by business investment companies (*Kapitalbeteiligungsgesellschaften*) of banks, savings banks, other credit institutions, and insurance companies, which are more interested in capital gains and invest little or no capital in NTBFs. Furthermore, there is a small number of independent German venture capital companies that coordinate the interests of industrial firms, banks, fund managers, and foreign venture capital firms (Wupperfeld, 1994).

In the United States, the use of venture capital to finance NTBF start-up is relatively commonplace. The American concept, combining equity as well as technical and management support, is not working that well in Germany, however. This is in part because of legal regulations, but also because traditional ways of doing business are difficult to change. Even if habit and mind-set problems are easy to identify, it is quite impossible to verify their impact and the scope of their influence on the difficulties faced by NTBFs. And while it is possible to modify habit and mind-set on an individual basis, broad structural change is taking place very slowly. Nevertheless, it is necessary to address these "soft" factors because they point to the limits of a sudden change in regulations.

Legal restrictions can be addressed more easily. Of particular relevance to NTBFs are the restrictive regulations concerning bankruptcy and liability. The bankruptcy law (*Konkursordnung* [KO]), dating back to 1877, states that every partnership and legal entity can apply for the commencement of bankruptcy proceedings and can be made personally liable. Some companies have no limits on their liability. If bankruptcy is declared, the partnership is automatically made personally liable in the event that the company's equity capital is insufficient to pay off debts. But even with limited liability, partners can be made liable beyond the level of their investment in the firm. In general, bankruptcy is closely connected to the securities offered to and demanded by banks: Debtors with a weak financial background, in particular, are often made personally liable. For the GmbH (*Gesellschaft mit beschränkter Haftung*), banks often demand personal liability when a limited partnership has liable equity capital of at least DM 50,000. The same standard can apply to legally formed corporations (*Aktiengesellschaft* [AG]), where the partnership can also be made personally liable. In this case, private assets are used to repay excess business debt by means of an attachment. According to the regulations, employee claims have first priority, followed by "ordinary" business creditors, employee pension plans, and public social insurance and pensions (§ 61 KO). A number of other public entities follow those four, but very often private assets are not sufficient to pay back all the claims.

This threat of personal liability in bankruptcy contrasts sharply with the situation in the United States. If somebody goes out of business in the United States, he or she faces almost no problems starting another business. In contrast, one failure in Germany almost always ends the dream of operating one's own business. Therefore, the risk inherent in establishing an NTBF is higher in Germany than it is in the United States. The apparent risk-averse mentality of founders of German NTBFs can be connected directly to these legal restrictions.

Two-thirds of American venture capital is administered by independent funds, one-fourth is handled by corporate venture capital firms, and the remaining portion is held by small business investment companies (SBICs). By contrast, the German venture capital and equity stock market is dominated by subsidiaries of banks. Some general characteristics of banks, their goals, and attitudes may hamper their supposed supportive function.

- Banks do not possess the necessary technological knowledge to assess accurately the risks of investing in NTBFs. Their risk-diversifying portfolio strategy limits the financing of uninsurable and risky operations.
- The universal banking system in Germany and the different inherent functions of saving, lending, and issuing bonds means banks can lose their reputations if a business they finance fails, and this loss can have a negative impact on their saving and lending functions. Therefore, banks do not want to take the risk of issuing bonds for NTBFs. Banks are also concerned about a possible loss in reputation and the profit margin, which is low for a small firm compared with a large one, when private businesses go public. For these reasons, banks prefer to issue credit to NTBFs without any further commitment to the start-up.
- Banks follow the strategy of constant returns. Therefore, rather than reinvesting their capital gains, NTBFs are required to pay dividends or interest to the bank or the equity stock company. This policy limits the growth potential of NTBFs.

One of the main barriers to success of the American venture capital model in Germany is the virtual impossibility for SMEs to go public. An equivalent to the U.S. over-the-counter market, which allows investors to sell off their shares in a start-up company, does not exist in Germany, but will be established in spring 1997. Venture capital companies therefore have faced a relatively low rate of return when investing in NTBFs. Going public in Germany is only possible for corporations (i.e., the legal form of AGs) and can only be done following strict and conservative financial requirements. An attempt has been made to remedy this problem by creating the "small corporation" (*Kleine AG*), which is linked to significantly fewer financial and bureaucratic requirements. In November 1996, a European stock exchange, the European Association Securities Dealers Automated Quotation (ESDAQ), came into being for NTBFs and small technology-based firms. Its purpose is to promote the concept of venture capital companies, as the individual national markets of the EU are too small to match the supply and demand of the relevant actors on the market. Another barrier is that there are no tax privileges for share capital and capital share gains, which are major incentives in the United States.

Resistance to venture capital investment is not encountered only on the supply side. Venture capital and the underlying concept are not widely accepted by the German founders of NTBFs. Founders vehemently oppose equity stock capital and venture capital because the investors are accorded executive rights (Kulicke, 1993). When faced with the financial difficulties associated with launching an NTBF, a majority of individuals change their minds about wanting to start their own companies. Most of those who do attempt to form NTBFs favor remaining independent (Kulicke and Wupperfeld, 1996). Even if NTBFs accept managerial help and agree to share the executive right of decision making, the limited competence of German venture capital and equity stock companies in anything but financial matters can create problems. NTBF founders recognize the financial expertise of these firms but lament their lack of other supporting competences (Kulicke and Wupperfeld, 1996).

The main current impediments to a functioning venture capital market in Germany, according to Kulicke and Wupperfeld (1996), are

- a lack of attractive "exit routes" for venture capital companies to achieve high rates of return from NTBFs going public;
- no favorable tax treatment for investors in venture capital companies or for venture capital companies themselves;
- avoidance of risk by investors and venture capital and business investment companies; and
- aversion to loss of independence on the part of NTBF founders and entrepreneurs.

There are a variety of steps that could be taken to increase the usefulness of the venture capital option for German NTBFs. These include:

- allowing a tax reduction for investors' contributions to special funds and reducing the applicable capital gains tax rate;
- strengthening the pan-European stock exchange for NTBFs;
- teaching managerial skills in natural sciences and engineering schools; and
- improving the competence of venture capital and equity stock companies to assess financial and technological risks and to deepen their knowledge of technology.

To sum up, in Germany, NTBF formation is discouraged by an unfavorable financial, legal, and social environment. As a result, this important instrument of technology transfer is used insufficiently. However, various steps are being taken to adapt this means of technology transfer, which has proved very successful in the United States, to the specific conditions in Germany and Europe.

Impact of European Research

RESEARCH PROGRAMS OF THE EUROPEAN UNION

The Single European Act, ratified in 1987, formulated a European research and technological development policy. Its most important aim was to strengthen the international competitiveness of European industry in technology-intensive sectors such as information and communication technologies, the biosciences, and materials research.

The policy's main instruments are the Framework Programs of Community Activities in the Field of Research and Technological Development. Practical realization of the framework programs takes place in so-called specific programs, which describe in detail the scientific topics and the procedures for carrying them out. These programs last for 4 years. The Fourth Framework Program, started in 1995, is the most recently initiated, although some specific activities of the Third Framework Program (1990–1994) are still in operation. The latter established three focal areas for R&D: basic technologies, management of natural resources, and management of intellectual resources. These areas were in turn broken into six sections and a series of specific programs:

- Information and communications technologies
- Industrial and materials technologies
- Environment
- Life sciences and technologies
- Energy
- Human capital and mobility

The Fourth Framework Program extended the specific programs within the existing sections and added another two sections, in transportation technologies and socioeconomic research; however, those two programs account for only 4 percent of the EU budget for research and technological development. Information and communication technologies clearly dominate with more than 36 percent of the budget.

With a volume of European Currency (ECU) 5,700 million, or just under 5 percent of the total EU budget, the Third Framework Program is relatively modest compared with other EU operations. The significant increase in budget for the Fourth Framework Program, to ECU 9,432 million, is a further indicator of the growing importance of EU funds. In absolute terms, EU support for R&D is becoming increasingly important, especially considering the expected decreases in the flows of national R&D funds. The growing importance of the EU in science and technology becomes even more apparent if one looks at the substantial efforts that have been made since the late 1980s to strengthen the research and technology base, particularly of the less-developed regions of the EU, with so-called structural (regional, social, and agricultural) funds.

EU support for research and technological development is awarded without regard to national proportional representation or quotas. The success rate of project applications is influenced mainly by the number and quality of applicants. In fact, the number of applications has risen substantially in the past few years, and application approval acceptance rates have dropped continuously. The increase in applications can be explained by a number of factors. The relatively high number of applications from British institutes of higher education, for example, is due to a severe cut in the national research budget for universities (Figure 3.6).

EU support primarily takes the form of contracted research with cost sharing.

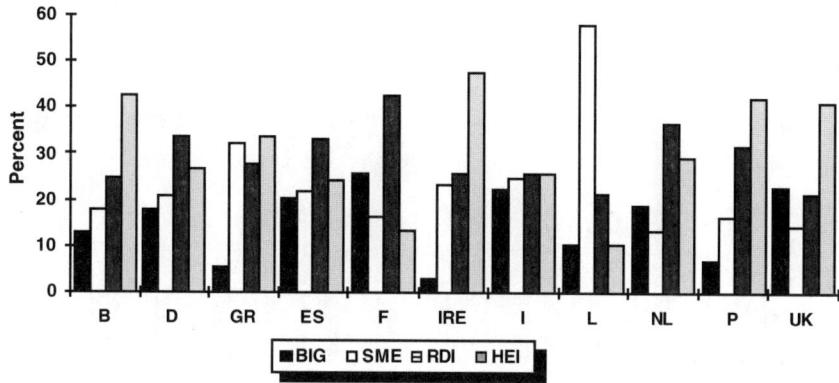

FIGURE 3.6 Participation structure in the Second Framework Program, by country, 1987–1991. NOTE: BIG = large enterprise; SME = small and medium-sized enterprise; RDI = nonuniversity research institute; HEI = higher education institution. B = Belgium; D = Germany; GR = Greece; ES = Spain; F = France; IRE = Ireland; I = Italy; L = Luxembourg; NL = Netherlands; P = Portugal, UK = United Kingdom. SOURCE: The database CORDIS.

The selection of projects is based on the following general criteria (see Kommission der Europäischen Gemeinschaften, 1990):

- Precompetitive character of the proposed R&D activities
- Transnationality of the project
- Scientific and technical quality and originality of the project proposal
- European dimension of the proposal (value added through European cooperation that could not be attained at a purely national level)
- Technical and economic usefulness
- Exploitation possibilities for the expected results

However, the EU is currently redirecting its technology policy from precompetitive research toward market-oriented projects (Klodt, 1995). Therefore, the precompetitive character of proposed projects is no longer a formal prerequisite—and actually was not strictly applied in former programs.

Compared with what is contributed by industry and the federal and state governments, the importance of EU financing is still minimal from the German perspective. From 1987 through 1991, a total of DM 1.3 billion (ECU 653 million) was received by German institutions from the EU. This represented only 0.4 percent of total German domestic expenditures on R&D. EU funding represented about 1.8 percent of R&D expenditures by the federal government and about 5.9 percent of direct project support by the government (not including R&D expenditures by the Ministry of Defense).

In some fields of research and technology, however, EU financing has gained

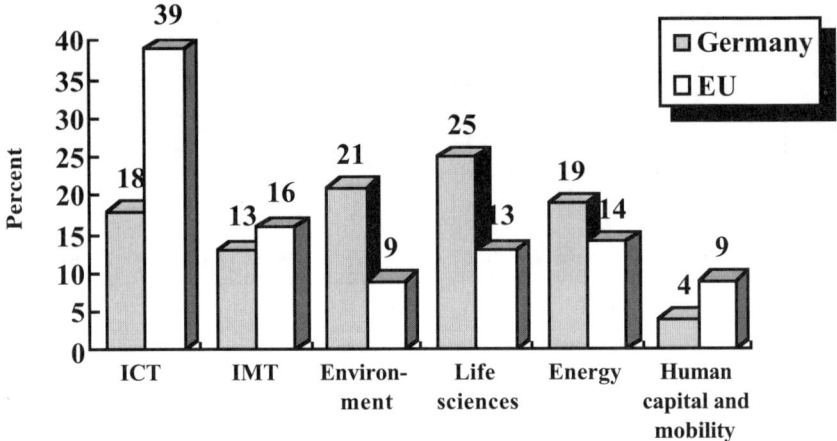

FIGURE 3.7   R&D expenditures of Germany (1992–1993) and the EU by sections of the Third Framework Program.   NOTE: ICT = information and communication technology; IMT = industrial and materials technologies.   SOURCE: Reger and Kuhlmann (1995).

considerable significance in Germany. The EU is relatively more active in information and communications technologies (ICT) than is Germany (Figure 3.7). In absolute terms, EU financing of German R&D activities in this field is equivalent to approximately one-fifth to one-quarter of what the German government spends on R&D in this area.

As to the four focal areas of this report, the research and technological development activities of the EU are particularly relevant for information technology, microelectronics (included in information and communications technologies), and biotechnology. Production technology is supported under the heading of industrial technology.

In 1991, German participants received 22 percent of available EU funding, a quite significant percentage. However, the allocation of those funds among different sectors of the R&D systems is uneven. EU funding for predominantly industrially oriented programs goes chiefly to German industry. Even in the relatively science-oriented programs, contractors from industry predominate. However, nonacademic German R&D institutes are significantly underfunded compared with similar institutions in other countries that receive EU support. German institutes of higher education are also underrepresented compared with the average of all EU countries (see, for example, Figure 3.6 and Reger and Kuhlmann, 1995, p.25).

The impact of EU-funded R&D varies significantly among recipient countries (Figure 3.8). Whereas EU funds play a minor role in the German R&D system, in other, structurally weaker countries (especially southern countries),

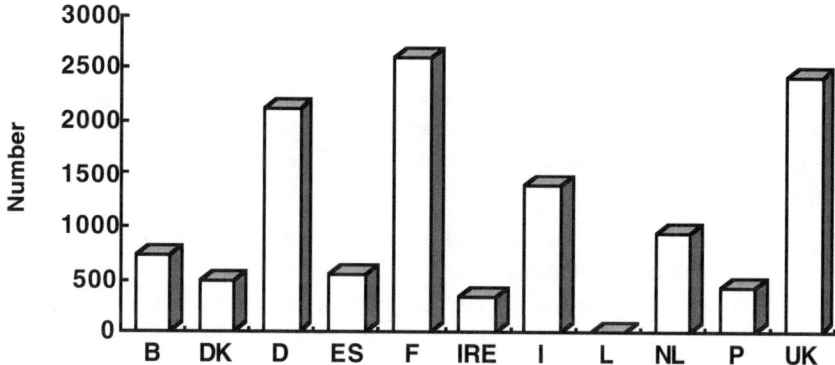

FIGURE 3.8 Participants in the Second Framework Program, by country, 1987–1991. NOTE: B = Belgium; DK = Denmark; D = Germany; ES = Spain; F = France; IRE = Ireland; I = Italy; L = Luxembourg; NL = Netherlands; P = Portugal; UK = United Kingdom. SOURCE: The database CORDIS.

EU funds support a significant portion of national R&D programs. Thus, if a country's gross domestic product is taken into account, the dominance of Germany, France, and the United Kingdom is diminished. From Figure 3.8, of course, one can only imagine the relative importance of EU support for "weaker" countries like Ireland, Portugal, and Spain.[10]

German R&D institutions have a number of concerns about EU R&D support (see, for example, KoWi, 1992), including

- that there is inadequate representation of some research fields among those that gain EU support;
- that there is excessive amount of bureaucracy involved in the application procedure and the management of the project (see, for example, Schmoch et al., 1996b);
- that the dominant role of the English language can be a hindrance in the running of EU projects and in the work of transnational project consortia; and
- that low rates of approval for project proposals waste resources if the application fee is high.

Despite these problems, German R&D institutions will likely become more interested in receiving EU funding. This is in part because they are becoming increasingly aware of the growing importance of international—and in this context, European—cooperation in R&D. In addition, the "years of affluence" in German national R&D support are over. Therefore, research institutions must search for other sources of support to compensate for the reductions in government funding.

THE EUREKA INITIATIVE

General Structures

The EUREKA initiative was launched in 1985 as a reaction to the American Strategic Defense Initiative. It is not a program of the EU, but it has provided a framework for international collaboration among firms and research institutes in the fields of advanced civil technologies. Its aim is to

- strengthen the productivity and competitiveness of European industry,
- develop a common infrastructure, and
- solve problems, especially environmental ones, affecting more than one country.

EUREKA is not intended to harmonize European R&D policy, but rather to use available potentials for common goals. In contrast to the generally precompetitive EU programs, EUREKA projects are market oriented. EUREKA projects are intended to complement existing programs of the EU. Members of EUREKA are the countries of the EU, the European Free Trade Area (EFTA) countries, Turkey, and the European Commission.

Two keys to the EUREKA concept are its bottom-up approach for setting an R&D agenda and its flexible structure. This means that, in contrast to EU programs, there are no predetermined technological areas. It is left to participating companies, universities, and other public- or private-sector research bodies to determine their particular areas of interest. In principle, there are no limitations to the type of projects undertaken. However, nine focal areas have been identified: communication technology, information technology, lasers, transportation, energy technology, robotics, biotechnology, new materials, and environment (Figure 3.9). Each EUREKA project is conceived and managed independently.

There are no limitations to the size or scope of EUREKA projects. Although governments may play a role in setting standards and norms (e.g., in the environmental area), the particular R&D approach is left to the participants. A special

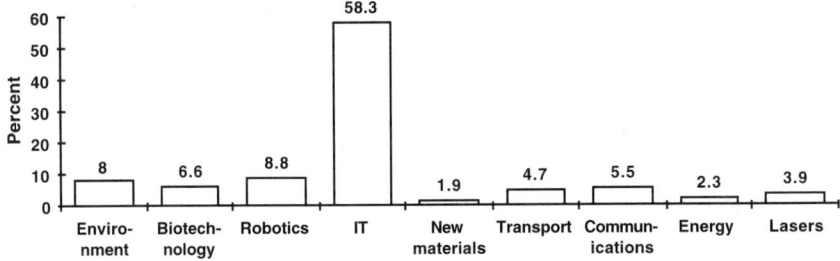

FIGURE 3.9 Volume of research conducted in areas of technology, as a percentage of total EUREKA financing, status as of 1995. SOURCE: EUREKA (1995).

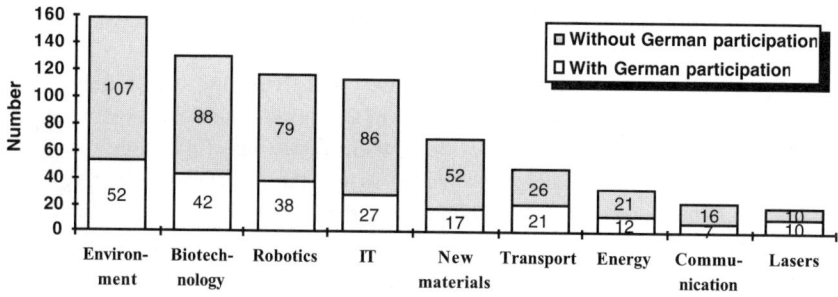

FIGURE 3.10 EUREKA projects, including those with German participation, according to technology, status as of 1995. SOURCE: EUREKA (1995).

form of cooperation has arisen with so-called umbrellas. These serve as framing projects in which single projects are organized and carried out in a flexible but coordinated manner. Results are shared among participants as a way of promoting awareness (EUREKA, 1995). Important to EUREKA's flexibility is its decentralized structure. Each member nominates a National Project Coordinator to assist participants from that country.

As of June 1995, 711 EUREKA projects were in progress, and 226 had been completed. The budget for current projects is DM 19.6 billion (including the contribution of participants). A total of 3,591 participating institutions were counted. Figure 3.10 shows the considerable German participation. Large companies are participating in about 43 percent of EUREKA projects, SMEs are involved in 24 percent (Figure 3.11). The budget figures for EUREKA and EU programs are not directly comparable, since they relate to different periods of

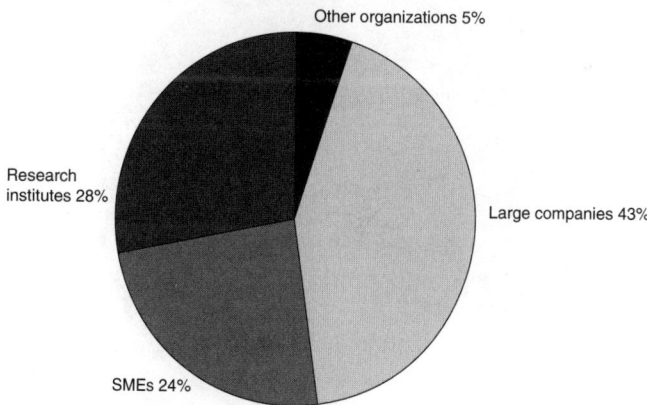

FIGURE 3.11 Involvement of EUREKA participants by major organization type, status as of 1995. SOURCE: EUREKA (1995).

time and involve different contributions by the participants. In any case, the budgets have the same order of magnitude.

A prerequisite for EUREKA projects is sound financing, which may come from either national or EU sources. In most cases, if German participants apply for public support, the BMBF will allocate them funds out of its programs.

All in all, the participating companies assess the EUREKA initiative very positively. Almost two-thirds considered that they have improved their international technological competitiveness, nearly 90 percent expect to produce new or improved products, and about 40 percent expect to achieve an increase in sales (EUREKA, 1993).

The Impact of JESSI

JESSI became a EUREKA project in 1989 and was scheduled to end in 1996. Its goal was to enhance the competitiveness of Europe in the areas of information technology and microelectronics. Financing sources for JESSI are shown in Figure 3.12.

More than 180 partners from 16 countries contributed to the JESSI program, providing approximately 3,100 person-years of effort annually. The estimated cost of this work was ECU 460 million in 1994 and will probably turn out to be the same in 1995. Approximately 50 percent of the work is carried out in France and Germany.

All of the JESSI projects are funded on a cost-shared basis. The partners pay 50 percent and either national public authorities or the EU pays the remaining 50 percent. The total budget for 1989 through 1996 was ECU 2,560 million.

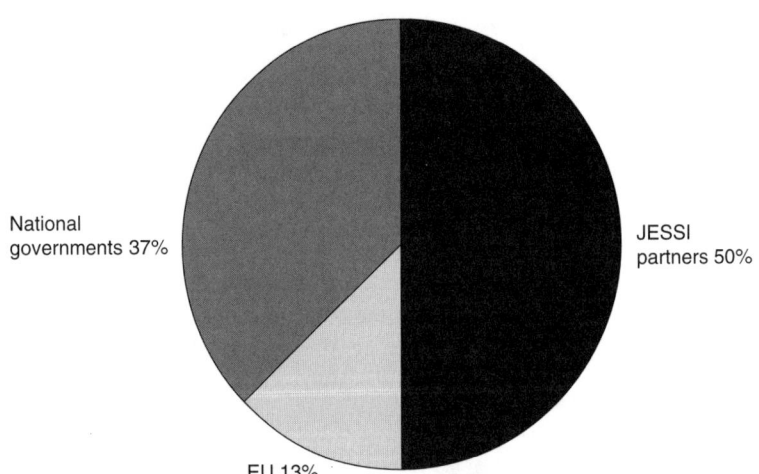

FIGURE 3.12 Financing sources for JESSI, 1989–1996. SOURCE: JESSI (1995).

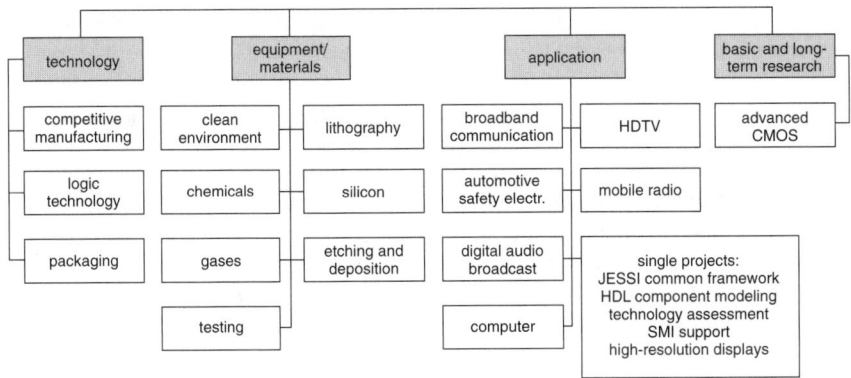

FIGURE 3.13   Program structures of JESSI. SOURCE: JESSI (1995).

The program was divided into four subprograms (see also Figure 3.13):

- *Technology*: Development and testing of the basic flexible competitive manufacturing technology for advanced system applications, to be available by the mid-1990s.
- *Application*: Devising flexible, competitive system-design procedures and tools for the development of highly complex integrated circuits.
- *Equipment and materials*: Development by the European supply industry of manufacturing equipment and materials for selected areas of microelectronics.
- *Basic and long-term research*: Basic and complementary applied research with the long-term perspective.

JESSI was successful in establishing a pan-European platform for collaborative research. Europe leads the field of digital audio broadcast thanks to relevant activities in the application subprogram. The digital audio broadcast project provided all the necessary components for planned field tests. Transmissions are in progress in 21 areas (10 more are planned), and 8 million people can already receive digital audio broadcasts.

Significant results have been achieved in the important JESSI subprogram of technology. Technological competence has been attained in the area of microelectronics. For example, a close cooperation between all major European integrated circuit companies has been established. Outcomes of this collaborative research include:

- the 0.5-micron CMOS technology of Crolles (jointly developed by SGS-Thomson, CNET, and PHILIPS), which is being transferred to the new PHILIPS Waferfab. Siemens and ES2 have signed an agreement allowing ES2 to produce chips with the Siemens 0.5-micron CMOS process.

- a marked increase in the production of European suppliers of computer memories like SGS-Thompson, which became the leading supplier worldwide for EPRON in 1993.
- the specification for the 0.35-micron CMOS logic process. For this project, industry, institutes, and universities were joint partners.

In the field of equipment and materials, one can see the benefits even for smaller equipment manufacturers. Various companies have strengthened their global competitive position through international cooperation with users, co-producers, and research institutes and through new innovative technologies. They include the following:

- ASTI, which introduced its all-plastic ultraclean chemical pump with a favorable global market response;
- AST, which is now accepted as one of the leading suppliers of rapid thermal processing equipment;
- Plasmos, whose share of the worldwide ellipsometer market has increased to about 32 percent; and
- Successful sales of GEMATEC's ELYMAT machine, which is used for mapping on silicon.

JESSI cooperation with SEMATECH in the field of minienvironment and mask technology resulted in internationally accepted standards, an increased understanding of U.S. market requirements, and increased European access to U.S. markets.

Even though JESSI's funding ended in 1996, the program has accepted a variety of new projects and decided to continue others. These focus on important application-oriented topics like digital audio broadcasting and also on new developments in integrated circuit technology and equipment for integrated circuit manufacturing.

The main achievement of JESSI is that the major suppliers in microelectronics and information and communications technologies have been brought together, forming a critical mass for large-sized research projects.

TECHNOLOGY TRANSFER FROM UNIVERSITIES

Universities

HISTORY OF TECHNOLOGY TRANSFER

The development of German universities in the nineteenth century was influenced by the idealist philosophers as well as the growing industrial sector's need for well-trained personnel.

Philosophers like von Humboldt, Fichte, and Schleiermacher influenced decisively the organization and orientation of the German university system. The idealists, who were involved in the founding of the Berlin University (1809–1810), viewed research at universities as an important element of teaching. At the outset, however, German academic research was focused primarily on areas such as philosophy, mathematics, and humanities; empirical research had to fight for recognition.[11] Nevertheless, by the end of the nineteenth century, German university research had achieved world leadership in several major fields of science, including medicine, chemistry, and physics. Due to rapid increases in student enrollments, particularly since 1870, many universities created separate departments and institutes with laboratories for natural sciences.

Through the mid-1800s, the idealist orientation of German professors and administrators led them to elevate the natural sciences and neglect the "less-dignified" engineering sciences. Ultimately, it was the demand of German industry for skilled engineers that led the German states to establish special polytechnical schools outside universities. In the 1870s, the polytechnical schools were elevated to higher status, becoming technical higher education schools (*Technische Hochschulen*). Initially, the efforts of these new institutions to achieve academic recognition led them to overemphasize theory and neglect research targeting industrial needs. At the end of the nineteenth century, however, the establishment of engineering schools in the United States induced German technical higher education schools to begin introducing research laboratories. The main benefit to industry of universities and technical higher education schools was the provision of trained personnel. Even at this early stage of the development of the German academic research system, professors had consultancy arrangements with industry. In other words, the first forms of technology transfer appeared.

Universities and technical higher education institutes focused on education, whereas the central government and the states established a variety of research institutes in applied areas. A prominent example of the latter is the Imperial Institute for Physics and Technology (*Physikalisch-Technische Reichsanstalt*), which served as a model for the National Bureau of Standards in the United States. In addition, some smaller research institutes were financed jointly by government and industry. Finally in 1911, the Kaiser Wilhelm Society, the predecessor of the Max Planck Society, was founded, at that time with a strong focus on applied science and nearly totally financed by industry.

The increasing engagement of industry in government or industry institutes outside universities was stopped by the economic problems caused by the two world wars. Following World War II, the government and the states assumed important roles in the national innovation system through such institutions as the Max Planck Society, the Fraunhofer Society, and the National Research Centers, today called Helmholtz Centers, which are described in more detail in the following sections. Increased public investment in R&D, beginning in the 1970s, was motivated by a perception that Germany was lagging technologically compared with the United States.

In the educational sector, the main development after World War II was the official recognition of technical higher education schools as equivalent to universities. By integration of other nontechnical disciplines some of them officially became "Technical Universities" or even normal universities.

In the 1970s, German universities began to consider seriously their role in technology transfer, and university-industry relationships grew. Between 1970 and 1980, industry support of universities increased by 25 percent, and between 1980 and 1990, such support grew by 44 percent. In addition, universities' needs for external funds increased with student enrollments and, in recent years, because of a scarcity of public funds. Both factors have created pressure on academic research.

Beginning in the 1980s, initiatives on a number of fronts document Germany's increasing interest in technology transfer. Early in the decade, German universities established a special working group of university chancellors to look into the topic (Selmayr, 1987), and the former Ministry of Education and Science (BMBW) initiated a research project called *Projekt Wissenschaft* (PROWIS) that had a strong technology transfer focus (see the publication list in Allesch et al., 1988). In the mid-1980s, the German Science Council issued a statement on technology transfer (Wissenschaftsrat, 1986). These efforts led to the easing of very strict regulations concerning the budget and personnel structures of universities, recommendations on how to handle technology transfer instruments (e. g., the establishment of external institutes), the establishment of technology transfer units at universities, and a generally more open-minded attitude in universities toward technology transfer.

STATISTICS ON GENERAL RESEARCH STRUCTURES

This section presents information about the development of research funding at universities and the distribution of money among research fields. It also analyzes the sources of external funds, which are good indicators of the major channels of technology transfer. Data are not available, however, on the four focal areas of this study.

A main characteristic of the German research system is the public funding of most universities; students do not have to pay tuition fees. It is the states, not the national government, that are responsible for education and hence for the support of universities. According to the principle of equality of research and teaching, the states assign general budgets to each university without dictating how the money should be used. Since universities are not required to report how much of their general budget they allocate to research and its associated overhead, there are no precise statistical data, only general estimates, on this base of institutional research funding. The latter are based on the assumption that a certain share of the total general budgets is used for research, with the share differing from discipline to discipline (Wissenschaftsrat, 1993a). In addition, research funds from

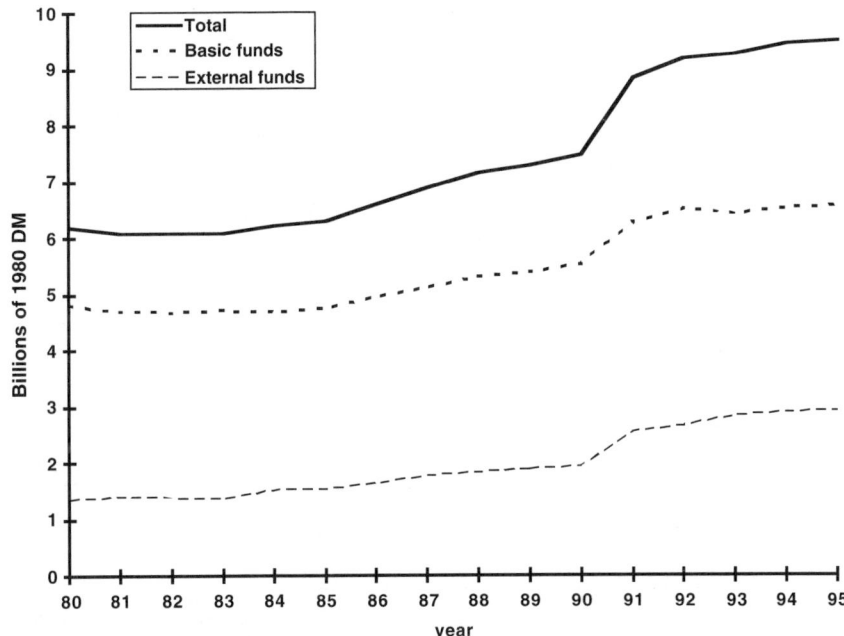

FIGURE 3.14 Research funds of German universities in constant 1980 DM. SOURCES: Bundesministerium für Bildung, Wissenschaft, Forschung und Technologie (1996); Bundesministerium für Forschung und Technologie (1993a); Wissenschaftsrat (1993b); calculations of the Fraunhofer Institute for Systems and Innovation Research.

third parties (*Drittmittel*), called external funds (contracts and grants), have to be taken into account.

Between 1980 and 1990, the overall research budget of universities increased nominally by 59 percent; in constant 1980 DM, the increase was 21 percent (Figure 3.14 and Table 3.2). The sharp growth that followed in 1991 and 1992 was mainly due to the inclusion of the new states in the former East Germany. At about 12 percent, however, the share of research conducted by former East German universities is quite small. In 1990, 49 percent of the total research budget of German universities was related to natural sciences and engineering (Figure 3.15). This has to be taken into account when a direct comparison between the total R&D budgets of industry and universities is made, because industrial research focuses primarily on these two areas. If only the natural sciences and engineering are considered, the relative share of universities in the German R&D system is much lower than suggested in the general comparison presented in the "General Structures" section, above. (See in particular Figure 3.2.)

In real terms, university institutional research budgets grew by 15 percent and the external funds by 42 percent between 1980 and 1990. Hence, the share of

TABLE 3.2 Research Funds of German Universities in billions of DM

Year	Nominal values			Real values (1980)			Indexes (real, 1980)		
	Basic	External	Total	Basic	External	Total	Basic	External	Total
1980	4.82	1.36	6.18	4.82	1.36	6.18	100	100	100
1981	4.92	1.47	6.39	4.69	1.40	6.09	97	103	99
1982	5.08	1.51	6.59	4.68	1.39	6.07	97	102	98
1983	5.26	1.54	6.79	4.71	1.38	6.09	98	101	99
1984	5.31	1.73	7.04	4.69	1.52	6.21	97	112	100
1985	5.51	1.78	7 29	4.75	1.54	6.29	99	113	102
1986	5.86	1.95	7.81	4.95	1.65	6.60	103	121	107
1987	6.19	2.15	8.34	5.11	1.78	6.89	106	131	112
1988	6.53	2.26	8.78	5.31	1.84	7.15	110	135	116
1989	6.83	2.40	9.23	5.38	1.89	7.27	112	139	118
1990	7.29	2.56	9.85	5.53	1.94	7.47	115	142	121
1991	8.64	3.53	12.17	6.27	2.56	8.83	130	188	143
1992	9.33	3.83	12.16	6.50	2.67	9.17	135	196	148
1993	9.59	4.25	13.84	6.41	2.84	9.26	133	208	150
1994	10.06	4.48	14.53	6.53	2.91	9.44	136	213	153
1995	10.31	4.56	14.90	6.56	2.93	9.49	136	215	154

SOURCES: Bundesministerium für Bildung, Wissenschaft, Forschung und Technologie (1996); Bundesministerium für Forschung und Technologie (1993a); Wissenschaftsrat (1993b); calculations of the Fraunhofer Institute for Systems and Innovation Research.

external funds within the total budget became more important, increasing from 22 percent of the total in 1980 to 26 percent in 1990. These figures have to be interpreted with care, however, because the apparent growth in external funds is partly due to more complete publication of financial sources (Wissenschaftsrat, 1993b). Furthermore, the method of calculation used by different statistical sources varies, leading to different results (Selmayr, 1989). The following analysis of external funds is based primarily on data compiled by the German Science Council (Wissenschaftsrat, 1993b), which seems to be the most consistent source.

The external funds come chiefly from semipublic agencies, federal ministries, foundations, and industry (Figure 3.16). The major semipublic agency (*Förderinstitutionen mit überwiegend staatlicher Finanzierung*) is the DFG, which provides about 90 percent of the funds in this category. The DFG is the most important central organization for science promotion and is, to a certain extent, comparable to the National Science Foundation in the United States. The largest part of its budget comes from the central government and the states, each of which usually makes an equal contribution for the support of individual projects (see Meyer-Krahmer, 1990; Wissenschaftsrat, 1993b). The DFG supports all areas of science, including the humanities and social sciences, and is generally, but not exclusively, oriented toward basic research. The coordination of university research at the federal level is one of its major statutory tasks. In this context,

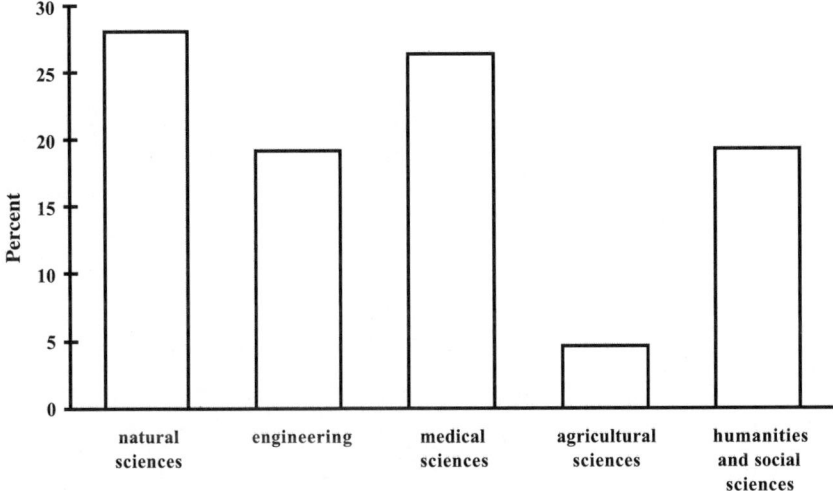

FIGURE 3.15 Distribution of research funds at universities, according to major areas, 1993. SOURCE: Bundesministerium für Bildung, Wissenschaft, Forschung und Technologie, 1996.

an effective instrument is the special research areas (*Sonderforschungsbereiche*), representing about 25 percent to 30 percent of DFG's budget (Bundesministerium für Forschung und Technologie, 1993a). The special research areas are temporary institutions at selected universities, established for a period of 12 to 15 years, where scientists from different disciplines cooperate in joint research programs (Deutsche Forschungsgemeinschaft, 1993). Focal programs (*Schwerpunktverfahren*) are another instrument for supporting the supraregional cooperation of scientists of different universities.

In the case of external funds from federal ministries, the former Ministry for Research and Technology (*Ministerium für Forschung und Technologie* [BMFT]), now the BMBF, contributed the largest share, about 86 percent. The BMFT support of university research increased by about 110 percent between 1980 and 1990. In other words, the general increase in external university funds is due largely to the increase in BMFT support. A major reason for this growth was the introduction of collaborative research projects in 1984, whereby several industrial partners as well as university institutes work together (Bundesministerium für Forschung und Technologie, 1993a; Lütz, 1993). The projects of BMFT/BMBF are generally quite application oriented, but they also support many basic research projects, for example in the area of marine science. BMFT funding of university research in 1990 equaled about 67 percent of what the DFG invested in this area. Thus, BMFT/BMBF became a second major force in the external funding of university research.

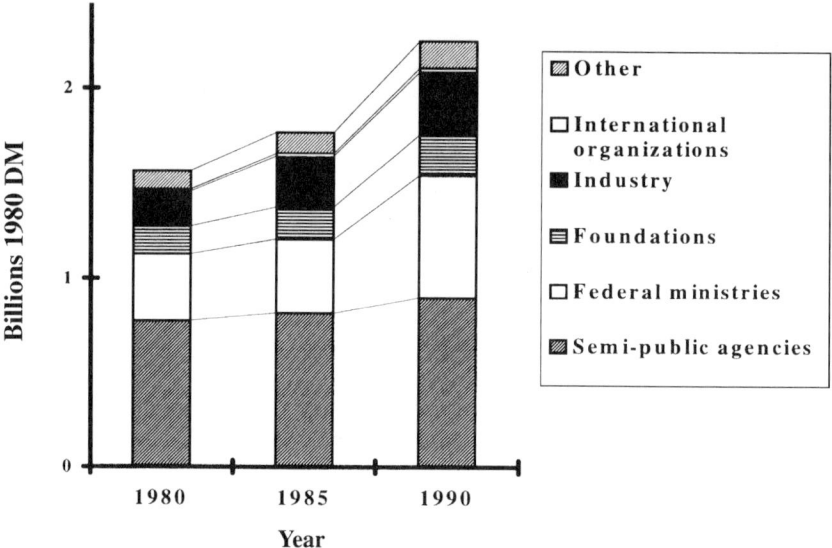

FIGURE 3.16 External research funds of universities, according to major sources, 1980, 1985, 1990. SOURCE: Wissenschaftsrat (1993b).

The Ministry of Defense invests relatively little in university R&D. In 1990, the ministry's contribution represented just 10 percent of all federal funds supporting university research.

The Volkswagen Foundation (*VW-Stiftung*) accounts for 70 percent of all foundation financing of research at universities. Although it was founded by a private company, the foundation generally supports basic research projects.

According to data of the German Science Council, in 1990, industry contributed 15 percent of all external university R&D funds, an 80-percent increase (in real values) from 1980 and a 115-percent increase from 1970 (Wissenschaftsrat, 1993b). This means that funding from industry has become both absolutely and relatively more prominent, particularly since 1980. Compared with the total research budget of universities—including institutional funds—industry support represents a mere 4.4-percent share. The industrial funds can be divided into donations, money for collaborative research, and money for contract research. Most industry support (81 percent in 1990) went to contract research. In addition, 11 percent of industrial funding went for "cooperative research" linked to projects of the AiF (see "Federation of Industrial Research Associations," below). Donations from industry or industrial associations accounted for a modest share, 8 percent, of all industrial funds for universities in 1990.[12]

Recent data of the BMBF based on a more complete survey than that of the German Science Council provide different figures for the industrial contribution to the university R&D funds.[13] According to the BMBF, in 1990, industry support represented 7.7 percent of the total research budget of universities or 25 percent of the universities' external R&D sources. In 1995, these figures increased to 8.7 percent of the total funds or 28 percent of the external sources (Bundesministerium für Bildung, Wissenschaft, Forschung und Technologie, 1996). Without including the contribution of industry-financed foundations, in 1995, the industry support represents about 7.5 percent of the total funds. To sum up, the BMBF data indicate that in recent years, industrial funding of universities has reached a significant level.

Among the international organizations that contribute to university research, the EU is the most important (about 85 percent of total international funding). According to the Science Council, EU funding for universities amounted to DM 23 million in 1990, or 0.8 percent of all external university funds. In contrast to that figure, Reger and Kuhlmann (1995) estimated, using data from the European Commission, that EU funding of German universities came to DM 170 million in 1991. Compared to the average situation, this value may be artificially high, since in 1991 the Second and Third Framework Programs of the EU overlapped. But even if EU contributions came to roughly DM 100 million, this is still a relatively small amount compared with total external funding for German universities (see "Impact of European Research," above). According to recent BMBF data, the EU funding amounted to about DM 130 million in 1995, or 2.7 percent of all external university funds. This means that the contribution of the EU to university funding has increased considerably in recent years (Bundesministerium für Bildung, Wissenschaft, Forschung und Technologie, 1996).

The only other significant external source of academic research funding is project-related funding from the states, which represented about 4 percent of the total external funding in 1990. Finally, it should be noted that not all external funds are linked to research activities: Only 86 percent of those funds were so linked in 1990. This reduced share has already been considered in Figure 3.14 and Table 3.2.

For the analysis of technology transfer, it is important to note that the external funds are not equally distributed across disciplines. For example, in the humanities and social sciences, the absolute and relative volume of external funds is rather low; in law, the external funds are about 4 percent of the institutional research funds; and in economics, they are about 9 percent (Wissenschaftsrat, 1993a). As Figure 3.17 shows for selected areas, the level of external funding in the natural sciences and engineering is much higher. The greatest amount of external funding, DM 266 million, or 41 percent of the total research funds, is apparent in mechanical engineering. This high proportion of external funds can be taken as a strong indication of considerable industrial funding of technical disciplines; the proportion is much higher than the overall 4.4 percent share of university research funding contributed by industry. In physics and electrical engineering, external funding represents about 29 percent of university research budgets. In absolute terms, external support for physics R&D is second only to that for mechanical engineering. This outcome is quite remarkable because of the generally basic orientation of physics research; unfortunately, more detailed statistics on the industrial funds in physics are not available. External funds amount to about 25 percent of the total research budget in computer science, chemistry, and biology. Thus, all focal areas of this report have above-average levels of external funding and probably relatively high contributions from industrial sources.

These budget statistics, however, can lead to an underestimate of research activities supported by external funds. In the German system, external sponsors pay only direct personnel costs and, to a limited extent, costs for facilities. They generally do not pay any overhead costs (e.g., for buildings, administration, central services). This is true for public and semipublic as well as private sponsors. Therefore, research with external funding has to be cofinanced, or matched, by infrastructure funds of about the same amount. Universities must take these infrastructure funds from their base-institutional support. These infrastructure funds can be considered cross-subsidies to external funds. To sum up, the share of research supported by external funds is about twice as high in terms of time and personnel than it is in terms of budget. This relationship is indicated in Figure 3.17. For example, in terms of time and personnel, the real share of research supported by external sponsors is equivalent to 82 percent of university research funds in mechanical engineering, 58 percent in physics and electrical engineering, and 50 percent in computer science, chemistry, and biology.[14] The German delegation is of the opinion that the shares of external support in those areas have increased since 1990 and in many technical institutes have reached 100 percent. Many of these institutes often have the opportunity to acquire additional external funds but cannot take advantage of them because of insufficient infrastructure funds; this insufficiency generally is manifested by a lack of space (see also Hochschulrektorenkonferenz, 1996).

Against this background, the overall figure of 4.4 percent of industrial funds within the research budget of universities, according to the data of the German Science Council, has to be adjusted to at least 8 percent in terms of research personnel and time. In other words, the industrial share can be considered equivalent to the U.S. share of 6.9 percent, because in the United States, industrial support generally covers the full cost of research, including overhead. If the more realistic share of 7.5 percent of industrial funds, according to the BMBF data, are taken, the German level including related infrastructure funds is even substantially above 10 percent.

On the basis of available statistics, it is quite difficult to assess the growth rates of external funding for specific disciplines because the disaggregated figures for 1980 and even 1985 are quite incomplete. According to a survey of the Science Council (Wissenschaftsrat, 1993a) and the German delegation's own estimates, computer science shows the highest increase in external funding, and the

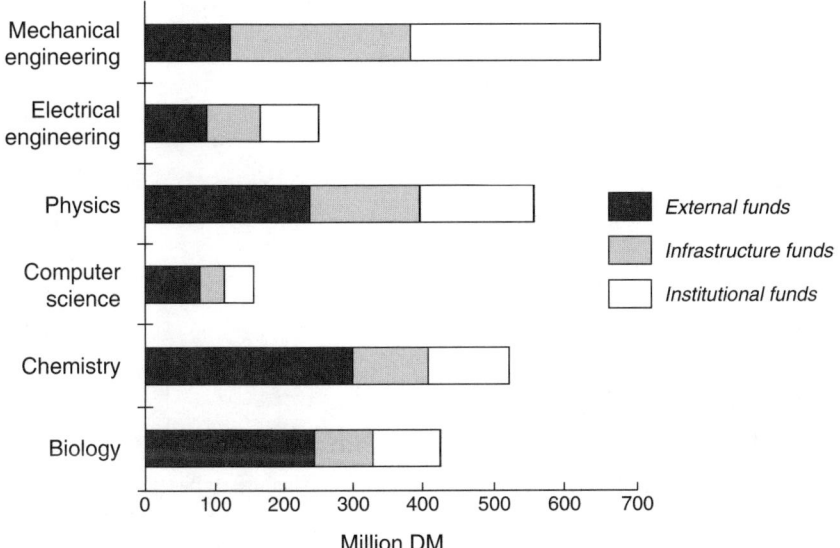

FIGURE 3.17 Relation of external, related infrastructure, and institutional base R&D funds of universities in selected areas in 1990 in current DM. SOURCE: Wissenschaftsrat (1993a).

increase in electrical engineering seems to be considerable, too. It is, however, not possible to say to what extent the increase in electrical engineering is related to (traditional) energy technology or (modern) electronics. The external funding in mechanical engineering shows only a moderate growth rate, probably because the rates at the beginning of the 1980s were already quite high.

In order to provide at least a rough estimate of industrial funding in different disciplines, the data for the University of Karlsruhe, one of the largest technical schools in Germany, are presented in Figure 3.18. The school of mechanical engineering receives the largest volume of industrial funds, but the growth rate in the 1980s was quite modest. These findings support the general results for external university funds. Electrical engineering, computer science, chemistry, and biological sciences (including geography)[15] occupy the next positions, whereas physics is quite low on the scale. This can be taken as an indication that the high general level of external funds in physics does not necessarily reflect a high share of industrial funds. The high absolute level of industrial funding for computer science is due to specialization in this area at the University of Karlsruhe. The highest growth rates can be observed for computer science, electrical engineering, and biological sciences (including geography), which confirms the upward trend found in the general data for computer science and electrical engineering for the total amount of external funding.

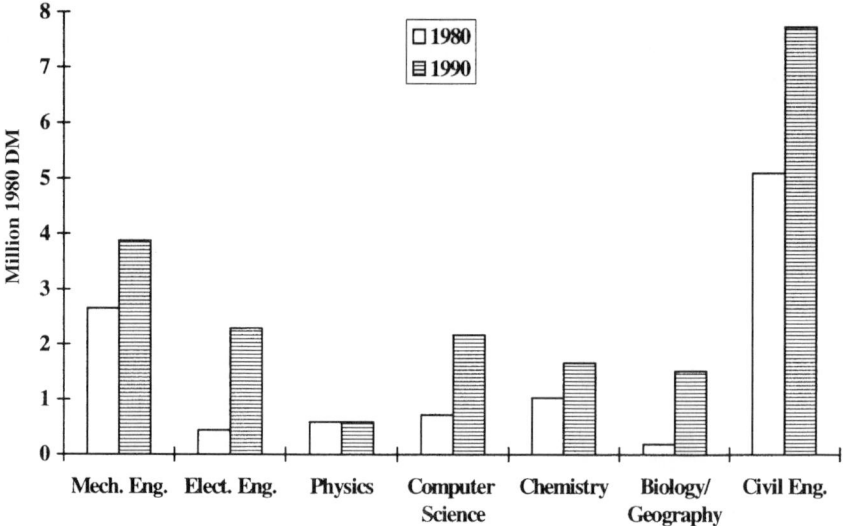

FIGURE 3.18 External funds from industry at the University of Karlsruhe, for selected areas, 1980 and 1990. SOURCE: Universität Karlsruhe (1995).

A special finding at the University of Karlsruhe is the extremely high level of industrial funds in the school of civil engineering. (The level is twice as high as that in mechanical engineering.) However, the majority of these funds are raised as a consequence of the activities of two departments, which act as official certifying institutions in the field of building materials. Also, according to the chancellor of the university, many construction companies expect universities to conduct the bulk of needed civil engineering research. The general statistics of the Science Council for "other engineering," however, indicate that the dominance of civil engineering as a recipient of industrial funds cannot be generalized to Germany as a whole.

To sum up, the major external sources of funds linked to technology transfer between universities and industry are collaborative research funded by BMBF and, at a much lower level, contract research paid for by industry. These activities are concentrated in the natural sciences and engineering, especially the latter, leading to distinctly higher rates of industry funding than the average rates suggest. The funds for collaborative research as well as for research contracts increased considerably during the 1980s; the greatest growth was in the areas of electrical engineering and computer science.

ADMINISTRATIVE STRUCTURES

The different means of technology transfer at German universities are largely determined by the public status of these institutions. This obliges the universities to manage their budgets according to rules established by public law. As public institutions, universities have to follow strict guidelines for budgetary planning and must balance revenues with expenditures. They are not allowed to make a profit. Funds must be clearly linked to projects or well-defined tasks and planning must be done on an annual basis. This system leads to a certain rigidity and can hamper industry-oriented projects, where greater flexibility is needed. The universities have recognized this problem and—especially in the 1980s—introduced more flexible ways of managing external funds (see Selmayr, 1987). For example, universities are now allowed to make profits from externally funded projects and to use them to fill in the financial gaps between different contract projects. However, some problems still exist, like the inflexible handling of travel and related expenses for invited visitors. Furthermore, the administrative procedure for achieving more flexible solutions is quite complex. Most universities, however, have sufficiently experienced administrative staff to cope with these problems. As a consequence of their public status, universities are not allowed to engage in entrepreneurial activities. In particular, they cannot work with new technology-based firms.

Furthermore, full professors are generally permanent civil servants (*Beamte*), and the other scientific staff are salaried public employees (*Angestellte*). Professors and the scientific staff are employed full-time over the whole year. Projects financed by external sources are taken on in addition to teaching and "regular" research and do not lead to additional income. The external money becomes part of the university budget and the related research activities are considered regular activities (*Dienstaufgaben*). The major advantage for the professor in participating in such research is the possibility of obtaining additional staff and equipment and enlarging the scope of his or her R&D activities.

However, there is a second way to carry out projects for private clients. Professors have permission to take on secondary activities (*Nebentätigkeiten*), as long as their regular work is not restricted in a decisive way. Most states assume that the upper limit for these secondary activities is about one-fifth of the total work time (Hartl and Hentschel, 1989). Professors can retain the money from their secondary activities for their private use; hence, there is a strong financial incentive for this type of external activity. Secondary activities are subject to official approval, but in the case of professors, this is usually routine. In the case of scientific assistants, approval of secondary activities is rarely granted, however. It is almost impossible for professors and scientific staff to engage in secondary activites together.

Another important aspect for technology transfer is the organizational structure of German universities. The major bodies in charge of the distribution of institutional funds for teaching and research are the various schools (*Fakultät, Fachbereich*). The schools comprise a number of chairs (*Lehrstühle*) responsible for different areas of teaching. In the natural sciences and engineering, several chairs often establish joint university institutes. The institutes are the most interesting partners for industry because they have a sufficient critical mass and integrate professors from several disciplines. The additional staff financed on the basis of external funds is generally linked to institutes, not to chairs. Many institutes employ 50 or more people and are the main users of external funds, especially industrial funds. In the case of the University of Karlsruhe, the external funds are raised chiefly by such institutes, with a few large ones generating the greatest share of university income that comes from industry. This is true for other technology-oriented universities like those in Stuttgart, Aachen, Darmstadt, and Hannover. These large university institutes have a variety of links to industry (e.g., seminars for industrial experts, supportive associations that include industrial members).

TRANSFER CHANNELS

At German universities, the major channels of technology transfer are collaborative research with industrial partners funded through BMBF projects and contract research for industrial clients. This statement is based on statistics for external university funds (see "Statistics on General Research Structures," above), interviews with university professors conducted in the context of this report, and a special survey on the four focal areas (see "Technology Transfer in the Four Focal Areas," below).

From the perspective of the universities, a special advantage of contract research for industrial clients is the possibility of using the funds more flexibly than is possible with institutional funds or funds from public projects. Thus, paradoxically, the rigidity of the public status of universities is actually a direct incentive for getting money from industry.

Universities are not obliged to price contract research services on the basis of total costs. They can—but need not—exclude general overhead costs already covered by institutional funds. In consequence, they can offer these services at relatively low prices, which is a special advantage for industrial clients (see Püttner and Mittag, 1989, and "Statistics on General Research Structures," above). Of course, other motives, such as the attainment of new research results, also play an important role. These will be discussed in detail in the section "Technology Transfer in the Four Focal Areas."

According to interviews and a survey, university professors view collaborative research with industrial partners as a very positive channel of technology transfer. It provides interesting insights into industrial research results and needs, and the resulting scientific independence is greater than in the case of contract research for industry. However, collaborative research seems to be less effective than is often assumed. First, the involved companies primarily expect to receive public funds and are often less interested in university-industry relations. Second, each partner linked in a common project generally works in its own laboratory; the outcomes are presented solely at a few meetings. Third, the model of cooperation involving several industrial enterprises and one or more university institutes leads to an orientation toward precompetitive research; hence, technology transfer mainly occurs in the early stages of the innovation cycle. Rarely are collaborative research projects followed by contract research projects, according to those interviewed for this report.

In contrast to the American situation, in Germany, grants from industry to universities without clearly determined tasks and deliverables are exceptions. The main reason for this situation is obviously the German tradition of university-industry relations. In addition, the tax regime is not very favorable: Grants can be deducted from taxable income (tax deductions) as normal donations, or special expenses (*Sonderausgaben*), with upper limits. Therefore, companies contribute to the base funds of university institutes only in few cases of special common interest.

One instance where industry contributes to university base funds is in the definition of focus projects in the area of biotechnology, in which research institutions, industry, and the BMBF cooperate. For this purpose, eight so-called "gene centers" (*Genzentren*)[16] were established for a limited period of about 10 years (Bundesministerium für Forschung und Technologie, 1993a). In these centers, university institutes, Fraunhofer institutes, and Max Planck institutes cooperate. The BMBF pays the major part of the project costs, and some large enterprises contribute to the base funds as well as to the project costs.

In the area of chemistry, the relations between industry and university are particularly close and have existed for many decades (see Herrmann, 1995). Chemical companies often have permanent consultancy contracts with professors or university institutes. Furthermore, the chemical industry has established a special fund (*Fonds der Chemischen Industrie*), which was set up as early as 1950, with the purpose of supporting university research. The university scientists personally get financial assistance for research purposes, but without clearly defined projects. Thus, the aid helps to enlarge the scope of university research, especially in basic activities. For the period from 1995 through 1997, the fund plans to grant a total of DM 21.7 million for these purposes (Fonds der Chemischen Industrie, 1995). All in all, the chemical industry provides a good example of how to improve stable long-term relations and cooperation with universities.

Collaborative and contract research represent only a part of a broad range of technology transfer mechanisms. Universities present the results of their research in scientific articles, at fairs, and at conferences. Especially at conferences, university researchers meet with people from industry to discuss the applicability of research to industry. Furthermore, academic and industrial researchers often meet informally or discuss their problems in telephone conversations. Knowledge exchange is broadly supported by a variety of scientific associations with academic as well as industrial members, which organize conferences and publish journals (Schimank, 1988b). As to the different channels of technology transfer, Allesch et al. (1988) asked university professors about the different forms of their "contacts with industry" (*Praxiskontakte*). The professors reported that informal consultancy and the provision of personnel (e.g., graduates) are by far the most frequent type of contact with industry. Of course, the intensity and length of these informal contacts are hardly comparable to what happens with formal contract or collaborative research, but the significance of informal channels of technology transfer should not be overlooked.

Consultancy by university professors for private clients is in general carried out as a secondary activity, because professors can obtain income in addition to their regular salary. In addition to consultancy and expert evaluations, professors also can conduct research for private companies as a secondary activity. If the professor uses staff or equipment from the university, he or she has to pay the related costs. According to Allesch et al. (1988), professors carry out 40 percent of their contracts with industry as secondary activities. (The present situation is discussed in the section "Technology Transfer in the Four Focal Areas.")

When contract research is a secondary activity, the legal requirements are quite complex and sometimes represent a barrier to cooperation with private enterprises (Püttner and Mittag, 1989). The major constraint in this regard seems to be the regulation that contracts from third parties cannot be split into regular and secondary activities (Püttner and Mittag, 1989). Thus, it is not possible that in such a project, the professor carries out his or her part as secondary activity and the scientific staff is engaged therein as a regular activity. Since the scientific staff generally is only allowed to have regular activities, the professor must do his or her part as a regular activity, too. The professor may then no longer be interested in the additional work because he or she receives no additional payment as in secondary activities.

In the context of consultancy for industry, the role of polytechnical schools/ technical colleges (*Fachhochschulen*) has to be mentioned. In contrast to universities, polytechnical schools are less science oriented. Their primary missions are education and the development of practice-oriented capabilities. Their research activities are rather limited; therefore, they are not explicitly mentioned in the statistical discussion above. Nevertheless, professors at polytechnical schools do considerable consulting for industry, especially regarding the solutions for problems that emerge from the daily business on the companies' shop floors. Furthermore, students at polytechnical schools are obliged to prepare their master's theses on subjects relating to industrial enterprises. In the state of Baden-Württemberg, the Steinbeis Foundation has established an effective network with polytechnical institutes for supporting technology transfer (see "Technology Transfer to Small and Medium-Sized Enterprises," above).

The major form of transfer of personnel to industry includes the provision of graduates and the permanent transfer of scientific staff from university to industry. In contrast, the temporary transfer of professors or scientific staff remains a rare event. Due to the requirements of public law, the short-term transfer of personnel is linked to a variety of conditions and is not easy to put through (Püttner and Mittag, 1989). In recent years, university administrations have become more open to this kind of technology transfer; but up to now, only minor changes can be observed.

German universities, especially technical faculties, have a long-standing tradition of appointing high-level researchers from industry as professors (Wissenschaftsrat, 1986). This leads to practice-oriented education and close research ties between universities and industry. In some cases, industrial firms endow professorships for a limited time period.

As previously discussed, technology transfer from universities to industry was a major issue in science policy in the early 1980s. A highly visible outcome of the belated initiatives is the formation of technology transfer units at all universities with schools of engineering or natural sciences. According to Kuhlmann (1991), these units serve primarily a so-called window function. In other words, they provide information to industry on the research capabilities of the university. Technology transfer units serve also a catalyst function by bringing industrial clients and university institutes or individual professors together. The units often help companies find the appropriate institute or professor to address specific problems. Other, less dominant functions are the systematic monitoring of industrial needs, the negotiation of contracts with industrial partners, and the supply of services (e.g., business consultancy). To sum up, the transfer units have only a limited supportive function and cannot replace the transfer activites of the university professors discussed above (see, for example, the criticism of Reinhard and Schmalholz, 1996). The transfer units, being responsible for the university departments altogether, can only assist firms in finding appropriate professors. However, the latter can present their specific research capacities more precisely than a general unit can, and have to build up the actual industry contacts. Nevertheless, the transfer units play a decisive role in making contact with SMEs, which are their major clients. Professors often do not have sufficient time to actively address this heterogeneous group and generally work with big companies. All in all, the transfer units have become an indispensable instrument of technology transfer.

AN-INSTITUTES AND OTHER EXTERNAL INSTITUTIONS

Professors can establish, as secondary activities, private institutes, as long as the legal limitations on work time are observed (Hartl and Hentschel, 1989; Tettinger, 1992). These institutes vary widely, from being completely independent to having close links to a university. A major advantage of external institutes is that they make it easier to carry out applied research and development, which in general goes beyond the scope of the usual research activities of universities. Further advantages are the simpler administrative procedures concerning contracts and the employment of scientific staff. To a certain extent, external institutes can help to enlarge the personnel and equipment capacity of universities. Since the mid-1980s, most universities have supported the establishment of external institutes.

In this context, it is worth noting again the quite inflexible regulations concerning employment of professors and university staff. These rules hamper both the temporary transfer of personnel from universities to industry and the establishment of part-time employment contracts. The latter could in theory be a means of accommodating parallel activities of scientific staff at universities and external institutes. In practice, this instrument is rarely used.

A special type of external institute is the so-called An-Institute. An-Institutes are legally defined as independent bodies in order to achieve sufficient administrative flexibility (Krüger, 1995; Tettinger, 1992). They may have a completely private or semipublic status. In most cases, they are nonprofit institutions and thus pay reduced taxes. Important common characteristics of all An-Institutes are that they are officially acknowledged by a university and operate under a cooperation agreement. Some states (*Länder*) have official rules and regulations for An-Institutes.

The main goals of An-Institutes are to

- foster technology transfer and application-oriented research and development;
- perform research in areas that are the focus of university research; and
- perform research that does not fit into the administrative structures of universities.

To summarize, An-Institutes are "mediators" between universities and industry. Because of their legal independence, they have short decision paths and can react to market demands and opportunities in a flexible way. Furthermore, they can establish a business-oriented budgeting and accounting system. For example, they can freely use their budgets for special remunerations of their staff, for public relations activities, or for the further professional training of their researchers. For interested companies, especially SMEs, the research areas and competences of An-Institutes are more transparent than those of large universities with a variety of faculties and internal institutes. This is a special advantage that helps An-Institutes integrate themselves into regional commercial activities.

At the same time, An-Institutes have close relations to universities and thus good access to basic research. In most cases, the directors of An-Institutes are also regular (part-time) professors at universities and are engaged in teaching. Hence, they can employ the brightest students.

Some critics fear that university research activities are being shifted to the An-Institutes and that universities are losing external funds from industry. In reality, universities generally profit from the industrially oriented activities of An-Institutes and acquire additional funds through the cooperation agreements.

The various An-Institutes differ not only in their legal status, but also in the scope of their research. Some An-Institutes have narrow markets linked to a special industry, for example VLSI design for the microelectronics industry (*Institut für Mikroelektronik* [IMS], Stuttgart). Others have broad markets, for example, software systems for the manufacturing industry (*Forschungszentrum Informatik* [FZI], Karlsruhe) or office technology for all industries (*Oldenburger Forschungs- und Entwicklungsinstitut für Informatik-Werkzeuge und -Systeme* [OFFIS], Oldenburg). The institutes with broad markets normally have multiple directors. As a general rule, An-Institutes carry out research in strategic areas, such as information technology and microelectronics.

The various legal statuses of An-Institutes correspond to their diverse budget structures. In some states, like Baden-Württemberg, the An-Institutes receive one-third of their institutional funds from the state, one-third from contract research for industrial clients, and one-third from projects for public clients, such as the BMBF, the European Commission, the states, and so on. In this regard, the model of the An-Institutes is comparable to that of Fraunhofer institutes (see "The Fraunhofer Society," below). However, many An-Institutes receive no public contribution to their institutional base and so depend almost totally on private and public contracts. In some cases, industrial partners provide some institutional funds.

The main problem for An-Institutes is survival in a market that is dominated by competitors from large institutions with superior organization and connections (e.g., Fraunhofer institutes), more generous basic funding (e.g., national research centers), or hidden overheads (e.g., universities). Therefore, only An-Institutes with a special competence profile, close linkages to industrial partners, and dynamic structures have a potential for long-term survival.

The activities of private institutes and particularly An-Institutes represent a considerable portion of technology transfer. According to a recent official survey, the R&D-related expenditures of An-Institutes amounted to DM 580 million in 1994, equal to 4 percent of the R&D expenditures of universities and about 50 percent of those of the Fraunhofer Society (Bundesministerium für Bildung, Wissenschaft, Forschung und Technologie, 1996).[17] All in all, they have developed into an important institutional sector that supports technology transfer.

Technology centers and science parks are a further source of technology transfer. These facilities aim to establish technology-oriented enterprises in the vicinity of universities (Eberhardt, 1989; Wissenschaftsrat, 1986). Collaboration between universities and such enterprises may include the use by firms of the expertise of universities or the paid use of university equipment. From the perspective of the German authorities, a clear legal distinction between universities and private companies is necessary to avoid any dependence of university research on the private sector. For instance, universities are not allowed to hold shares in industrial enterprises in technology parks. The companies in technology centers are often already well-established firms, which use the special facilities at universities. But the centers also support the establishment of spin-off companies. In the case of the technology center in Karlsruhe, 11.5 percent of all companies are spin-offs from universities, and 20 percent are spin-offs from other research institutions.[18]

TECHNOLOGY TRANSFER IN THE FOUR FOCAL AREAS

Results of a German Survey

Because available statistics and surveys do not contain any specific information about technology transfer in the four focal areas of this report, FhG-ISI conducted its own survey of German university institutes (not including An-Institutes and other external institutes). The survey included institutes in the focal areas of production technology, microelectronics, software, and biotechnology[19] and was conducted in May 1995. The addresses of presumably relevant institutes were determined with the help of a manual on universities and research institutions wherein the research area of each institute is briefly described (Vademecum, 1993).

In all, 783 questionnaires were sent out, and 332 questionnaires with valid responses were sent back (Table 3.3). This response rate of 42 percent has to be considered very high, particularly since the description of the institutions in the manual often was quite poor, so that the selection of really appropriate institutions was difficult. The high response might have been due to the user-friendly design of the questionnaire, which had a limited number of questions, the majority of which could be answered by simple multiple choice.

The first group of questions concerned the volume and composition of external funds.[20] A striking result was the very high proportion of external research funds in all of the focal areas. The average share of external funds was 62 percent (Table 3.4), which is considerably higher than the average figures cited in official statistics for superordinate areas (e.g., in 1990, the figures were 41 percent for mechanical engineering and 33 percent for electrical engineering; see Figure 3.17).

TABLE 3.3 Response Rate of Survey Sent to German Universities, by Focal Area

Area	Questionnaires Sent Out	Questionnaires Sent Back	Response Rate
Production technology	185	97	52%
Microelectronics	155	60	39%
Software	175	68	39%
Biotechnology	268	107	40%
Total	783	332	42%

TABLE 3.4 Percent Share of University External Funds in Four Focal Areas, 1995

Area	Share of External Funds Within Total Budget	Share of Industrial Funds Within Total Budget	Share of Secondary Activities Within Industry Contracts
Production technology	68	25	11
Microelectronics	63	18	10
Software	43	13	16
Biotechnology	69	12	25
Total	62	17	15

Two reasons may help explain these differences:

- Due to the increasing number of students, the relative share of research funds from institutional sources has diminished since 1990, and the universities have become more active in the acquisition of external funds.
- Since the questionnaire was clearly oriented toward technology transfer, primarily institutions with a high level of external funds answered (respondent bias).

However, the main reason for the disparity is probably different. Expert interviews revealed that many professors are not aware of the real cost structures and do not sufficiently take into account the contribution of institutional funds to universities' overhead costs (see the related discussion in "Statistics on General Research Structures"). Only some respondents answered in terms of money. In consequence, the results presented in Table 3.4 are a little bit lower than the real values in terms of personnel and time.

Among the focus areas, the high share of external funds in production technology is closely related to a high share of industrial income. It seems to be easier in biotechnology research than other technology areas to acquire external funding through BMBF, EU, DFG, and other sources.

The average level of industry-related research within the total research activities is (as explained above) probably a little higher than 17 percent.[21] In any case, it is far above the average level of about 8 percent for universities altogether. The industrial budget does not include collaborative projects with industrial partners funded by public sources (e.g., BMBF, EU), so the actual rate of industry-related activities is even higher.

Production technology, microelectronics, software, and biotechnology are ranked first through fourth, respectively, in terms of the percentage of industrial funds that make up their total budgets. This ranking results because the focal areas are at different stages of their technology cycles, reflected by different degrees of concentration on basic research. For example, in production technology,

TABLE 3.5 Orientation of University R&D Activities, by Percent, 1995

Area	Basic Research	Applied Research	Experimental Development
Production technology	29	53	18
Microelectronics	41	47	12
Software	50	38	12
Biotechnology	66	27	7
Total	47	41	12

29 percent of R&D activities can be labeled as basic research; the amount of basic research is much higher in biotechnology (66 percent; Table 3.5).[22] Thus, biotechnology is still at an early stage of development, whereas production technology has already matured. It is interesting to note that in all areas, the universities do not restrict their activities to basic and applied research but devote some effort to experimental development work.

A rather interesting result shown in Table 3.4 is the relatively low level of secondary activities within industrial contracts (average 15 percent) compared with the results of Allesch et al. (1988), who found an average share of 40 percent. A partial explanation is that the relative share of regular activities has increased since 1984, when Allesch et al. conducted their survey. The greater flexibility of university administration has made the integration of industrial contracts into regular work easier (Wissenschaftsrat, 1993a). Second, Allesch et al. focused on individual professors, whereas the present questionnaire included whole research teams. Thus, with respect to professors, the level of secondary activities is more important than Table 3.4 suggests. Secondary activities are still a relevant incentive for technology transfer.

Survey respondents also were asked to assess the importance of different channels of technology transfer. As it is not possible to measure and compare the various channels using common quantitative units, the respondents could choose from among the statements "very important," "important," "somewhat important," and "not important." These assessments were meant to reflect the specific importance of the industrial contacts for the institution, not general opinions. For the analysis of the results, the statements were arranged in an ordinal scale from 1 (not important) to 4 (very important). In Table 3.6, the assessments of the different channels and the overall mean scores are recorded.

The respondents regarded collaborative research as the most important transfer channel, with a mean score of 3.2. Despite the various points of criticism raised in accompanying interviews, this type of technology transfer, which is primarily supported by BMBF programs, seems to be very effective. Informal channels (e. g., telephone conversations or informal meetings; see also Rappa and Debackere, 1992) are second in importance, with a score of 3.0. Thus, the

TABLE 3.6 Channels of University Technology Transfer by Percent and Mean Score

	Very Important	Important	Somewhat Important	Not Important	Mean Score
Cooperative research	53	25	12	10	3.2
Contract research	35	25	18	22	2.7
Consultancy	21	31	36	12	2.6
Informal contacts	34	39	20	7	3.0
Industry-related committees	10	23	32	35	2.1
Workshops, conferences	24	35	28	13	2.7
Organization of seminars	14	30	32	25	2.3
Exchange of scientists	16	25	30	29	2.3
Provision of personnel for industry	27	31	24	17	2.7
Exchange of pubications	10	25	36	28	2.2
Industrial participation in master's and doctoral theses	29	33	20	17	2.7

establishment of appropriate conditions for the arrangement of informal meetings is important (e.g., the availability of travel funds and meeting rooms). With scores of 2.7, 2.6, 2.7, 2.7, and 2.7, respectively, contract research, consultancy, workshops and conferences, provision of personnel for industry, and industrial participation in master's theses and doctoral dissertations are quite important, too. In contrast, participation in industry-related committees and the organization of seminars for people in industry generally are viewed as only somewhat important. The same applies to the exchange of publications, which is a major instrument for information exchange in academia but obviously is less important for industry contacts. The exchange of scientists was given a low score, a result that confirms the outcome of other studies. This low score means that the temporary exchange of scientists is rarely used. But according to the interviews with professors, when used, the exchange of scientists has been very effective.

The results, disaggregated according to the four focal areas, are similar to those for the total sample, but not completely uniform. For instance, contract research has a high score in the application-oriented area of production technology, and a low score in biotechnology with its distinct focus on basic reasearch. A detailed discussion of these differences, however, lies beyond the scope of this study.

Not suprisingly, university researchers saw the availability of additional funds as the most important advantage of industry contacts (Table 3.7). However, the opportunity to confer with industry had almost the same impact. Thus, technology transfer does not only flow from universities to industry, but aca-

TABLE 3.7 Benefits to University Researchers from Contacts with Industry, by Percent and Mean Score (percent total sample), 1995

	Very Important	Important	Somewhat Important	Not Important	Mean Score
Additional R&D funds	66	22	8	5	3.5
Flexibility of industrial funds	51	25	14	10	3.2
Additional facilities	31	30	26	13	2.8
Opportunity to confer with industry	54	33	11	3	3.4
References for acquisition of public funds	22	33	27	18	2.6

demic researchers receive new intellectual input from industry as well. This finding was confirmed by interviews, in which university scientists emphasized the relevance of information from industry for their research and for improved, practice-oriented teaching. As already explained in the context of administrative structures, the flexibility of industrial funds compared with public funds is a major incentive for German universities to undertake contract research for industry.

As to the barriers to industry contacts (Table 3.8), university researchers regard only the short-term orientation of their industrial partners as relevant (mean score of 2.9). All of the other reasons were "somewhat important" or even "not important" (scores between 1.8 and 2.3). The low score for administrative barriers confirms interview results indicating that today's university administrations cope better with the problems of industrial contracts than in the 1980s (Selmayr, 1986). University researchers' assessment of a limited indigenous industrial base as a barrier (mean score of 2.3) showed interesting differences in the four focal areas. (This internal differentiation is not indicated in the tables.) In biotechnology, the mean score was 2.6 (still an "important" barrier); in microelectronics, the

TABLE 3.8 Barriers to Industry Contacts, by Percent and Mean Score, 1995

	Very Important	Important	Somewhat Important	Not Important	Mean Score
Less interesting topics	8	23	34	35	2.0
Industry's short-term orientation	35	32	21	12	2.9
Restrictions of publications	10	26	39	25	2.2
Administrative problems	7	17	38	38	1.9
Unfair contracts	4	14	38	44	1.8
Limited industrial base in Germany	20	28	17	35	2.3

TABLE 3.9 Reasons for Industry Interest in University Research, by Percent and Mean Score, 1995

	Very Important	Important	Somewhat Important	Not Important	Mean Score
Observation of scientific development	40	42	16	3	3.2
Solution of technical problems	39	36	20	6	3.1
Personnel recruitment	26	43	25	5	2.9

mean score was 2.5; in software, it was 2.4; and in production technology, it was 1.9 ("somewhat important"). These results reflect the strong focus of German industry on all areas of mechanical engineering and a lower level of specialization in information technology, microelectronics, and biotechnology.

The university researchers were asked to describe what they believe to be the reason for industry's interest in their research (Table 3.9). It is interesting that they ranked "observation of scientific development" even higher than "solutions to technical problems." This ranking confirms once again researchers' belief in a scientific dialogue between universities and industry on mid- and long-term questions and an acknowledgment that industry needs solutions for its immediate technical problems. In addition, the provision of qualified personnel—a basic function of universities—plays an important role.

As mentioned above, the relative importance to universities of different transfer channels, industry contacts, and barriers to working with industry are generally the same in all selected areas. Nevertheless, differences in the absolute values of the scores can be observed (Table 3.10). To demonstrate this effect, the mean scores of all responses to a group of questions were combined and then averaged. In the case of transfer channels, those working in production technology generally saw the different channels more positively than did those in biotechnology. The similarity of this result to the differences in industrial funding in these four areas is obvious. The same phenomenon emerges with respect to the

TABLE 3.10 Average Mean Scores in Major Question Groups

Area	Channels	Benefits	Barriers
Production technology	2.8	3.3	2.1
Microelectronics	2.8	3.2	2.2
Software	2.6	3.0	2.3
Biotechnology	2.3	2.9	2.2
Total	2.6	3.1	2.2

benefits of industrial contacts, but to a lesser extent. In the group of questions dealing with barriers to industrial contacts, the differences between the areas are negligible although the scores are generally low.

All in all, the contacts between universities and industry in the selected areas are above average, and universities are more engaged in technology transfer to industry than generally assumed. Of course, the differences between the analyzed institutions are large, and technology transfer could be improved in many cases. Nevertheless, the potential for a further increase in technology transfer seems to be limited—at least in the selected focal areas. It is important to take the different stages in the technology life cycle into account. In biotechnology, for instance, a great increase in applied research and a corresponding reduction in basic research would be detrimental to the quality of research given the present stage of the area's technology life cycle.

Comparison with the American Situation

The results presented above give interesting insights into how technology transfer occurs at German universities. It is informative to compare this with the situation at American universities. A direct comparison is not possible, because an equivalent U.S. survey does not exist. But Cohen et al. (1994) conducted a survey of the University-Industry Research Centers (UIRCs), which are in many respects comparable to German university institutes. For the purpose of the present study, Cohen et al. (1995) prepared a special analysis for the four focal areas.

UIRCs are research centers at U.S. universities that get base funds from the federal government, mostly the National Science Foundation, on the precondition that they also raise money from industry. In most cases, the industrial funds are base funds, too, and are not linked to contracts with clearly determined deliverables. The funding companies, however, are involved in the general planning of research activities and have early access to research results.

With respect to the four focal areas, Cohen and his colleagues received input from 411 UIRCs (Table 3.11), a magnitude of response comparable to the Ger-

TABLE 3.11 Responses to the Survey of UIRCs, 1990

Area	Number of UIRCs
Production technology	109
Microelectronics	64
Software	129
Biotechnology	109
Total	411

SOURCE: Cohen et al. (1995, Table 3.1).

TABLE 3.12 Industrial Contributions to UIRCs, Percent Share by Area, 1990

Area	Share
Production technology	41
Microelectronics	30
Software	33
Biotechnology	21
Total	31

SOURCE: Cohen et al. (1995, Table 2).

man survey (Table 3.3). Only in software was the German absolute response rate distinctly lower, but the remaining sample is still sufficiently large.

A revealing outcome of the U.S. survey is the share of the industrial contribution to the funds of the UIRCs (Table 3.12). Like in Germany, U.S. centers devoted to production technology receive the highest share, those for biotechnology the lowest, and the area of microelectronics falls in the middle. The share of U.S. industrial funds for software R&D are comparable to, or even a little higher than, that for microelectronics, whereas in Germany funding for microelectronics research is near the level of funding for biotechnology. This difference may be due to closer university-industry relations in U.S. software development.

The level of industry contributions to the UIRCs is generally higher than the average level of industry contributions to German universities, because the special mission of UIRCs is to improve technology transfer.[23] In contrast, the German survey sample covered all types of university institutes and also included institutes with few industrial relationships. The U.S. survey, like the German one, asked respondents about the distribution of their R&D activities in basic research, applied research, and experimental development (Table 3.13). The differences between the four focal areas are less distinct in the United States than they are in Germany (Table 3.5), but the level of basic research in production technology is lowest in both Germany and the United States.

The distribution of the three types of R&D activity in the United States is comparable to that in Germany for production technology and microelectronics. But the U.S. orientation toward basic research is clearly less pronounced in software and biotechnology. Of course, such comparisons are of limited usefulness, because the German and American interpretations of the different R&D types might be different. The higher U.S. level of applied R&D in software, however, correlates to the higher share of industrial contributions in this area. In the case of biotechnology, the difference between Germany and the United States is so large that it cannot be explained by a methodological bias. To summarize, the application orientation in German academic R&D is apparent in production technology

TABLE 3.13 Orientation of R&D Activities at UIRCs, Percent Share, 1990

Area	Basic Research	Applied Research	Experimental Development
Production technology	32	46	22
Microelectronics	44	42	14
Software	38	44	18
Biotechnology	44	41	15
Total	44	43	18

SOURCE: Cohen et al. (1995, Table 3.3).

and microelectronics and just as it is at U.S. UIRCs (Table 3.13). In contrast, the German university research in software and microelectronics appears to have a distinctly basic orientation. Unfortunately, the available U.S. data do not shed light on the extent to which other academic research outside UIRCs is oriented toward more basic activities.

Like the German survey, the UIRC survey asked about the relevance of different transfer channels (Table 3.14). Because of the different structures of German university institutes and American UIRCs, responses to the UIRC survey do not always have a counterpart in the German survey. Nevertheless, some comparisons can be made. The U.S. scores, however, seem to be generally higher than the German ones (Table 3.6). This is due to different ways of analyzing the questionnaires. According to the German approach, all questionnaires are included as long as the respondents assessed the importance of some channels of technology transfer. The channels not marked by respondents were considered to be "not important." According to the U.S. approach, however, only questions with a definite answer were included. If the German questionnaires are dealt with according to the U.S. method, the scores of German respondents rise and become comparable to the U.S. figures (Table 3.14). The only distinct difference concerns the temporary work of UIRC/university personnel in industry laboratories, where the U.S. score is clearly higher; in other words, the movement of personnel is less often a mode of technology transfer in Germany.

The approach of the UIRC survey to assessing the benefits of industry contacts was different than that of the German survey, and the U.S. data are combined rather than separated out by the four focal areas (Table 3.15). The U.S. questionnaire asked whether or not the UIRCs see a benefit, without further differentiation, so that only the percentages of positive answers are available. The outcome, however, indicates that the U.S. and German respondents gave similar rankings to the value of "R&D funds," "opportunity to confer with industry," and "equipment." In other words, the U.S. survey, like the Germany survey, revealed the importance of dialogue with industry for the advancement of academic research.

As to the barriers to industry contacts, the U.S. survey asked only about restrictions on publication. Thirty-nine percent of UIRCs reported that partici-

TABLE 3.14 Channels of U.S. UIRC and German University Technology Transfer, Mean Score in the Four Focal Areas

	U.S. Mean Score (1990)	German Mean Score (1996)
Collaborative R&D projects	3.4	3.5
Seminars, workshops, symposiums	2.9	3.0
Research papers, technical reports	2.8	2.6
Telephone conversations	2.9	—
UIRC personnel in industry labs	3.3	2.8
Industry personnel in UIRC	3.5	—
Informal meetings with industry people	3.3	3.2
Delivery of prototypes or designs	3.4	—

NOTE: German mean scores are calculated according to the method used in the U.S. survey.

SOURCES: Cohen et al. (1995, Tables 18 to 22); survey by the Fraunhofer Institute for Systems and Innovation Research.

pating companies can require information to be deleted from research papers before they are submitted for publication; 58 percent said that companies can delay the publication of research findings, and 34 percent indicated that companies are able to both delay publication and have information deleted. The data do not indicate the actual frequency of these interventions. In Germany, the problem of publication restriction exists, too, but is generally less important (see Table 3.8). However, the Geramn and U.S. data sets are not really comparable due to the different types of questions asked.

TABLE 3.15 Benefits of Industry Contacts at UIRCs, by Percent, and at German Universities, by Mean Score

	Percent Share of UIRCs	German Mean Score
R&D funds	91	3.5
Opportunity to confer with industry	70	3.4
Equipment	68	2.8
Information on industry needs	56	—
Operational funds	49	—
Access to industrial facilities	45	—
Practical experience for students	38	—
Research direction	36	—
Industry personnel loaned to academic programs	22	—
Other	6	—
None of the above	1	—

SOURCES: Cohen et al. (1994, Table 3.29); survey by the Fraunhofer Institute for Systems and Innovation Research.

All in all, the results of the U.S. survey confirm the German outcome. They emphasize the importance of collaborative research and informal contacts for technology transfer and highlight shortcomings of the German system with respect to the difficulty of temporarily moving academic researchers into industrial laboratories. The data seem to indicate that German universities have less of an orientation toward applied research in software technology and biotechnology than their U.S. counterparts, a result that might be due to a lack of complete data for all types of research units of U.S. universities (i.e., not only UIRCs).

PATENTS AND PATENT STATISTICS

Intellectual property rights, especially patents, play an important role in technology transfer. The particular situation at German universities is characterized by the privilege of professors to exploit for their own benefit inventions created during their work on institutional base funds at the university (*Verwertungsprivileg*). The consequences of this policy for technology transfer are contradictory. On the one hand, the private holding of patents can be an incentive, if the invention is generated within the framework of existing ties to industry. In this case, the patent is licensed or transferred directly to the industrial partner, leading to a generally moderate extra income for the professor. On the other hand, if no industrial partner is directly available, the professor has to pay the patent application fees at his or her own risk. Therefore, many inventions at universities are not patented. Later on, as a result, companies may not be interested in investing in further development because the basic idea has not been protected.

If the research is funded by external sources, especially the BMBF, the university, not the professor, is responsible for patent protection. Due to the increasing relevance of external funding, the significance of the exploitation privilege is diminishing. However, the incentives for patenting by the universities themselves are low due to various factors. Among the most important are that most universities have neither funds nor infrastructure to support patenting and licensing activities; inventions resulting from federally funded academic research generally can only be licensed on a nonexclusive basis to industrial partners; and a portion of any licensing income earned from developments with federal government funds must go back to the funding agency.

In recent years, the University of Karlsruhe and the University of Dresden established patent and licensing offices comparable to those at American universities. These offices offer professors advice on patent affairs and, if the invention seems to be marketable, provide financing for the patent application and search for potential licensees. (For more details, see Schmoch et al., 1996a.) Some federal states plan to start similar programs, with the aim of better supporting inventors at universities. The states do not wish to abolish professors' exploitation privilege, but rather to offer institutional support.

It is not possible to directly track German academic patents. However, the

German database PATDPA allows one to search for the title "Professor" among inventors or applicants. Such a search turns up not only university-related patents, but also inventions by former professors now working in industry. Thus, the search sample is somewhat too broad and does not include inventions by scientific assistants at universities. Nonetheless, it can be assumed that the largest part of the search sample adequately reflects university patents.[24] From 1980 to 1990, the number of patent applications registered for professors jumped by 46 percent (Figure 3.19). This rate of increase is comparable to that for external university funds (42 percent), lower than that for industrial funding (80 percent), but higher than that for overall university research budgets (21 percent). Obviously, the number of patents is linked primarily to the share of external funds. Remarkably, 54 percent of university patents are applied and owned by companies (Becher et al., 1996). These patents are obviously sold directly by the professors who have taken advantage of the exploitation privilege.

It is interesting to note that the number of patent applications by German professors in 1992 was about 1,000, whereas the number of patent applications originating in American universities was about 2,500 (Association of University Technology Managers, 1993; Schmoch et al., 1996a). Despite the absolute difference in patent activity between the two countries, the relative number in Germany in relation to the gross domestic product seems to be quite high. However, it has to be taken into account that U.S. universities reported about 8,000 inven-

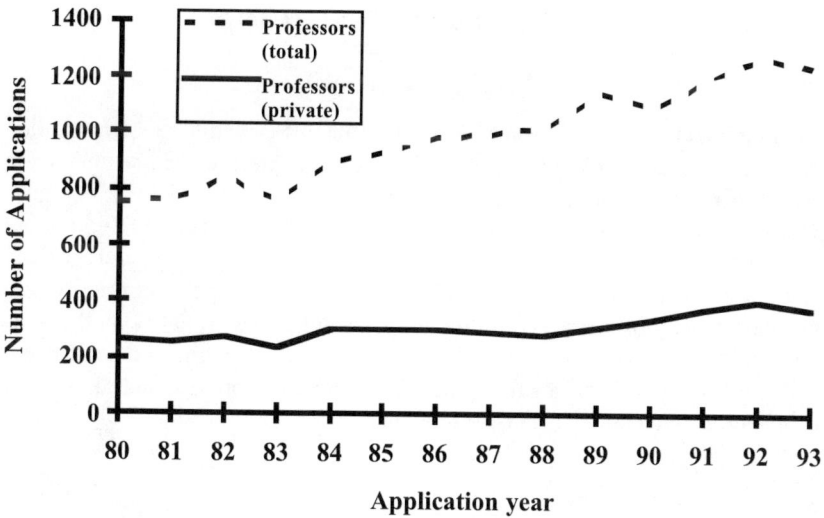

FIGURE 3.19 Patent applications to the German Patent Office by German university professors. NOTE: private = application by the professor; total = includes applications by firms or other institutions. SOURCE: Schmoch et al. (1996a).

tion disclosures; that means that not all invention disclosures resulted in patent applications. The U.S. technology licensing offices at universities have established an effective system to select inventions with sufficient economic prospects. A similar system does not exist in Germany, so that the reletively high number of patent applications from German universites can, at least partly, be taken as indicator for an insufficient quality selection.

University-related patents do not reflect the general orientation of academic research but can be used as an indicator for transfer-related activities. For analysis of these activities, differentiating university patents according to technology areas is quite revealing (Figure 3.20; for methodological details, see "Research Programs of the European Union," above). With reference to the general international distribution, patents of German professors are primarily in the field of chemistry, including pharmaceuticals and biotechnology. In mechanical and construction engineering, the specialization indexes are mostly negative for patents of German professors, but at a moderate level. Compared with the large volume of external funding for mechanical and construction engineering, the outcome in patents is quite modest, and the question arises whether this finding can be taken as an indicator for less effective technology transfer. In all fields of electronics and information technology the specialization indexes of the patents of German professors are distinctly below average, which has to be interpreted against the background of a low level of industrial activity in this area.

TECHNOLOGY TRANSFER FROM PUBLIC INTERMEDIATE R&D INSTITUTIONS

Max Planck Society

Complementary to German universities, the MPG is the major institution performing outstanding basic and long-term applied research. The MPG's main areas of focus are physics, biology, and chemistry. Many Max Planck institutes perform research in areas of strategic interest to industry. The most important channel of knowledge transfer is the exchange of scientific personnel. However, collaborative research with industry plays a modest but increasing role. Up to now, the intensity of contacts with industry has depended primarily on the willingness and interest of individual MPG scientists. With declining public funding, the usefulness and achievements of the MPG have to be proved, and the society has to approach technology transfer more actively.

GENERAL ORIENTATION

Reestablished in 1948 as the successor to the Kaiser Wilhelm Society, founded in 1911, the MPG basically has the same role today that it had in 1948. In the German landscape of scientific research, the MPG is a prominent research

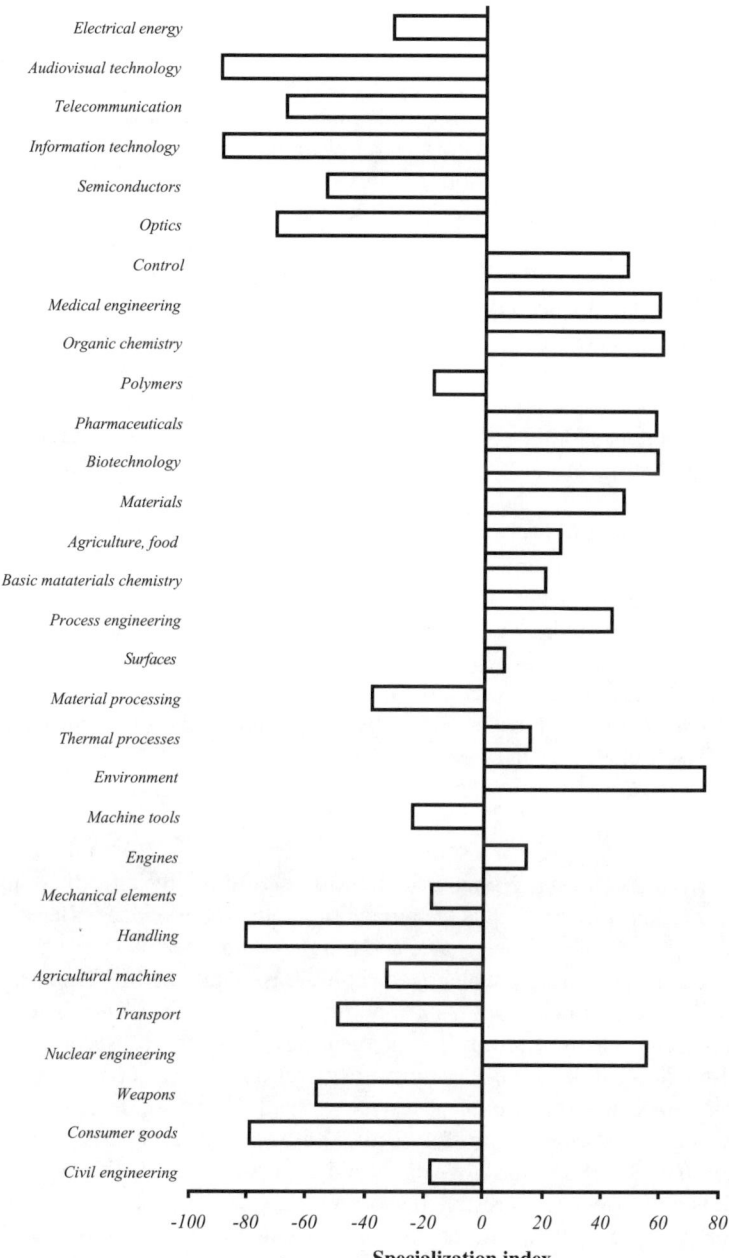

FIGURE 3.20 Specialization of German Patent Office patents of German university professors, in relation to the average distribution at the EPO for the period 1989 to 1992. SOURCE: Schmoch et al. (1996a).

body with a focus on promoting science and conducting basic research in various areas of natural and social sciences in the public interest.

Whereas industry and other nonprofit institutions such as the FhG and the AiF are involved chiefly in the field of applied R&D, universities and especially the MPG are almost completely oriented toward basic and long-term applied research. This balanced structure of the institutes and their respective "output" is seen to legitimate the existence of the MPG—including its public funding.

The definition of research areas and even the establishment of the society itself can be seen as a reaction of the federal government to an established situation in which universities fall under the jurisdiction of the federal states. With their priority of educating a broad array of students, universities are not in a position to focus on specific research-intensive topics. Prior to 1948, the central government had practically no way to promote areas of research thought to be of strategic importance for the country's international competitiveness. The formation of federal scientific institutes, through the MPG, was a solution. These institutes

- conduct research in important or strategic fields of science with an adequate concentration of personnel and equipment;
- quickly enter newly developing fields, especially those outside the mainstream, or fields that cannot be covered sufficiently at the universities; and
- conduct research that requires special or large equipment, or research that is so costly that it cannot be undertaken at universities (see Max-Planck-Gesellschaft, 1994a).

RESEARCH AREAS

Whereas the Kaiser Wilhelm Society focused primarily on promoting the natural sciences, the MPG adds the humanities and social sciences. Because the MPG aims to be a pioneer in science and tries to complement research at universities, it cannot do research in all conceivable areas. Thus, the MPG concentrates on fields that contain extraordinary opportunities for science. The society's research is focused in three areas: the chemical-physical-technical section, the biological-medical section, and the humanities section. These sections cover, for example, biochemical and clinical research, metal research, astrophysics, comparative law, education, and history—all with a strong focus on basic research (Table 3.16). The MPG has not established an institute devoted to engineering, since it is not seriously interested in short-term applied research.

In recent years, there has not been much change in the research priorities of the Max Planck institutes (MPIs); only research activities in biology have increased to any significant extent. So far, the principal areas supported have been physics and biology research, amounting to almost 60 percent of total expenditures (Figure 3.21, Table 3.17).

TABLE 3.16 Average Number of Permanent Staff and Scientists at Max Planck Institutes, Main Sections, 1993

Area of Research	Full-Time Staff	Full-Time Scientists	Percent Scientists
Chemical-Physical-Technical Section	208	556	29
Biological-Medical Section	127	39	31
Humanities-Social Sciences Section	57	20	34

SOURCE: Max-Planck-Gesellschaft (1994b); calculations of Fraunhofer Institute for Systems and Innovation Research.

ORGANIZATION

The society's main units are its institutes. In 1994, research was conducted at 62 institutes, 2 laboratories, and 3 independent research groups (figures for West Germany only). The size of the institutes differs widely; only a small number contain fewer than 50 or more than 800 permanent staff members (Table 3.16). In 1994, about 11,050 persons were employed full-time in MPG units. Beside the senior scientists (about 3,050 people), technicians, and other regular employees, there have been an increasing number of visiting researchers, fellows, and junior scientists (doctoral candidates); in 1994, there were a total of 5,500 individuals in this latter group. The average period of stay of visiting researchers and fellows was about 7 months.

In the Kaiser Wilhelm Society, institutes were designed around an outstanding scientist (Harnack principle). Today, given the complexity of research at MPIs, this principle is being applied at the departmental level. This personality-centered form of organization can explain the rise and fall of individual institutes

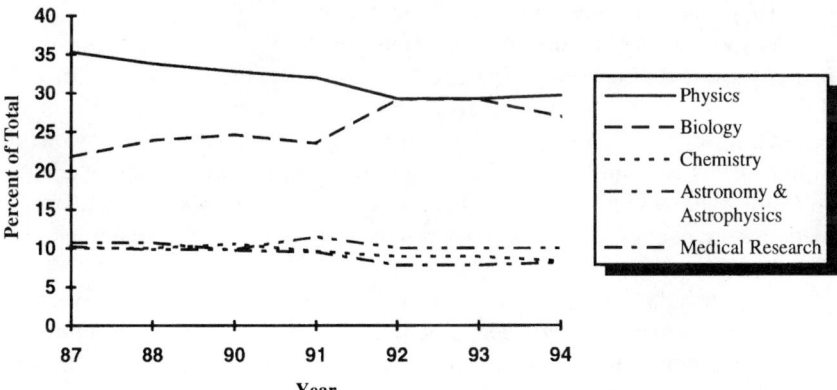

FIGURE 3.21 Max Planck institutes' expenditures in main supported areas, percent of total. SOURCE: Max-Planck-Gesellschaft (various years).

TABLE 3.17 Areas of Research at Max Planck Institutes, Percent by Expenditures and Scientists, 1994

Section and Area of Research	Expenditure	Scientists
Chemical-Physical-Technical Section		
Chemistry	8.3	7.7
Physics	29.7	29.2
Astronomy and astrophysics	10.0	10.1
Atmospheric and geological sciences	4.1	4.0
Mathematics	0.6	0.6
Information technology	1.4	1.4
Biological-Medical Section		
Biological research	26.9	22.3
Medical research	8.1	6.2
Social Sciences-Humanities Section		
Law	3.6	5.0
History	3.0	6.5
Sociology	1.1	1.5
Psychology	1.4	1.5
Linguistics	1.3	2.7
Education	0.3	0.5
Economics	0.1	0.6

SOURCE: Max-Planck-Gesellschaft (1994b).

or departments. If a departing head scientist is not replaced by an equivalent successor, the research focus of the institute or department might be changed (depending on the new leader) or even dissolved. The Harnack principle is considered to be an important basis for scientific excellence. In recent years, however, strategic considerations about relevant and declining research areas increasingly supplement this personality-centered principle.

As a rule, a board of directors is responsible for the entire institute; the members of the board elect a managing director, who serves for a set period. In addition to the board of directors, an advisory board (*Fachbeirat*), consisting of experts from different local and nonlocal scientific institutions, functions as an evaluating and advising body, submitting its reports to the president of the MPG. At many of the institutes, there are boards of curators (*Kuratorium*) as well, whose members are public authorities or interested scientists, including representatives from industry.

The chief administrative bodies of the Max Planck Society are the Executive Committee (*Verwaltungsrat*) and the Senate (*Senat*). The Executive Committee comprises the president, four vice presidents (three from each section and one from industry), the treasurer, and up to four senators. Together with the secretary-general, who heads the general administration, the Executive Committee forms the Board of Trustees (*Vorstand*).

The central decision-making body is the Senate. In addition to its supervisory role, the Senate assumes functions crucial to MPG existence, such as the

- establishment, closure, or reorganization of institutes and independent departments, including decisions on the incorporation of new areas of research;
- appointment of scientific members, directors, and heads of independent departments;
- election of the president, the vice presidents, and the members of the executive committee;
- assessment of the budget and other decisions concerning the use of funds; and
- approval of the guidelines of the institutes (see Meyer-Krahmer, 1990).

The Senate, comprising approximately 60 members, contains various representatives of the Executive Committee and of the three sections. The federal government can appoint two ministers or secretaries of state (*Staatssekretäre*) as official MPG senators, and the federal states can appoint three. Other senators are elected for a period of 6 years and represent other scientific institutions, industry (most of them members of the board of leading German companies), government, banks, employer and employee associations, public media, and other institutions of public interest. Permanent guests of the Senate include presidents or chairmen of the main science-promoting organizations in Germany. All in all, about 50 percent of the senators are scientists, most of them representing the MPG.[25]

The Scientific Council (*Wissenschaftlicher Rat*), which includes about 270 scientists (all scientific members of the society) and 1 scientific staff member for each institute elected by the institute's scientists, is the most important advisory body with a role in guidelines for scientific research.

In the context of technology transfer, it is interesting to assess the influence of industrial representatives and other nonscientific groups on the orientation of the MPG. As mentioned, about one-quarter of the Senate is made up of industry representatives and another one-quarter of government representatives. These nonscientific officials have a non-negligable influence on the general policy of the Max Planck Society. However, nonscientific groups have a rather marginal influence at the level of the institutes. Advisory board members of the institutes are highly reputed scientists, not lobbyists for a particular interest. All in all, the organizational structures reflect the general aim of the MPG to pursue independently basic research (see also Max-Planck-Gesellschaft, 1994a,b).

BUDGET AND FINANCE

Like universities, the MPG conducts chiefly basic research and is financed largely by public funds. Whereas financing of the universities is a duty of the

TABLE 3.18 Budget Structure of the MPG, 1994

Type of Funds	Million DM	Share in Percent
Public institutional[a]	1,534	88.5
Project[b]	199	11.5
Total	1,733	100

[a]Includes special allowances, general revenues, and transfers from 1993.
[b]Includes transfers from 1993 and additional project support.

SOURCE: Max-Planck-Gesellschaft (1994c).

states, the MPG was initially financed primarily by the federal government. Gradually the share of state funding increased to 50 percent. Since a long-term commitment of financial resources is needed to ensure the continuity of basic research and to generate new technical knowledge, the so-called institutional financial support, or promotion, of scientific bodies has been established.

Out of MPG's budget of DM 1.73 billion in 1994, DM 1.53 billion (88.5 percent) were public institutional funds. This money covered expenditures like wages, building maintenance, investment in equipment, and other payments. DM 199 million (about 11.5 percent) were noninstitutional allowances designated for individual research projects (Table 3.18). A further breakdown of project funds, for 1993, can be seen in Table 3.19. The project-specific money came primarily from the Ministry of Science and Technology and the EU. With the decrease of project funding by the federal and state governments, the allowances for individual research projects by the EU have become increasingly significant.

The importance of externally funded scientists is clearly demonstrated by comparing their numbers with the number of regular scientists (i.e., those paid within the institution-funded part of the budget). In the biomedical section, externally funded scientists comprised 53.5 percent of the total number. In the physical-chemical section, their share was 25.5 percent, and in the social science sec-

TABLE 3.19 Structure of Project Funds, 1993

Source	Million DM	Share in Percent
Federal government and states	105.2	62.1
EU, other public institutions	35.3	20.9
Foundations, industrial contracts, endowments	23.6	13.9
MPG assets	5.2	3.1
Total	169.3	100.0

SOURCE: Max-Planck-Gesellschaft (1994b).

tion, it was 16.6 percent. Overall, about 35 percent of all scientists are sponsored by external funds.

One indicator of the amount of applied research being done is the number of contracts or direct grants to MPIs by industry. Not surprisingly, this figure is very low. Only between 5 and 6 percent of project funds are the result of such contracts. In 1994, about 2,000 contracts brought in DM 37 million to the institutes; for 1995, this revenue was estimated to be between DM 37 and DM 40 million, or the equivalent of 0.5 percent of the overall budgets of the MPIs. Nonetheless, a few institutes support a considerable number of their scientists with industrial grants. These institutes conduct research in the fields of biochemistry (e.g., *MPI für Biochemie*), synthetic polymers (*MPI für Polymerforschung*), and material analysis (*MPI für Metallforschung*). Still, the scientific community within the MPG prefers to obtain grants from foundations and public agencies.

TECHNOLOGY TRANSFER

The MPG emphasizes its identity as an organization for basic research. But, especially in the late 1940s and 1950s, the society carried out a large volume of applied research. The current strong orientation toward basic research occurred over time (Mayntz, 1991), in particular against the background of the growing relevance of the FhG. Today, the main function of the MPG inside the German framework of science is to perform basic research.[26] The assumption is that basic research provides an important stimulus for more applied R&D in industry (Dose, 1993); therefore, the work of the MPG pays off.

How is the transfer of basic research findings accomplished and assessed by the MPG? The "classic" type of transfer, through the exchange of research personnel, may be the most effective. Most MPI directors are at the same time honorary professors at a local university. Thus, there is close contact with the other institution promoting basic research. Some of the expensive MPG facilities—especially for research in astronomic and solid-state physics—are used by university research groups as well. Another important factor is the number of recipients of doctoral degrees, an estimated 80 percent of whom will be employed in industrial R&D departments. In 1993, the mean number of recipients of doctoral degrees for the institutes in the chemical-physical-technical section was 13[27]; the biomedical section graduated an average of 7.2; and the social sciences section graduated an average of 1.9. Several institutes were well above the average, like the MPI for Polymer Research (*MPI für Polymerforschung*), which graduated 42 Ph.D.'s, and the MPI for Psychiatric Research (*MPI für Psychiatrische Forschung*), which graduated 19. Although many of these graduates will work as scientists in industry, those scientists who prepare a habilitation thesis[28] tend to become professors at universities. Again, as professors, they educate dozens of students and junior scientists and are an important means of knowledge transfer.

The MPG allows its scientists to take a sabbatical term for doing research in industry. This temporary transfer has to be approved by the MPG and is as yet quite underdeveloped. Consultancy contracts with industry and the supply of expert reports are additional means of knowledge transfer. Recently, MPG scientists have been allowed to engage actively in the development of spin-off companies.

The already-mentioned decrease in public funds and increase in public pressure toward a stronger and more active technology transfer to industry has forced even the MPG to document its capabilities and achievements for a broader public. Because basic research is the main focus of the MPG, long-term applied research, which is of greater interest to industry, is pursued only by certain institutes. It is helpful to concentrate on the examples of more industry-oriented institutes and thereby explain different technology transfer mechanisms.

The MPI for Polymer Research belongs to the chemical-physical-technical section. In 1993, the institute had an average size staff: 167 full-time employees (including 51 scientists), 31 externally funded employees (including 18 scientists), 25 visiting researchers, 26 fellows, 118 doctoral candidates, and 24 master's candidates. Partly due to a high percentage of chemical research (high even for an industrial laboratory), this institute has an above-average number of contacts with industry. These contacts, which include domestic and foreign companies of all sizes in chemistry or chemistry-related areas, are established by means of publications, exhibitions, and conferences. A considerable amount of collaborative research with industry takes place in several projects of joint interest. Generally, there is no cash flow from industry to the institute; the major interest is in a mutual exchange of knowledge. Sometimes, a company is acquainted with the spectrum of topics dealt with by the institute and wants to contract for certain research services. However, the MPG accepts research contracts very restrictively. Such work will be undertaken only if free publication of all research results is guaranteed. Another prerequisite is that the contract research be formally approved by the society. Another type of contact arises when the institute needs to perform experiments but does not possess the equipment or facilities. In these cases, the experiments are performed in industrial laboratories. The exchange may occur in the other direction, too: The institutes are permitted to offer their facilities to industry (Wegner, 1995).

The MPI for Biochemistry (*MPI für Biochemie*), located near Munich, provides another example of active knowledge transfer. With more than 800 employees, half of them scientists, this institute is one of the largest, as it was formed by combining three formerly independent institutes. It is located next to a large medical clinic and the Center of Genetic Research of the University of Munich. Interdisciplinary research and applied clinical research are carried out, as is basic research, depending on the specific work group or department. This institute will function as the nucleus for a biotechnology incubator that is currently being established there. The concentrated settlement of companies with the core business of biotechnology is being funded by the Bavarian state and managed by the

Fraunhofer Management Society. This form of state-promoted science, which integrates applied and basic research institutes, universities, and industry, will be a major achievement, as it is not yet well developed in Germany.

The MPG always claims to be an advocate for pure basic research, but at least 19 institutes in the biological-medical section (out of a total of 24) and approximately 16 institutes in the physical-chemical-technical section (out of 26) conduct research in areas that are generally interesting for their industrial application. By the definition of the *Frascati Manual* (Organization for Economic Cooperation and Development, 1994a), they perform basic research. These activities are primarily carried out in two major areas: biotechnology and materials (see Bild der Wissenschaft, 1994b). In biotechnology, there are MPIs for biochemistry, biophysics, molecular genetics, and brain research; in materials, there are institutes for solid-state physics, microstructure physics, and metal research.

The MPG has made a major effort to make it easier for its institutes to undertake technology transfer to make the benefits of technology transfer more apparent. In a recent publication, the MPG stated that its institutes contribute to 9 strategic areas with 70 subgroups of strategic technologies like new materials, cell biotechnology, and nanotechnology (the definition of the strategic areas is from Grupp, 1993). Only a few subgroups are not represented by the MPG (see Max-Planck-Gesellschaft, 1995).

A major indicator of the extent of application-oriented MPG research is patents. Between 1989 and 1992, most MPG patents were registered in biotechnology or in related areas like organic chemistry and pharmacy. In terms of registered European patents, the MPG heads the field of genetic engineering in Germany; it is ranked number seven among the leading patent assignees worldwide (Bild der Wissenschaft, 1994a). In addition, MPG research that requires new tools and advanced equipment leads to spin-offs and a certain number of patents in measuring and control technology. As to the four focal areas, there have been a small number of MPG patents related to semiconductor devices; MPG has no patents in either production technology or information technology. Few information-technology-related patents have been awarded because that particular institute was established only recently.

Within the MPG, the Garching Innovation GmbH is responsible for intellectual property rights. Garching Innovation was established in 1969 as the central institution for technology transfer from MPIs and serves as its mediating agent for the industrial use of research findings. If the results of basic research carried out at an MPI can be exploited technically, an attempt is made to transfer the findings to industry through licensing or, in the case of collaborative research, through direct transfer of patents. MPG scientists are free to publish or apply for patents, so not all of the research findings are reported to Garching Innovation first. Garching has to cope with the very necessary, but sometimes hindering, attitude of scientists: They want to publish their results as soon as possible. They are often not aware that with intelligent timing, patents and publications do not hinder each other and can exist in parallel. In 1994, Garching Innovation completed 45 license agreements and had license revenues of DM 7 million, with a trend toward growth. It received about 90 new inventions for exploitation and managed about 600 domestic patents and 860 patents in foreign countries.

To sum up, the MPG always emphasizes the value of technology transfer, but it never views the success of transfer as a criterion for excellence. Technology transfer seems not to be a priority; rather, it is seen as a by-product or spin-off of the institutes' research activities. Up to now, the question of whether there are strong ties to industry has depended primarily on the willingness and interest of each individual scientist. Some scientists tend to work in more applied research fields and are ready to maintain contact with industry. Because collaborative research, applied research, and technology transfer are not considered to be priorities, but rather depend completely on the willingness of individual scientists, much industrially applicable research is probably undertaken by the institutes but is forgotten before industry becomes aware of its relevance. It will be a challenge for the MPG to overcome this apparent gap without losing its independence and focus on basic research.

Helmholtz Centers

Helmholtz Centers conduct primarily research on long-term problems entailing considerable economic risks in areas of public welfare and in fields requiring large investments. Besides the classic instrument of scientific publications, the major mechanisms of technology transfer are the participation of industry in advisory boards and committees and collaborative research uniting industry and the centers on large projects or programs. The centers are funded primarily with public money, but industry and the federal government are striving to increase the share of industrially relevant research these centers conduct. This can be achieved by reducing institutional funds in favor of project support and broader participation of industry in the centers' research planning procedures. It is not clear to what extent these different measures suggested will be implemented. In any case, the centers will go through a process of considerable structural change within the next few years.

INSTITUTIONAL STRUCTURES

The first Helmholtz Centers were founded in the late 1950s, when the allied forces gave Germany permission to perform nuclear research, then called Large Research Centers (*Großforschungseinrichtungen*). At that time, the federal government was struggling to establish a role for itself in technology policy, which was generally the province of the states as part of their responsibility for education and science. Federal technology policy was limited to special federal purposes. In this situation, the establishment of Helmholtz Centers opened a way for the federal government to increase considerably its influence in this area. Following the pattern of U.S. and British national laboratories, all Helmholtz Centers worked initially in various areas of civilian nuclear research. Since the late 1960s, other areas of research have been added such as aeronautics, computer science, and biotechnology (Meyer-Krahmer, 1990; Schimank, 1988a, 1990). It is not possible to describe the research orientation of Helmholtz Centers in terms of simple categories like basic or applied. Their activities include

- basic research requiring large research facilities;
- large projects and programs of public interest, sometimes undertaken with international cooperation, requiring extraordinary financial, technical, and interdisciplinary scientific resources and management capacities; and
- long-term technology development, accompanying the whole innovation cycle from basic research to applied research to development, including preindustrial fabrication (e.g., nuclear fusion, magnetic railway).

Helmholtz Centers are institutionalized as private companies, associations, or foundations. The autonomy of science in Helmholtz Centers is constitutionally comparable to the situation in universities (Meusel, 1990). Each center defines its research program independently of government or industry, but program implementation requires the agreement of a Supervisory Board (*Aufsichtsrat*), on which the federal government and the states hold dominant positions. Furthermore, each center has a Scientific Advisory Board (*Wissenschaftlicher Beirat*), which evaluates scientific quality and regularly makes recommendations on the future orientation of research. In contrast to the situation of universities, the MPG, or the FhG, which can freely decide on the use of institutional funds, the institutional funds of Helmholtz Centers are linked to program tasks determined by the government. Thus, the actual political intervention is more important in Helmholtz Centers than it is in most other research institutions.

The umbrella organization of the Helmholtz Centers is the HGF. It represents the interests of the Helmholtz Centers and has the major task of coordinating their research activities. For that purpose, about 20 HGF committees have been established to deal with technical, economic, and administrative questions.

BUDGET AND RESEARCH AREAS

In 1993, the HGF comprised 16 Helmholtz Centers with about 24,000 employees located throughout the old and new states of Germany (Arbeitsgemeinschaft der Großforschungseinrichtungen, 1994; Bundesministerium für Forschung und Technologie, 1993a). In 1994, these 16 centers received a total of about DM 4.1 billion. Of this amount, about 80 percent was institutional funds. Ninety percent of the institutional funds were contributed by the federal government; the remaining 10 percent came from the states. The support for the Helmholtz Cen-

TABLE 3.20 Spending, Percent Share of Total Budget, and Trend for Major Research Areas of the Helmholtz Centers, 1993

Area	Budget (million DM)	Share of Total Budget (%)	Trend
Energy	518	18	+
nuclear energy	418	15	+
Transport, traffic	253	9	+
Aerospace	294	10	+
Geophysics, polar research	122	4	++
Environment	367	13	+
Health	302	11	++
Biotechnology	80	3	+
Information, communication	254	9	O
New technologies, materials	176	6	O
Basic physical research	498	17	+
Total	2,864	100	+

NOTE: + = increasing; ++ = increasing considerably; O = stagnating.

SOURCE: Arbeitsgemeinschaft der Großforschungseinrichtungen (1994).

ters amounted to two-thirds of all grants awarded by BMBF to research institutions and about one-fourth of BMBF's total budget in 1994.

The other sources of support for the Helmholtz Centers include funds generated by the Helmholtz Centers themselves, institutional funds from nonpublic sources, and external funds linked to specific research projects or programs. On the basis of the available publications of the Association of Large Research Centers (*Arbeitsgemeinschaft der Großforschungseinrichtungen* [AGF]), it is not possible to determine the exact volume of these project-related funds and consequently, the share of contract research for industrial clients.

Table 3.20 shows the overall distribution of the R&D activities in different areas. At first sight, the distribution seems to be quite balanced and stable. In reality, the Helmholtz Centers' research has gone through a process of dramatic reorientation as several areas have reached maturity; nuclear energy, especially, is no longer seen as a major strategic field. By the beginning of the 1990s, the federal budget for nuclear energy research had fallen by about one-third compared with its level in 1985 (Bundesministerium für Forschung und Technologie, 1988, 1993a). Seen in this light, the current 15 percent share of the budget devoted to nuclear energy research is still considerable.

When one looks more closely at specific research programs and individual Helmholtz Centers, one sees that the situation is characterized by an enormous restructuring process. Three new Helmholtz Centers and eight affiliations have been established in East Germany, so that the overall budget figures hide stagnation or even cutbacks at Helmholtz Centers in the old federal states. On the program level, only some areas, like polar research and cancer research, are expected to grow substantially. The need for reorientation is magnified by the financial constraints stemming from the reunification of Germany. Stagnation in key areas like information and communication and modest share in biotechnology may be interpreted as a signal that, in the face of public financial restrictions, technology-related R&D requires additional contributions from industry.

As to the four focal areas of this study, the Helmholtz Centers play a significant role in biotechnology (although the share for biotechnology in the total budget of the Helmholtz Centers is modest) and in information technology. Research in microelectronics is subsumed in the official statistics under information technology and is performed at several Helmholtz Centers to a significant extent. Production and manufacturing are generally not explicit topics of Helmholtz Center–related research. The one major exception is the development of chemical and physical processes for environmental purposes at the Helmholtz Center Gesthacht.

TECHNOLOGY TRANSFER

The Helmholtz Centers see their mission regarding technology transfer largely according to a science-push approach. According to this view, they develop the scientific and technological basis for future applications that have great public relevance (Bundesministerium für Forschung und Technologie, 1993a; Meusel, 1990; Schimank, 1990). They consider high-quality research and the publication of research results to be the most effective means of technology transfer. They do not follow a demand-pull model, as this orientation would not be compatible with the autonomy of scientific research (a prerequisite for scientific excellence). This approach is in many ways similar to that of universities and the MPG. Furthermore, Meusel (1990) emphasized the division of labor between, on the one hand, application-oriented institutions like the FhG and institutes of industrial research associations and, on the other hand, the Helmholtz Centers.

Since many research programs have long-term relevance for industrial applications, Helmholtz Centers often invite industry to collaborate. The dialogue with industry on specific projects or research areas can be mediated by industrial members on the advisory boards of the individual center or program committees of the AGF. Furthermore, Helmholtz Centers and industry conduct collaborative research in areas of common interest (e.g., energy, information technology, biotechnology). In these cases, the division of labor is fixed by formal cooperation contracts, which also determine the conditions of the mutual transfer of results and the exchange of personnel. In general, each partner bears its own costs, and the Helmholtz Centers do not get any additional funds from industry. In very large projects, the Helmholtz Centers and companies involved often establish joint ventures for technology development and exploitation. In some cases, where

TABLE 3.21 Budgets and Staffing of Selected Helmholtz Centers that Emphasize Industrially Relevant Research, 1993

Institution	Major Areas of Research	Budget (million DM)	Staff (full-time equivalents)
Deutsche Forschungsanstalt für Luft- und Raumfahrt, DLR	Aeronautics, aerospace, energy	694	4,469
Forschungszentrum Gesthacht, KFA	Climate, materials, process technology, nuclear safety	132	845
Forschungszentrum Jülich, KFA	Materials, information technology, life sciences, environment, nuclear and other energy	682	4,263
Forschungszentrum Karlsruhe, FZK	Environment, nuclear technology, super-conductivity, micro-systems	956	3,790
Gesellschaft für Biotechnologische Forschung, GBF	Biotechnology	75	487
Gesellschaft für Mathematik und Datenverarbeitung, GMD	Mathematics, information technology, VLSIs	195	1,599

SOURCE: Bundesministerium für Forschung und Technologie (1993a).

Helmholtz Centers have special knowledge or facilities, they also carry out contract research for industry.

However, there are considerable differences in the research orientations of the different Helmholtz Centers. Examples of Helmholtz Centers with a primarily basic orientation are the *Gesellschaft für Schwerionenforschung (Darmstadt)* and the *Hahn-Meitner-Institut (Berlin)*, both working in the area of basic physical research. Table 3.21 documents the staff, budget, and major areas of activity for six selected Helmholtz Centers (among them the three biggest centers) whose orientation is particularly appropriate for technology transfer to industry. For example, the Research Center in Karlsruhe has collaborations with 25 large enterprises and 85 SMEs.

Helmholtz Centers also claim to trigger substantial technology transfer through their investment in research facilities, since industrial suppliers of these facilities often develop new leading-edge technology. Such technological developments could be transferred to other markets (Bianchi-Streit et al., 1984; Commission of the European Communities, 1992).

Some Helmholtz Centers have established spin-off-related technology transfer units for the active marketing of their own patents (Arbeitsgemeinschaft der Großforschungseinrichtungen, 1995; Wüst, 1993). For example, the Research Center in Karlsruhe receives about DM 2 million from license revenues; this is, however, less than 1 percent of its total budget. The transfer units actively address potential users of Helmholtz Centers' technology, visit exhibitions, and work in cooperation with other, generally regionally based technology transfer institutions. The transfer units were initiated in the early 1980s by introducing new regulations for the use of license revenues. Before that time, license revenues did not increase centers' budgets because their base funds were reduced by the same amount. At present, two-thirds of license income can be used for technology transfer projects, in particular the adaptation of research results to the needs of SMEs. Even so, the current stituation has some shortcomings. For instance, one-third of the license income has to be transferred to the government and cannot be used by the Helmholtz Centers themselves. Furthermore, license revenues cannot be used for purposes other than technology transfer (the Helmholtz Centers are actively trying to change this ruling). Since the department where the invention comes from does not get to use the license income, its incentives for patenting are limited. A further problem is that the exclusiveness of license is generally restricted to a period of 5 years, which is a decisive impediment for industrial partners. In practice, most exclusive licenses are extended. Nevertheless, a more industry-oriented policy would be desirable.

In the context of technology transfer, it should be emphasized that Helmholtz Centers cooperate intensively with other scientific institutions, particularly universities. In many cases, leading scientists of the Helmholtz Centers simultaneously hold chairs at universities, and Helmholtz Centers and universities cooperate directly in the recruitment of their scientific staff (Meusel, 1990). For example, the Research Center in Karlsruhe currently has 110 collaborations with German universities, 120 with other Helmholtz Centers and German research institutions, 125 with foreign R&D institutions, and 55 with foreign universities (Forschungszentrum Karlsruhe, 1996).

The technology transfer activities of the Helmholtz Centers have been criticized since about the mid-1970s. Because Helmholtz Centers' research is limited to a relatively small number of research topics, it is crucial that these topics be chosen appropriately. However, it is difficult to define the long-term problems that will be relevant for future technology transfer; it is equally difficult to negotiate between the sometimes different perspectives of different advisory groups in academia, industry, and government. Because it is necessary to find a compromise, decisions can easily lead to failures (Kantzenbach and Pfister, 1995; Schimank, 1990). The severe restructuring process of the last few years can be taken as proof of the validity of this very fundamental criticism. To solve this problem, the government tries to implement improved methods of technology foresight.[29]

Especially in recent years, the federal government and industry have demanded new mechanisms and structures for increasing and accelerating technology transfer to industry. Thus, BMBF has suggested new types of cooperation, including the temporary merging of the research capacities of Helmholtz Centers and industry for specific projects and the institutional separation of industrially relevant departments. This institutional autonomy of parts of a Helmholtz Center could be the precondition for their cofinancing by industrial partners. BMBF views as a necessity the stronger engagement of industry in supervisory and advisory boards of the Helmholtz Centers. R&D in technology areas not seen by industry to be useful should be stopped (Bundesministerium für Forschung und Technologie, 1992, 1993a).

In 1993, BMBF charged a committee of industrial experts, the Weule Commission (*Weule-Kommission*), to assess the potential for closer industry relations for Helmholtz Centers in Jülich and Karlsruhe (*Forschungszentrum Jülich* and *Forschungszentrum Karlsruhe*). The commission stated that only 30 percent of the research activities they examined were application oriented and industrially relevant. Among 20 analyzed research areas, only 9 were industrially relevant; very few activities were interesting for the specific target group of SMEs. The commission suggested that the centers increase the application-oriented share of their research from 30 percent to 75 percent within the next 5 years. In addition, it proposed that industry become more closely involved in the planning of new projects and programs at the centers (Management-Informationen, 1995; Weule-Kommission, 1994).

Simultaneously, a commission of the Central Association of the Electrotechnical Industry (*Zentralverband der Elektrotechnik- und Elektronikindustrie*) analyzed public research institutions in the area of information technology. It stated that the institutional funding of Helmholtz Centers is too high and should be partly replaced by a higher share of project funding. In addition, it concluded that the transfer of personnel should be facilitated (Management-Informationen, 1995).

The directors of the Helmholtz Centers refused to increase to 75 percent the share of application-oriented research they conduct, because they saw the need to maintain a sufficient level of basic research, long-term research, and research for public welfare. This reaction can be partially explained by the fact that the commission evaluated two Helmholtz Centers that are already highly industry oriented compared with most others. Hence, the requirement of this large share of application-oriented research could only apply to some selected Helmholtz Centers. The Helmholtz Center directors also did not agree with the suggestion to devote a higher share of their budgets to project research. They felt this would be detrimental for an orientation on strategic medium- and long-term goals. They also feared that the reduction of institutional funds would lead to a loss of scientific competence. In any case, with a greater diversification of research areas and stronger emphasis on application, the overlap and competition with other research institutions, such as the universities, the MPG, and the FhG, will grow.

Although the suggestions of the Weule Commission and the Central Association of the Electrotechnical Industry will probably be only partially adopted, the discussion shows that the Helmholtz Centers will go through a further dramatic structural change. The mechanisms of technology transfer will be strengthened, and industry will come to participate more intensively in the centers' planning processes. This structural change also implies a change in the legal framework of the Helmholtz Centers. Regarding implementation of new structures, a crucial problem is whether industry will be ready not only to assume a more intensive advisory function, but also to engage more in the funding of Helmholtz Centers.

In November 1995, the former Association of Large Research Centers adopted the new name Helmholtz Association of German Research Centers (*Hermann von Helmholtz-Gemeinschaft Deutscher Forschungszentren*); at the same time, a senate was established. This new decision-making committee is responsible for general strategic planning and cooperation with other research institutions and industry (Bundesministerium für Bildung, Wissenschaft, Forschung und Technologie, 1995b). Thus, the HGF will have a status comparable to that of the MPG and the FhG and will achieve greater autonomy with regard to the BMBF. What actual impact on technology transfer these new organizations will have remains to be seen.

Blue List Institutes and Departmental Research Institutes

The semipublic institutes of the Blue List and the departmental research institutes carry out numerous research activities. However, only some of these institutes have close relations to industry and perform technology transfer activities.

Besides the Helmholtz Centers, the MPG, and the FhG, the central government and the states jointly support independent research institutes with supraregional importance and specific scientific interests. These institutes are called Blue List institutes because the first list of them was printed on blue paper.

In 1992, 82 Blue List institutes existed, of which 48 were located in the old federal states and 32 in the new ones. They employ about 10,000 people (i.e., about as many as the MPG). In 1994, the overall budget for the institutes was about DM 1.2 billion. In general, the host state and the federal government pay equal shares of the budget. The institutes have different legal forms but generally a semipublic status.

The structure and the technical orientation of the Blue List institutes are very heterogeneous. The research areas comprise the social sciences and humanities, economics, education, biomedicine, biology, other natural sciences, and information services. Examples of technology-oriented institutes are the *Heinrich-Hertz-Institut für Nachrichtentechnik* in Berlin (telecommunications), the *Institut für Halbleiterphysik* in Frankfurt/Oder (microelectronics), and the *Institut für Molekulare Biotechnologie* in Jena. Because of their heterogeneous structure, however, the institutes have no common research policy and especially no common policy of technology transfer. In 1991, the Blue List institutes established a common association, called *Wissenschaftsgemeinschaft Blaue Liste*, to represent the interests of members and to achieve a more coherent research policy.

Because the budget of the Blue List institutes is almost totally covered by public institutional funds, the incentives for technology transfer are low. However, some technical institutes (e.g., the Heinrich Hertz Institute) do some contract research for industrial clients. The regulatory regime concerning intellectual property rights is comparable to that for the Helmholtz Centers. However, only a few institutes have begun to engage in a more active patenting.

Many public agencies, which carry out official tasks for specific ministries of the federal government, also perform some research. They are called departmental research institutes (*Ressortforschungseinrichtungen*). Because of the large size of some of these institutions, their research activities are not negligible. Some of these institutions even have exclusive research missions. The overall volume of this research cannot be estimated precisely. In any case, institutions such as the *Physikalisch-Technische Bundesanstalt* in Braunschweig (measuring and testing, about 1,800 employees) and the *Bundesanstalt für Materialforschung* in Berlin (measuring and testing of materials, 1,600 employees) document the broad potential for technology transfer. Other examples are the *Biologische Bundesanstalt für Land- und Forstwirtschaft* in Berlin/Braunschweig (agro-biotechnology, 700 employees) and the *Paul-Ehrlich-Institut* in Langen (vaccines and serums, 350 employees).

Because the institutions have primarily an official mission for a special department of the federal government, an explicit policy of technology transfer does not exist. Technology transfer is generally considered a spin-off effect. However, some departmental research institutes cooperate closely with industrial enterprises within the framework of their official missions, e.g., the approval of technical products, so that de facto considerable informal technology transfer takes place (Bierhals and Schmoch, 1997). Because of their public status, income derived through research contracts or patent licenses cannot be used for the institutions themselves but must be transferred to the federal government. All in all, the incentives for an active licensing policy are low. In recent years, first attempts to formulate a more deliberate transfer and patent policy have been undertaken.

Fraunhofer Society

ORGANIZATION STRUCTURES

Founded in 1949, the Fraunhofer Society originally coordinated and controlled research projects that the Federal Ministry for Economic Affairs assigned to industry. In the mid-1950s, the FhG began to perform contract research financed only by two federal states and the Ministry of Defense. On this basis, it grew slowly during the1960s. It was not until 1973 that the FhG obtained the status of a federal research institution and received institutional funds from the BMFT, now the BMBF. This decision has to be seen in the context of the intense discussions that were taking place at that time about the technological gap between Europe and America and the more active technology policy being implemented by the German federal government (Schimank, 1990).

Today, the FhG is the major German nonprofit organization in the area of applied research, running 46 institutes in Germany—36 consolidated institutes in the old German states and 10 newly established institutes in the new states, supplemented by 12 subsidiaries of consolidated institutes in the new states. The FhG employs 7,800 people, of whom 2,600 are scientists and engineers. In 1994, the FhG budget amounted to DM 1.1 billion, or roughly $700 million.[30]

The FhG is organized as a registered society (*eingetragener Verein, e.V.*) whose principal statutory task is the furtherance of applied research. The FhG is instrumental in keeping up with worldwide technology developments and making new research results usable for industry and public needs (Schuster, 1990). Its roughly 700 members come from federal and state governments and other political, scientific, industrial, and economic institutions. The BMBF and state ministries are dominant members (Fraunhofer-Gesellschaft, 1985).

The society is managed autonomously according to its statutes. There are two principal management levels: the society and the institute. Decision making on the society level is in the hands of the Members' Assembly, the Senate, the Board of Directors, and the Scientific-Technical Council. The members elect the Senate, which is responsible for long-term decisions and general policy (i. e., budget and finance, opening and closure of institutes, major investments, and consensus management). Senate membership includes representatives of the scientific, economic, political, and public sectors in the German R&D system. The Board of Directors, composed of the FhG president and two full-time directors, carries out policies as determined by the Senate. The Board of Directors is supported by the central administration, which has a staff of more than 200. The Senate and Board of Directors are advised by the Scientific-Technical Council, which is made up of 102 members; 52 of these are institute directors and the rest are scientific and technical staff at the institutes. The council elects a Main Commission (*Hauptkommission*), which keeps in contact with the Board of Directors and thus is a major advisory body for consensus management between the society and the institute levels.

The organization and success of the FhG are based on decentralized initiative and responsibility. There are 40 civil research institutes and 6 defense institutes. The definition of research agenda and acquisition of funds, as well as personnel recruitment, are essential tasks of the institutes; the central administration is responsible for general planning, controlling, resource allocation, and business administration. The institutes have an average staff of 170, including part-time employees and students (the number varies greatly among the institutes), and are organized internally as profit centers according to the same concept of decentralization. Project and division managers have major responsibility for the acquisition and execution of research, including personnel recruitment.

Formal contact between the institutes and their sponsoring and cooperating partners in science, policy, and industry is fostered through advisory boards (*Kuratorien*) that usually meet once a year to exchange general information and discuss the institutes' activities and progress. In total, the advisory boards of Fraunhofer institutes (FhIs) have 450 members.

All in all, industry has only an advisory function at the central and institute levels through representatives in the Senate and the institutes' advisory boards. Thus, the FhG research orientation is largely independent and primarily determined by the institutes in a decentralized way.

BUDGET AND FINANCE

The typical FhG financial structure is best exemplified by the civil contract research activities of the consolidated 30 institutes in the old states of Germany (leaving aside civilian contract research in the new states, defense research, and investment expenditures). In 1994, the total budget for these institutes amounted to DM 603 million, of which about 70 percent were funds for contract research and 30 percent were institutional funds from the federal and state governments (Fraunhofer-Gesellschaft, 1994). Ninety percent of the institutional funding is contributed by BMBF; the remaining 10 percent as well as half the costs of establishing new institutes are paid by the state ministries hosting the institutes (sometimes, the states bear up to 100 percent of special investments).

A major characteristic of the Fraunhofer model is that the level of institutional funding is not stable but depends on the income from public and private contracts. In other words, for each institute, the level of institutional funding increases or decreases in relation to the institute's success in contract research (Imbusch and Buller, 1990). During the early years of FhG, the share of institutional funding amounted to about 50 percent and decreased later to about 30 percent. These funds were the basis for developing the FhG's reputation for high-quality applied research that thereafter allowed for successful expansion of research and technology transfer with considerably less institutional funding.

Figure 3.22 shows the contributions of base institutional funds, public projects, and industrial contracts to the FhG from 1976 to 1994. Each of the three sources contributed about one-third of the total budget, with so-called "other sources" not taken into account here. Income from private contracts showed a strong, steady increase over the period. Public project funding dominated FhG finances up until the economic recession that followed German reunification. There is still uncertainty as to whether industrial contract research will make up for the loss of public project funding. Contract research may fill the gap, because public programs for key technology research indirectly support industrial interests and thus contribute to technology transfer to industry. According to this

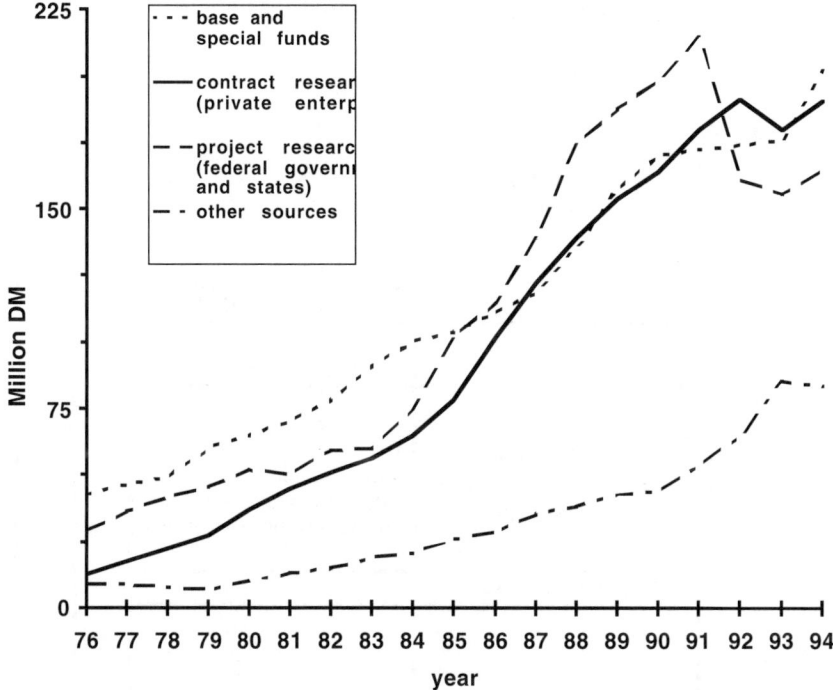

FIGURE 3.22 Budget structure of 30 consolidated Fraunhofer institutes in West Germany. SOURCE: Fraunhofer-Gesellschaft (1994).

perspective, 55 percent of civilian FhG contract research is relevant for technology transfer to industry (Figure 3.23). If only direct investiment is taken into account, the industry contributes about 30 percent of the total (DM 196 million in 1994).

It is interesting to note that the FhG is allowed to carry out contracts for foreign industrial clients. In 1994, DM 18.6 million, almost 10 percent of the industrial budget, came from foreign countries. The largest share of these contracts emanated from neighboring German-speaking countries (Switzerland 20 percent and Austria 10 percent); however the volume of contracts with U.S. enterprises is considerable (20 percent). These activities enable the FhG to monitor the international development of technology, not only on the supply side through communication with other foreign scientists, but also on the demand side. At the same time, the foreign clients profit from FhG competencies in applied research.

FhG activities account for about 1 percent of the German gross domestic expenditure on R&D. The FhG operates in the market of publicly funded technology programs that are partly relevant to private industry (key technologies)

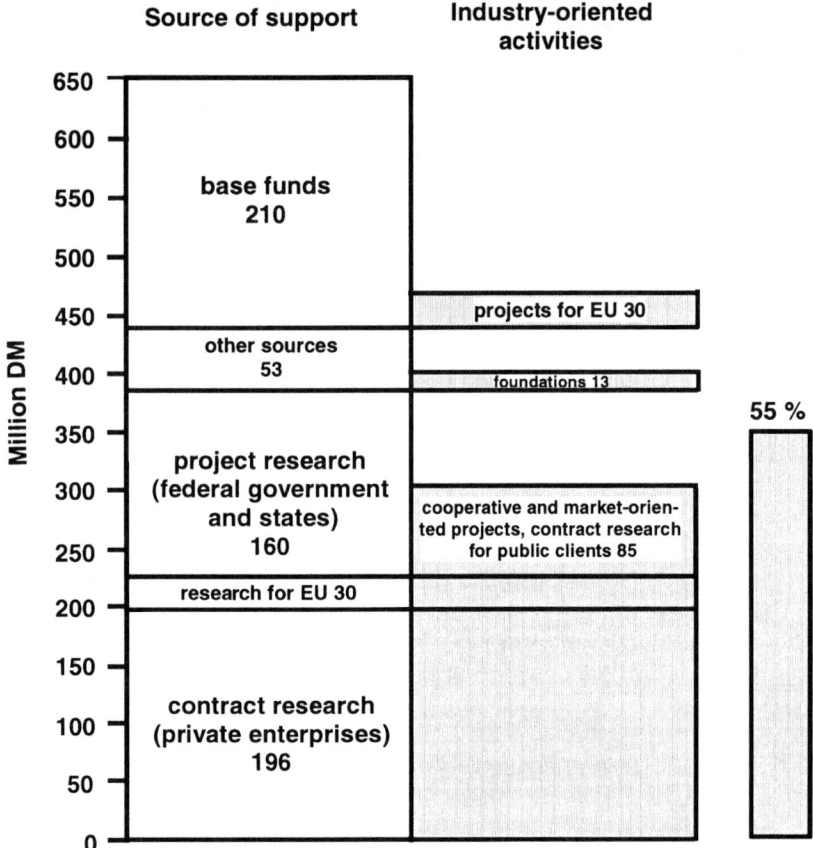

FIGURE 3.23 Industry-oriented activities of 30 consolidated Fraunhofer institutes in West Germany, 1994. SOURCE: FhG-Zentralverwaltung (1995).

and in the market of privately funded external R&D expenditure. The latter amounted to more than DM 6 billion in 1993. For research institutes, this market is actually much smaller, as most of the external industrial research is done by other companies in the private sector. A realistic level of industrial contract research in the publicly funded nonprofit sector would be in the region of DM 1 billion. In 1993, FhG institutes attracted about 20 percent of this market, second only to universities.

RESEARCH AREAS

Of the main research areas of the FhG, production technology is the largest and, when the materials area is included, shows the distinct focus of FhG on

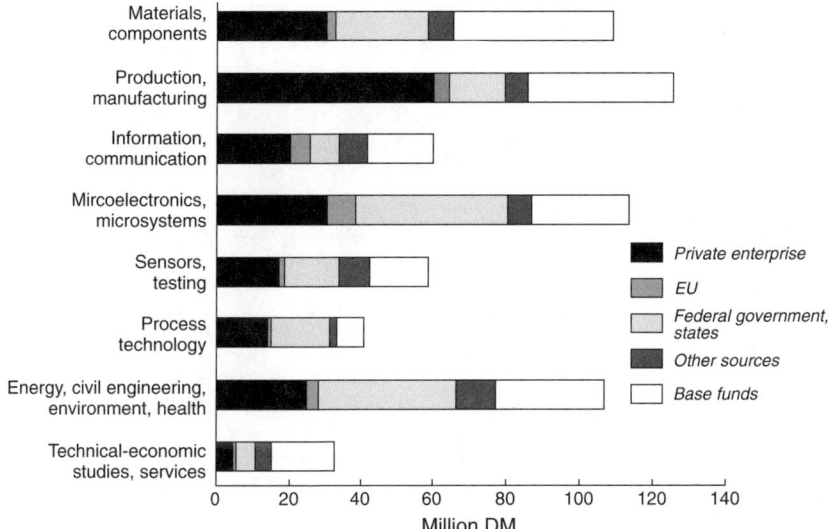

FIGURE 3.24 Budget structure of 30 consolidated Fraunhofer institutes in West Germany, by research area, in 1994. SOURCE: Fraunhofer-Gesellschaft (1994).

mechanical engineering (Figure 3.24). A second focus is microelectronics, in association with the related areas of information and communication and sensor technology. FhG activities cover application-oriented basic research (less than 5 percent of total expenditures), applied research, industrial product (process) engineering and prototyping (about 75 percent of expenditures), and technical and scientific services (about 20 percent of expenditures) (Imbusch and Buller, 1990; data for 1986). This special mixture of R&D types leads to a specific division of labor between the FhG and industrial enterprises (Figure 3.25). In this idealized scheme, small companies use the whole range of FhG activities up to prototyping, whereas large companies are interested primarily in more basic and long-term strategic research.

The average share of FhG industrial contracts varies greatly among institutes and technology areas. Figure 3.26 shows the major trends between 1989 and 1993. During this period, production technology received by far the strongest industrial support; 50 percent of the funding in this area came from industrial contracts. This corresponds to the traditionally close cooperation between industry and science in the field of mechanical engineering with Fraunhofer clients in important industrial sectors like the automobile industry. For material technologies, industrial support decreased from above average to average (around 30 percent over the 4 years). This may reflect economic difficulties in the German chemical industry and changes in R&D strategies (concentration on mid-term core competencies after a period of long-term diversification in R&D). The trends

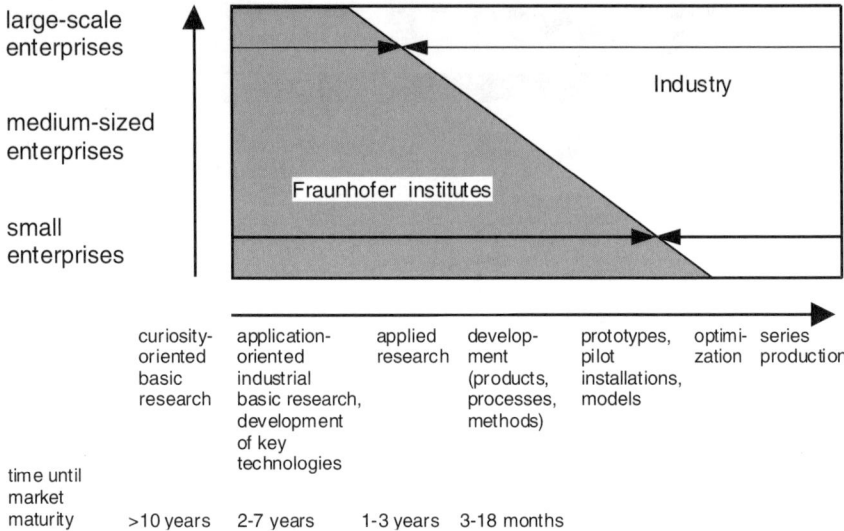

FIGURE 3.25 Typical division of labor between Fraunhofer institutes and industry. SOURCE: FhG-Zentralverwaltung (1995).

in sensor, process, and energy technology are relatively stable and the values are about average.

Trends in information and communication technology and microelectronics, two sectors characterized by a relatively weak industrial base, show perceptible changes. Whereas in information and communication technology the trend is significantly downward, possibly reflecting deep structural changes (decline in the information industry, privatization in the communication industry), the trend in microelectronics switched from a decrease to a significant increase after 1991. This may correspond to a strategic reorientation of Fraunhofer microelectronic institutes toward systems applications instead of devices in areas where U.S. and Japanese competition has grown.

TECHNOLOGY TRANSFER

In Germany, technology transfer is often seen to be either contract research or intermediary services of specialized transfer agencies (i.e., an institutional infrastructure added to R&D institutions like universities or national laboratories). Actually, the diversified sector of nonuniversity R&D institutes with its multiple levels of interaction with industry represents the major institutional framework for technology transfer. The FhG in itself can be regarded as an important transfer institution. It bridges the gap between basic research and industrial development, relying on a market-driven and demand-driven orientation to applied research.

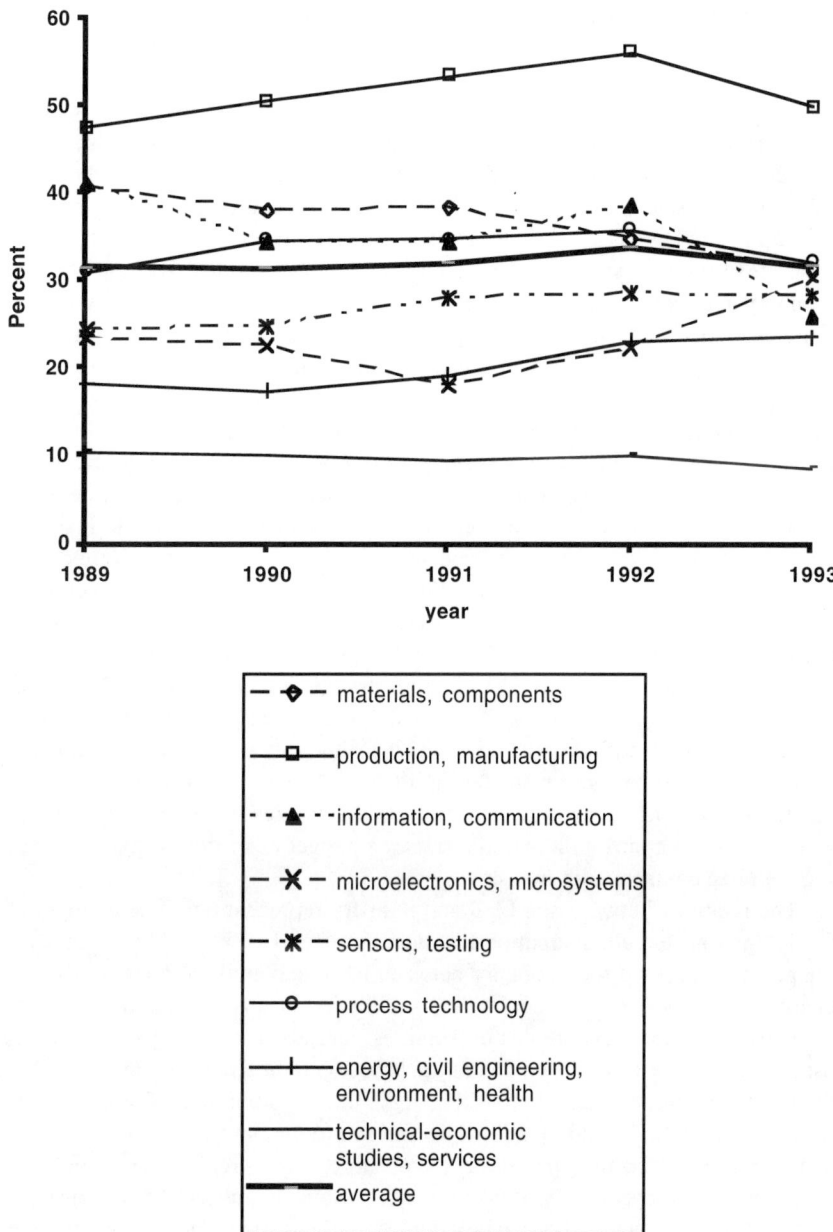

FIGURE 3.26 Share of FhG industrial contracts, according to research area. SOURCE: FhG-Zentralverwaltung (1995).

For the FhG, the most important channel for technology transfer is, as described above, contract research for industry. The Fraunhofer model assumes that contract research guarantees a research orientation geared to application. The targets of a research contract are defined by the sponsor; therefore, it can be assumed that the sponsor is highly interested in using the results for product (process) developments. The strong reliance of the FhG on industrial contracts implies that the research activities are closely related to market demand. The FhG philosophy implies taking the initiative in convincing potential sponsors of the relevance of research subjects. The acquisition of project funds gives the FhG the autonomy to allocate resources to particular research issues within the Fraunhofer institutes. This mechanism encourages in Fraunhofer scientists entrepreneurial behavior in terms of their strategic orientation toward future demand of the applied-research market.

The second major transfer channel of the FhG is contract research for public projects or programs related to government responsibilities like health care, environmental protection, energy and telecommunications infrastructures, defense, and so forth, as well as to German economic competitiveness in world markets. Public research programs that are relevant to industry focus on precompetitive research with the goal of improving national competitiveness in key technologies. Individual contract research projects allow for long-term, application-oriented research[31] with precompetitive prototypes as the typical transfer result. Public projects run collaboratively with industry are directly relevant for technology transfer.

Closely related to industrial contracts is technology transfer via consultancy or other services considered to be auxiliary. According to interviews with Franhofer researchers, the importance of these activities increases in relation to higher industrial contributions to the research budget of institutes; their purpose is to stabilize contacts with industry.

The relations between the FhG and industry represent only one element of technology transfer, albeit an important one. Another decisive step in the innovation process occurs at the boundary between basic and applied research. In this regard, the interaction between the FhG and universities is crucial. Most Fraunhofer institutes are located near universities, and about two-thirds have direct institutional connections based on contracts between the FhG and the university. The main element of such relationships is the joint appointment of a full professor as director of a Fraunhofer institute and to a university chair. The relevant faculty participate in the appointment procedure, but thereafter the Fraunhofer institute is run independently of the university. Some members of the faculty are elected to the institute's advisory board, thus getting full insight into its research activities. The knowledge transfer between the Fraunhofer institute and the university flows in both directions. At the university, the Fraunhofer director can carry out basic research funded by institutional funds of the university, and the director is in close contact with other academic researchers. At the same time, the university gets aquainted with the needs of applied research; the FhG director is a member of the faculty and can directly influence its research policy.

An important element of this close relation to universities is the direct access Fraunhofer institutes get to qualified students. This creates mobility of personnel, with more than 11 percent of the scientific staff annually moving from the FhG to other employers (Fraunhofer-Gesellschaft, 1993). Of the 11 percent, 41 percent join industry, thus accomplishing a process that begins when institute directors select qualified students for jobs that turn into regular employment at the institute after graduation from university. For doctoral theses, students are given the chance to participate in cutting-edge research with industrial applications. After 5 to 7 years, they may leave the FhG to start industrial careers. Many stay in contact with "their" FhG institute, thus stimulating further industrial cooperation. The level of personnel turnover is an indicator of successful technology transfer that is monitored continuously by the central administration of the FhG.

In addition to these formal means of technology transfer, the FhG also uses a variety of informal channels. For instance, the institute directors establish close contacts with industrial managers as well as with their academic colleagues. In addition, Fraunhofer scientists are expected to publish papers, attend conferences, and participate on academic and industrial committees. Through these activities, research results are disseminated to the technology and scientific communities, and at the same time, new scientific trends can be followed. These informal transfer activities are also a performance metric for the evaluation of an institute by the central administration. With regard to this kind of networking, the selection of members for the advisory boards of the institutes plays a decisive role.

A specific model of close cooperation with industry is the Microelectronics Alliance (*Mikroelektronikverbund*) of the FhG. This is an organizational union of the FhG's seven microelectronics institutes with a leadership composed of the directors of these institutes (Fraunhofer-Gesellschaft, 1988). In view of the often defensive position of the German and European microelectronics industry, the association was established to focus and coordinate the investment and research capacities within the FhG, and especially to coordinate its research orientation with the business policy of the German electronics industry and other research institutions. Cooperation with industry is organized by a special Technology Advisory Board (*Technologiebeirat*) in addition to the usual advisory boards of the institutes. The supporting ministries and the largest German electronics concerns are represented on the board. This institutionalized cooperation helps to concentrate the resource input for R&D according to the needs of industry, thus paving the way for future technology transfer. The Microelectronics Alliance represents the most direct form of industrial influence on the research policy of Fraunhofer institutes.

Within the chain of technology transfer, the present Fraunhofer model covers the range from basic research to prototyping. The final development of products or processes is left to the industrial partners. Some institute directors, however, see an increasing need to become involved even in this last stage. As a result, several institutes have established joint ventures with industrial partners or new technology-based firms more or less closely affiliated with the institute. Some of these new firms are spin-offs, run by former FhG researchers at their private entrepreneurial risk.[32] Since these initiatives are still young, it is not yet possible to evaluate whether these FhG-associated firms can become a standard element of the Fraunhofer model.

Recently, the FhG considered splitting in two, with a division of labor between the institutes and the innovation centers (*Innovationszentren*). The institutes would focus on applied research and keep their nonprofit status. The innovation centers would be associated with one or several Fraunhofer institutes, develop their results further to create industrial products, and introduce these products into the marketplace. The innovation centers would have a for-profit status and would be the basis for establishing spin-offs (FhG-Zentralverwaltung, 1995). Before the realization of this concept, a variety of administrative, financial, and legal problems have to be resolved. However, this approach seems to be a reasonable adaptation of the Fraunhofer model to the current needs of technology transfer.

The patent policy of the FhG is an important technology transfer tool. Institutes can decide whether patents are useful for their general contacts with industry. In most cases, inventions created within research projects are not given directly to industry but registered by the institute itself. An industrial partner generally gets an exclusive license, but only for the partner's special application; hence, the FhG is free to license the patented technology to another company for a different application. With more than 200 domestic patent applications in 1993, the FhG is among the most active patent assignees in Germany (Deutsches Patentamt, 1993).

The specialization of FhG patents may be distorted to a certain extent by the varying patent policy of the institutes. Overall, most focal areas are well represented (Figure 3.27). High index values in machine tools and handling (robotics) relate to production technology, as does the above-average value in optics (laser working). Other focuses are material technology (materials, surfaces) and microelectronics. The low index in data processing may be related to the fact that the research institutes involved have a strong software orientation and, according to the German and European patent laws, patent protection for software is limited. All in all, the Fraunhofer profile, to a certain extent, reproduces the general German profile (see Figure 3.3), because FhG activities must be close to the market demand. However, in several key areas such as semiconductors, optics, biotechnology, control, and materials, FhG patent activity is ahead of that industry.

MAJOR ELEMENTS OF THE FRAUNHOFER MODEL

The success of the Fraunhofer model, as reflected by steadily increasing budgets, is based on a variety of strategic elements, including the decentralized

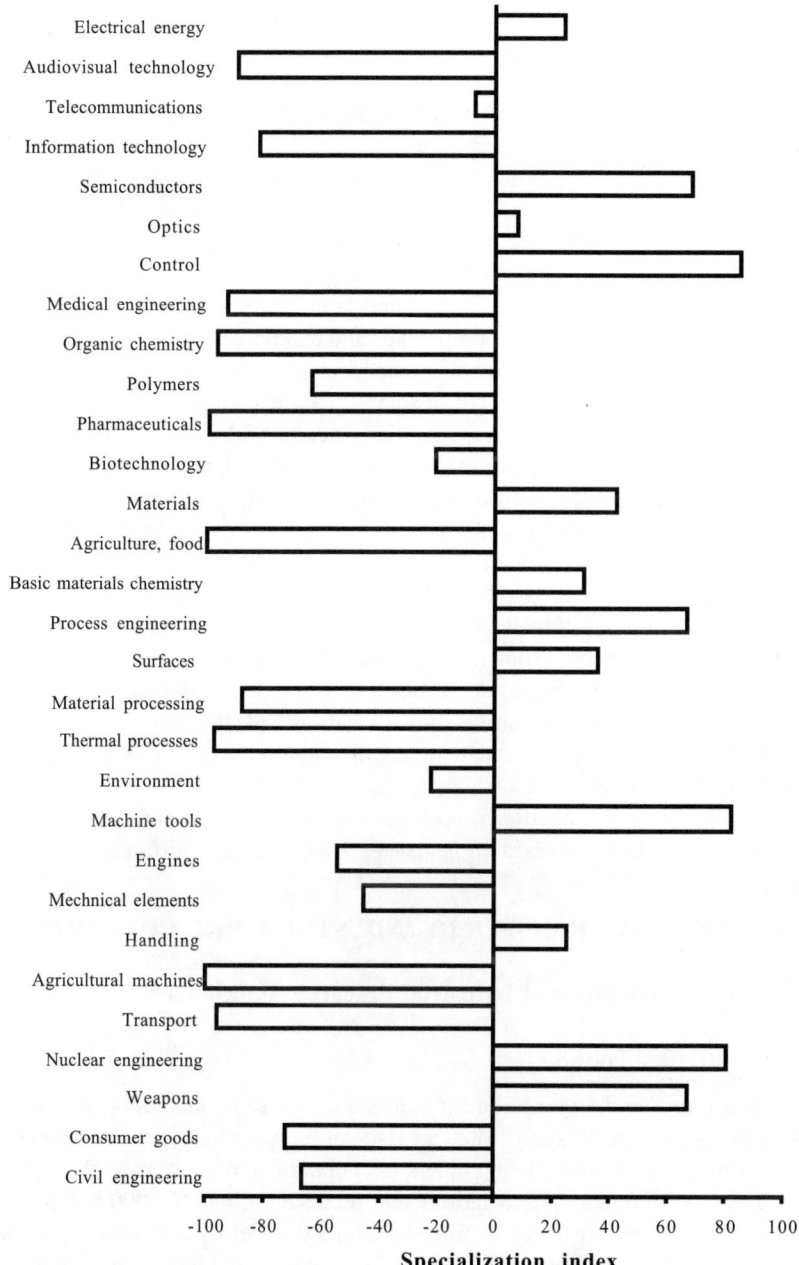

FIGURE 3.27 Specialization of German Patent Office patents held by the FhG in relation to the average distribution at the EPO for the period 1989 to 1992. SOURCE: The database PATDPA; Fraunhofer Institute for Systems and Innovation Research.

management and substantial autonomy of the institutes, which are prerequisites for flexible adaptation to the needs of the research market. Another element is the direct linkage of the level of institutional funding to success in contract research, which is a major incentive for market orientation and entrepreneurial behavior.

Furthermore, the Fraunhofer model builds on a balanced mixture of the three sources of support: institutional funding, public projects, and private contracts. On the one hand, a higher share of institutional funding would imply a decreasing interest of the institutes in industrial contracts, and thus a diminished orientation toward industrial needs. On the other hand, a considerable decrease of public funding would reduce scientific competence and call the institutes' transfer function into question. The institutional linkage to universities is another vital element in maintaining a high standard of scientific competence.

In the German debate on research policy, success with industrial contracts is often seen as the defining feature of the Fraunhofer model, and the close linkage to science is overlooked. Both elements, however, are important to guarantee effective technology transfer in the long run (see also Meyer-Krahmer, 1996). Therefore, managing the balance between scientific and technological competence is a major challenge for the FhG, which is met by regular control of all elements of technology transfer for each institute. In the present situation of scarce public funds, major problems could arise from further reduction of institutional funds, and public projects. At the same time, other public or semipublic research institutions such as universities or national research centers might be urged to carry out more contract research, which could lead to a growing competition for industrial funds. In such a situation, the FhG would not become obsolete because of its high specific competencies in many areas of applied research, but its role in the German research landscape would be quite different.

TECHNOLOGY TRANSFER BY INDUSTRIAL R&D CONSORTIA

Federation of Industrial Research Associations

STRUCTURE AND TASKS

It is a long-standing tradition for German SMEs to be linked together in an industrial research association. Such an association functions as the backbone of a special industry, enabling its members to cooperate and coordinate joint interests. Today, 106 research associations that represent around 50,000 companies and own 68 collective research institutes are joined under the umbrella organization of the AiF (*Arbeitsgemeinschaft industrieller Forschungsvereinigungen "Otto von Guericke"*). The AiF is a product of 1950s West Germany: a reemerging landscape of research units at universities, independent institutions, and industrial research associations, focusing on industrially applicable research. The

Ministry of Economics, which was very interested in efficient and applicable research, decided that a coordinating agency was needed. In 1954, private industry set up the AiF in order to fill this gap.

The federal government was interested in channeling public funds through one efficient, mediating umbrella organization. Therefore, in 1954, the AiF obtained responsibility for 20 industrial research associations. Industrial cooperative research has always been the core competence of the AiF, but the federation has broadened its authority by obtaining jurisdiction over the administration of the funds of three of the most important government programs for the promotion of R&D in SMEs, which include contract R&D, R&D subsidies, and subsidies for R&D personnel. The major aims of the AiF are as follows (Geimer and Geimer, 1981):

- to finance cooperative research projects originating in the member associations;
- to coordinate research projects;
- to promote personnel transfer between its members;
- to support its member associations in obtaining public funds;
- to advise on the establishment of new research associations in industry;
- to represent the members' general interests; and
- to act as a link between the members and the public administration.

AiF currently has 106 member associations that vary in size and structure. The AiF distinguishes three categories of members: type A comprises associations with individual companies as members; type B unites industrial organizations; and type A/B combines individual companies and industrial organizations. About 46 percent of member associations are type A; 41 percent are type B; and the remaining 13 percent are type A/B.

The different industrial associations comprise a broad range of industries and extend from textiles to mining and energy. One example of a member association is presented in Table 3.22.

ORGANIZATION OF COOPERATIVE RESEARCH

About 90 percent of all member companies of the AiF are considered small or medium sized; just 10 percent of the firms operate on a large scale (i.e., with 1,000 or more employees). With this structure, a special approach to technology transfer is needed, as SMEs are considered to have certain deficits in their capacity to finance research activities. An appropriate solution is cooperative research activities (*Gemeinschaftsforschung*), which are collectively initiated by several companies in an industrial research association, but conducted by separate research institutes. In this report, the term "cooperative research" is used primarily in the context of industrial research associations. It should not be confused with

TABLE 3.22 Structure of the Food and Beverages Sector and its Member Research Associations

Association (type)	Number of Companies (SME[a]/LE[b])	Research Institute(s)
Brewing (type A)	332 / 9	Experimental and Academic School of Brewing, at the Institute for Fermentation Industry and Biotechnology, Berlin.
Breweries (type B)	994 / 6	No special research institute; cooperation with universities teaching brewing.
Food-producing industry (type A/B)	3,836 / 313	Six research institutes of its own; cooperation with federal research institutes as well as with institutes at universities and polytechnical schools.
Yeast-producing industry (type A)	10 / 0	Research institute for baking yeast at the Institute for Fermentation Industry and Biotechnology, Berlin.
Plant growing (type A)	60 / 0	No special research institute.
Manufacturing of spirits (type A)	461 / 0	Experimental and Academic School of the Fabrication of Spirits, at the Institute for Fermentation Industry and Biotechnology, Berlin.

[a]SME = companies with fewer than 1,000 employees.
[b]LE = large-scale enterprises (1,000 or more employees).
SOURCE: Arbeitsgemeinschaft industrieller Forschungsvereinigungen (1991).

"collaborative research," in which different institutions collectively perform research within the context of joint projects.

Obviously, several SMEs competing in the same market will not want to work together on research applicable to their competitive position. Therefore, cooperative research projects are "strictly precompetitive" (Schiele, 1993) and the results that come out of such research leave room for individual companies to adapt the findings to their special needs. Typical areas of cooperative research are setting technical rules for safety purposes and standardization to reduce production costs. A variety of projects concern environmental problems, such as the elaboration of ecologically optimized products and processes. As all research topics are initiated by the companies, the former often emerge from daily business on the companies' shop floors; these are so-called bottom-up problems (Schiele, 1993). In this sense, industrial cooperative research is complementary to the top-down approach of research institutions like the MPG and of many programs of the BMBF.

Cooperative research is usually initiated by the SMEs themselves without any outside influence (Schiele, 1993). Next, the firms inform their research association of their particular research problem. The advisory board of the association discusses whether the problem is interesting for other members of the corresponding branch and whether it will be possible to define a suitable research project. In this first step of evaluation, about 50 percent of proposed research ideas are rejected. This first phase of project definition at the research association level must be viewed as crucial, since it is a collective, interactive process for determining common problems. The definition of appropriate problems is a major prerequisite for the success of the related research projects. In this regard, the projects of the industrial research associations are truly cooperative.

When a project is approved, an expert group plans its contents and its costs in detail and chooses an appropriate institute to implement it. In 1990, about 53 percent of the 1,102 projects were performed in one of the 68 institutes belonging to the research associations themselves; 43 percent were carried out by university institutes; and the remaining 6 percent were done by other public or private research institutions. Since then, the percentage undertaken by other institutions, primarily independent research institutes, has increased considerably, reaching 24 percent.

The individual research associations have to decide whether a project can be carried out exclusively with internal funds, or whether it requires public support. Getting public funding can be a rather difficult and time-consuming procedure. Only if the research association decides to seek public support will the AiF be involved. In this case, the member association submits an application to the AiF, where the project is again evaluated by an expert group. There are currently 8 groups with 140 experts from science and industry involved in this evaluation procedure. In this second evaluation, another 26 percent of the submitted applications are rejected. Next, the Authorizing Committee (*Bewilligungsausschuß*) of the AiF decides whether or not the accepted projects should be recommended to the German Ministry of Economics for public support. The whole evaluation procedure is illustrated in Figure 3.28.

If the project is recommended to the German Ministry of Economics, it gets complete financing from the ministry. The public funds are allocated to the AiF, which administers them. To receive public support, the particular research association must match the support it receives from the government by investing at least the same amount of its own resources in other cooperative research projects. During past years, funding by AiF member associations on average amounted to about twice as much as the public support.

To sum up, cooperative research projects can be financed either by industry, at the level of industrial research associations, or publicly, by the Federal Ministry of Economics. When a project is publicly funded, the AiF plays a decisive role in the evaluation of project proposals and in project administration. In this context, it has to be emphasized that the AiF is completely financed by its member associations and not by public funds.

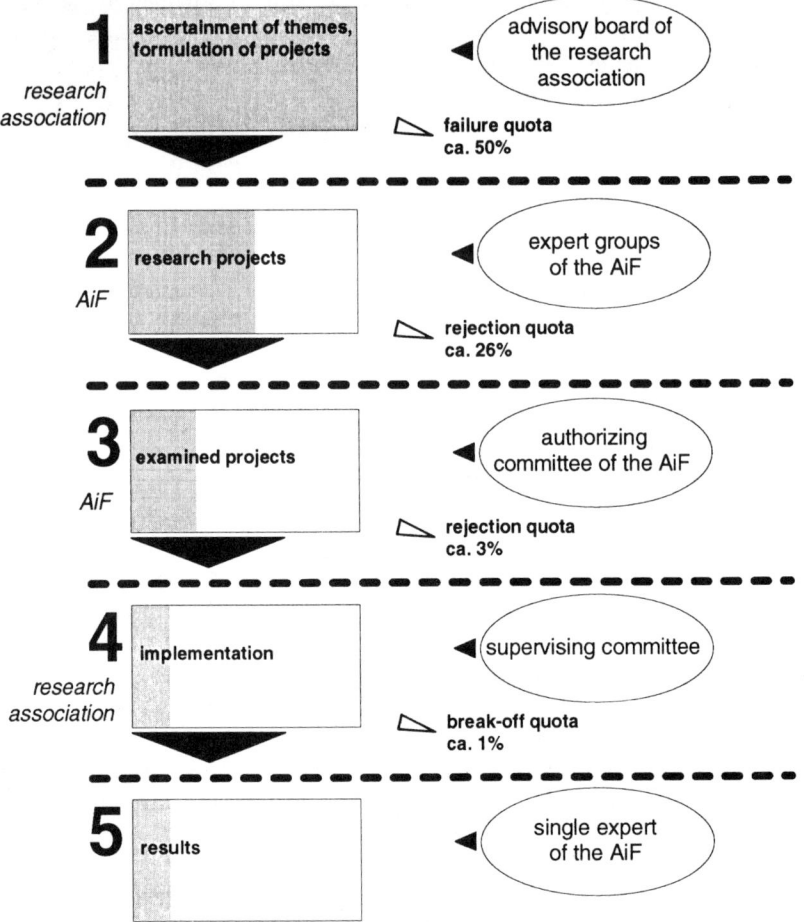

FIGURE 3.28 Evaluation steps for publicly funded projects involving industrial cooperative research. SOURCE: Schiele (1993).

BUDGET AND ORIENTATION

The history of the AiF until 1961 was marked by the financial aid of the European Recovery Program. From the time AiF was established, it was evident that the financing of cooperative research should be shared by the government, the participating companies, and the member associations. During the early years of the AiF, public funding exceeded contributions by AiF members, but a constant increase in membership brought in new financial resources. Public funding today serves as an incentive and allows the development of "riskier" research projects. In 1975, a remarkable decision enabled member associations to apply on behalf of single companies, even small firms without the required capital, in order to obtain research funds. This decision has resulted in the promotion of many small companies that otherwise have almost no chance of receiving public funds. The financial strength of their association not only serves as a guarantee but also contributes to the monetary support.

Figure 3.29 shows the development of public and private industrial funds for cooperative research. The striking increase of public support in 1991 and 1992 was due to a special effort made for SMEs in the former East Germany. At the same time, industrial funds increased as well, so that in 1993, they were again about twice as plentiful as public funds, just as was true in the 1980s. In 1991, public and private expenditures for cooperative research amounted to DM 470 million, or about 1 percent of total R&D spending in industry. This share seems to be quite modest. However, one should bear in mind that about 60 percent of industrial R&D is performed by very large companies with more than 10,000 employees. Only about 17 percent is carried out by SMEs, and SMEs are the target group of cooperative research (see SV-Wissenschaftsstatistik, 1994, 1996). About 6 percent of research conducted in SMEs is cooperative research within research associations.

The situation in different industrial sectors varies considerably. In the textile sector, the share of cooperative research represents about 42 percent of the total R&D expenditure of industry, whereas in electrical engineering it is only 0.1 percent (Table 3.23). Cooperative research is in general more important for indus-

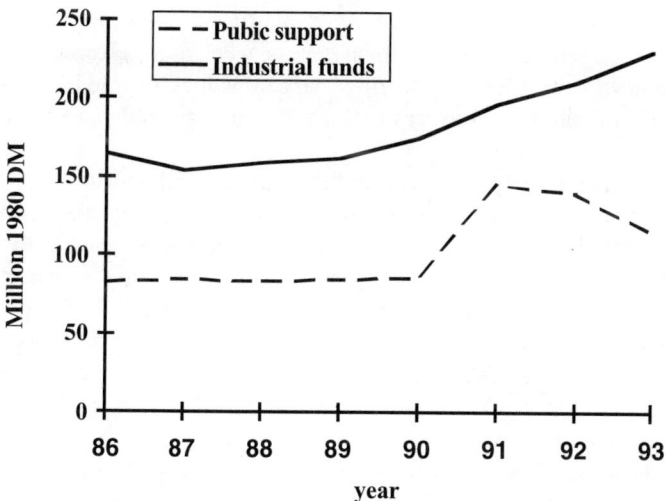

FIGURE 3.29 Public and industrial funds for cooperative research, 1986–1993, in constant 1980 DM. SOURCE: AiF-Verwaltung (1995).

TABLE 3.23 Importance of Cooperative Research in Different Industry Sectors in Germany, 1989

Sector	Sector's Total R&D Expenditure as Percent of Turnover	Sector's Percent Share of R&D Expenditures on Industrial Cooperative Research	Percent Turnover Represented by Sector's 10 Largest Companies
Aeronautics, aerospace	30.9	0.0	95
Electrical engineering	9.3	0.1	30
Chemical industry	6.3	0.1	30
Vehicle industry	4.1	0.0	74
Mechanical industry	3.6	1.4	12
Materials for construction	2.0	5.0	52
Wood, paper, printing	1.2	8.3	33
Textiles	1.0	42.2	12
Nutrition	0.7	7.0	11
Iron manufacturing	0.6	22.0	75

SOURCE: Schiele (1993).

trial sectors with a low R&D intensity. Sectors with a low level of concentration (indicated in Figure 3.30 by the share of turnover, or sales, of the 10 biggest companies) and thus a high share of SMEs also often rely to a great extent on cooperative research. However, in the highly concentrated sector of iron manufacturing, the impact of cooperative research is significant. All in all, cooperative research is focused primarily on traditional sectors, whereas in research-intensive areas like chemistry, electrical engineering, and aeronautics, its role is negligible. The distribution of public and private funds in different industrial sectors is shown in Figure 3.30.

As to the focal areas of this study, production and manufacturing are broadly represented in the activities of different research associations. For these kinds of industries, cooperative research seems to be a quite appropriate means of technology transfer. Biotechnology is not only an area of cooperative research unto itself but plays an increasing role in projects of the food industry. Cooperative research among microelectronics and software firms is rare, as the electrotechnical sector has traditionally not established relevant research associations. Cooperative projects in the two fields do occur, however, in the context of process and production automation.

An examination of research reports resulting from cooperative projects (e.g., Arbeitsgemeinschaft industrieller Forschungsvereinigungen, 1992, 1994), reveals that most such projects are highly application oriented. The project "Shortening of the Ripening Period for Hard Cheese" is a typical example. Conducted at a

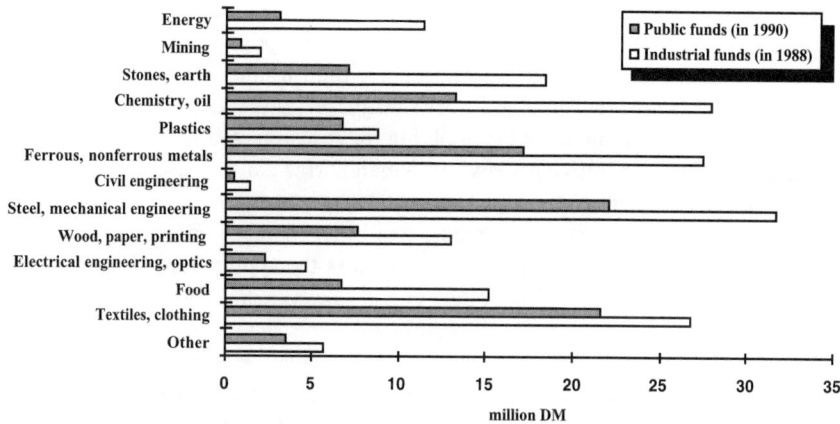

FIGURE 3.30 Volume of public funds and industrial funds spent on cooperative research. SOURCE: Böttger (1993).

research laboratory in the former East Germany, financed by the Ministry of Economics, and initiated by the Research Circle of the Food Industry (*Forschungskreis der Ernährungsindustrie e.V.*, Bonn), the project yielded the finding that adding cell-free extracts of lactobacillus to milk accelerates substantially proteolysis and therefore reduces the period needed to ripen cheese. This reduces the requirements for energy and space, and since adding the extracts actually improves the flavor of the cheese, it increases the quality of cheese.

Another example from the food and beverage industry is a project initiated by the Science Promotion of the German Breweries (*Wissenschaftsförderung der deutschen Brauwirtschaft e.V.*, Bonn) and carried out at the Institute for Brewery Technology at the Technical University of Munich. Funded by public sources, "Research on the Production of Volatile Aroma Essences for the Aging of Bottled Beer and the Possibilities of Technological Influence" resulted in techniques for the optimization of the brewing process. The development of an objective sensory evaluation technique made it possible to establish criteria for the stability of flavorings.

TECHNOLOGY TRANSFER

In the framework of industrial cooperative research, we have to distinguish between two different target groups for technology transfer. The first group consists of the enterprises that initiate projects and are directly involved in their definition and application. These firms are also closely involved in supervising the project's implementation, so that an intensive process of technology transfer takes place. The phase of project definition, especially, implies detailed consideration of the research problem, interaction with other companies, and discussions with research institutions; the submission of a qualified proposal usually requires an effort of several months (Arbeitsgemeinschaft industrieller Forschungsvereinigungen, 1992). Since the involved SMEs define the projects according to their special needs, they are interested in applying the research results. With regard to this target group, cooperative research is demand driven, a situation that is favorable for successful technology transfer.

The second target group is composed of other member companies of a research association that are not directly involved in the application and execution of a research project. According to the general rules of cooperative research, the research association has to make the results of its projects available to all of its members. The major means of disseminating information is the documentation of research results in regular journals of the association and research reports. In addition, meetings, seminars, demonstrations in companies, exhibitions, and similar transfer activities are organized. In many cases, the research results lead to the introduction or change of norms and standards (Arbeitsgemeinschaft industrieller Forschungsvereinigungen, 1992). The technology transfer to this second group is, of course, less efficient than it is to the first group. Companies have to become aware of the existence of the research results and must determine whether or not these results may be useful for their purposes. In the case of large associations, the activities of the member companies often vary considerably, so that the results of a research project are useful for only a limited group of members.

According to a recent study by Lageman et al. (1995), about 20 percent of member companies actively use the results of cooperative research. Of the companies that are involved in committees of their research association, almost 90 percent use the results of cooperative research. This latter subset of companies is largely equivalent to the first target group described above. Furthermore, the type of membership in the research association is important for the efficiency of technology transfer. If the companies are direct members, they frequently use the research results. If they have only indirect membership through an industrial organization (research association type B, see description above), about 75 percent never use research results and one-third do not even know that the possibility of cooperative research exists.

To sum up, technology transfer is very efficient for the group of companies that are directly involved in the definition and execution of projects. As to the rest, the research associations and the AiF undertake various activities to support technology transfer, but the effect is limited. Lageman et al. (1995) suggested a variety of measures for improving technology transfer to this second target group, such as establishing on-line databases and preparing more user-friendly documentation of research results. Probably the most important step for facilitating effective technology transfer is for associations to identify contact persons within their member companies to establish regular and stable communication structures. Despite all these improvements, the effect on the second target group will always be limited, because from their perspective, cooperative research has a supply-oriented character, which is a generally unfavorable precondition for technology transfer.

All in all, cooperative research can be a useful instrument of technology transfer for SMEs, and it has especially positive effects for the actively participating firms. Therefore, Lageman et al. (1995) judge its effect to be largely positive.

TECHNOLOGY TRANSFER IN SELECTED AREAS

Technology Transfer in Information Technology

The German delegation believes that German educational and scientific standards in information technology are excellent and internationally competitive. This assessment is confirmed by an above-average level of scientific publications as well as a high citation level. Furthermore, scientific publications in this area reveal a strong international orientation (Grupp et al., 1995).

In contrast, German information technology, comprising computer as well as software technology, is, with the exception of some internationally competitive companies such as Siemens and SAP, characterized by a low number of industrial technology providers. (In telecommunications, the German position is much stronger.) The German software industry is weak in prepackaged high-volume products; most such standardized software comes from the United States. German enterprises have strengths in information technology applications in all kinds of industries and in customized, value-added products, generally produced in low volume series.

A large percentage of NTBFs—25 to 30 percent of about 300 created annually in Germany (see "Conditions for New Technology-Based Firms," above)—are working in information technology (Kulicke and Wupperfeld, 1996). But only about 20 percent of the founders of NTBFs come from universities or other research institutions, so that the spin-out from scientific institutions is limited.

Due to the lack of industrial counterparts, the activities of German research institutions are oriented primarily toward projects and less toward products, a situation that is unfavorable for technology transfer. Conducting projects requires a stable institutional framework, so larger institutions dominate technology transfer.

One Helmholtz Center *(Gesellschaft für Mathematik und Datenverarbeitung,* with about 1,200 employees) conducts primarily basic research, mathematics, and computer science, and the Max Planck Society recently established an institute for computer science aimed at basic research. On the applied side, seven Fraunhofer institutes are partly or completely active in various areas of information technology such as software engineering, industrial automation, and business organization. Two of them were established only recently. Within these institutes, about 600 employees work in information technology. Also participating in the area of applied research are about five An-Institutes and some regional or state research centers for which statistics are not available. The An-Institutes generally focus sharply on a special application field such as industrial automation or office automation.

Information technology research at universities covers all types of R&D from basic research to experimental development (see "Results of a German Survey," above). Some 30 universities run programs in computer science, with about 150 to 200 mostly small institutes involved. Thus, universities contribute substantially to research in information technology. In the applied research market, there is a strong competition between universities, Fraunhofer institutes, An-Institutes, and regional/state institutes, leading to a high quality of research.

Even though the demand of information technology producers for scientific support is limited, there is a distinct need for the integration of information technology into other products, especially in mechanical engineering, chemical plants, and consumer goods. For this reason, many research institutions pursue an application- and system-oriented strategy.

The main channels of technology transfer are contract research for industry; collaborative research with firms participating in the European programs framework; and, first of all, collaborative research with projects on behalf of the BMBF.

Technology Transfer in Microelectronics

The situation of microelectronics, like software, is characterized by relatively little industrial activity apart from some long-established, internationally competitive companies like Siemens, TEMIC, Bosch, and the German subsidiaries of IBM, ITT Semiconductors, Philips, and Texas Instruments. Whereas the German market represents 7 percent of the world's demand for microelectronic devices, only 2 percent of the demand is met by German producers (see also "Orientation of Industrial R&D," above). Against this background, many German companies specialize in microelectronic applications for a wide variety of industrial branches, a situation that has a strong impact on the orientation of research institutions, too.

Of about 300 NTBFs created annually in Germany (an already low number), only about 10 percent are linked to microelectronic devices or applications (Kulicke and Wupperfeld, 1996), so technology transfer through this channel is modest.

The most important channel of technology transfer from universities is the provision of highly qualified personnel for industry. Each year in microelectronics, about 50 engineers with Ph.D.'s and 200 with diploma degrees move from universities to industry. At present, there are about 40 university research centers in microelectronics, many of which focus on microelectronic applications and microsystems (e.g., the integration of mechanics and electronics—so-called mechatronics). Many university professors come from industry, and many master's and Ph.D. theses are done in cooperation with industrial companies, mostly those by students in German engineering schools.

The major nonindustrial institution in microelectronics is the Fraunhofer Society, which supports eight institutes in microelectronics and microsystems. Six of these cooperate under the umbrella of the Microelectronics Alliance, coordinating research within the Fraunhofer Society with the needs of industry. The Microelectronics Alliance has about 700 employees with regular contracts and about 300 with short-term contracts.

Among the Helmholtz Centers, only the Jülich Research Center carries out significant amounts of—mostly basic—research on microelectronics. This situation applies also to the Max Planck Society and its MPI for Solid-State Research in Stuttgart. The Helmholtz Center in Karlsruhe conducts various activities in the closely associated field of microsystems. In addition to universities, Fraunhofer institutes, and Helmholtz Centers, some Blue List institutes, several An-Institutes, as well as a few state-run institutes are relevant actors in the field. The Association of German Electrical Engineers published a list of 17 institutes with a definite orientation toward technology transfer (VDE/VDI, 1994), including the Fraunhofer institutes, five university or An-Institutes, three state-run institutes, and four institutes run by industrial associations.

Remarkably, these application-oriented institutes have enlarged their scope of activity. Whereas previously their goal was to develop prototypes, they recently have started to include pilot production, too. Another decisive trend of the last 10 years is the increasing integration of European research in microelectronics through programs of the European Commission and the EUREKA initiative, especially JESSI. Although these European activities have shortcomings, in particular an enormous bureaucratic overhead, the German delegation acknowledges they have caused considerable change in the consciousness and international orientation of German researchers. Within this context, the German contribution is quite important, because it represents about 30 percent of all European activities in microelectronics.

All in all, the analysis of microelectronics shows various specific mechanisms, like the appointment of professors from industry and the coordination within the Microelectronics Alliance of the Fraunhofer Society. Nevertheless, institutionalized forms of technology transfer, especially contract and cooperative research, dominate.

Technology Transfer in Biotechnology

For the purpose of this study, we defined biotechnology broadly to include genetic engineering, cell cultures, and microbiology. In more general terms, biotechnology was defined as the use of living organisms or parts thereof for the production, modification, or the decomposition of substances, or the modification of organisms; or services such as analytical services.[33] At present, about 400 small and medium-sized enterprises and 30 large companies are engaged in the area of biotechnology. This estimate also includes firms in mechanical and process engineering as well as distribution and service companies (Reiss and Hüsing, 1992). NTBFs working in biotechnology play a negligible role; they represent barely 10 percent of the roughly 300 NTBFs created per year (Kulicke and Wupperfeld, 1996).

In the important application area of pharmaceutical products, the market is dominated by a dozen multinational, German-based concerns primarily in the chemical and pharmaceutical industry. (The description in this paragraph is largely drawn from Dolata, 1995.) For many decades, these firms have been very successful in pharmaceuticals and produce about 40 percent of the world's exports in this sector. Their products are primarily based on research in organic chemistry. Because of their undeniable success in traditional areas, pharmaceutical companies almost ignored the potential of genetic engineering for many years, although many German experts were already emphasizing its importance in the 1970s. Therefore, the first initiatives toward industrial applications for biotechnology came from the government. This general assessment does not apply to all companies; firms like Boehringer Mannheim and Bayer have an internationally competitive position. Furthermore, many companies broadly use "traditional" processes with cell cultures, microorganisms, or enzymes. In some application areas outside pharamaceuticals, such as environmental technology and analytic kits, German industry's performance is very good.

In contrast to this generally slow response of industry, German scientific performance has always been good (Eichborn, 1985; Kircher, 1993). This is also reflected in the number of publications and their frequency of citation (Grupp et al., 1995). In addition, in the early 1970s, German scientists contributed many discoveries, new methods, and processes to the world's knowledge of biotechnology. Many German scientists went to the United States to establish spin-off companies.

Today, the research landscape in biotechnology is quite diverse. About 30 percent of the staff of the Max Planck Society works in its biological-medical section with a strong focus on biotechnology. In addition, four Helmholtz Centers are partly or fully engaged in the biotechnological area (*Stiftung Deutsches Krebsforschungzentrum, Stiftung Max-Delbrück für Molekulare Medizin, Gesellschaft für Biotechnologische Forschung, Forschungszentrum Jülich*). So are several Blue List institutes and some departmental research institutes, especially those of the Federal Ministry of Agriculture and the Ministry of Health. Of course, universities also play an important role in biotechnology. The university survey turned up more than 200 institutes involved in biotechnology. The Ministry of Education and Research initiated the establishment of eight gene centers (*Genzentren*) or biotechnology centers where universities and Max Planck institutes cooperate (see also "Universities, Transfer Channels," above). The activities of the Fraunhofer Society in biotechnology are still limited; three institutes have partial involvement in this field.

Since about the middle of the 1980s, the large pharmaceutical companies began to acknowledge the potential of biotechnology and started a catch-up strategy. It is based on cooperation with external scientific institutions, the building up of internal research capacities, and the acquisition of SMEs abroad (Dolata, 1995). The cooperation with scientific institutions has a strong focus on American partners, so the industrial funds for German university centers in biotechnology are still low, as the results of the university survey show (see Table 3.4). Many experts believe that the stringent regulations affecting biotechnology and the debate on patentability of biotechnological products have been major reasons for this external orientation of the German firms. However, Hohmeyer et al. (1994) found recently that the legal approval procedures for biotechnology projects and facilities in foreign countries, especially America, are comparable to those in Germany and are no longer a decisive factor. Nevertheless, German regulatory and beaueaucratic requirements for the aproval and operation of facilites are considerable. The problems associated with the aforementioned regulatory requirements led to a new law on gene technology and encouraged the states to make the approval processes for biotechnology projects and facilities less bureaucratic and thus shorter. Furthermore, most biotechnology processes and products are now patentable in Europe (Knorr et al., 1996; Schmoch et al., 1992). Finally, the orientation of most German companies toward America and increasingly Japan has to be seen in the context of a general move toward internationalization.

All in all, the level of technology transfer between industry and research institutions is still quite low, but in the future, the intensity will increase. A recent study on the potential of contract research in biotechnology revealed a growing need on the side of industry, especially of SMEs (Reiss and Hüsing, 1992). In the 1990s, some states started initiatives for supporting biotechnology, primarily for attracting start-up companies and already-existing SMEs with a new focus on biotechnology. Recently, the BMBF started a new initiative, "Bioregio," to identify sites with a high transfer potential in biotechnology. These various initiatives are too recent to give a reliable assessment of their actual impact.

Technology Transfer in Production Technology

The international competitiveness of German industry in production and manufacturing technology is very high, not only in the specific field of machine tools, but also in related areas such as material processing and handling (see also "Orientation of Industrial R&D," above). The knowledge in this area is primarily generated by three sources:

- in-house company research
- university laboratories
- Fraunhofer institutes

The situation in industry is characterized by a high number of SMEs. In machine tools, about 94 percent of the firms are SMEs. This situation implies a special need for cooperation with external partners. One way to cope with this problem is the self-organization of industrial research associations under the umbrella of the AiF. The share of cooperative AiF-related research (e.g., in metal manufacturing and construction materials) is distinctly above average (see Table 3.23). For the success of AiF projects in this area, it is important that not only SMEs but also large companies are members of the association. For example, in the automotive industry, SMEs are suppliers for large companies; therefore, the large companies play a decisive role in the definition of production standards. Cooperative projects of the AiF are carried out not only by the institutes of the member associations, but also by external institutes such as university and Fraunhofer institutes. In addition, the diffusion of technology is supported by a dense network of supporting institutions such as the Chambers of Industry and Commerce, industrial associations outside the AiF, the *Rationalisierungs-Kuratorium der Deutschen Wirtschaft,* and the transfer centers of the Ministry for Economic Affairs.

The survey of German universities showed a high intensity of university-industry relations in production technology (see Table 3.4). At present, there are 30 university institutes in the area of machine tools. If the definition of production technology is broadened (e.g., to include materials processing, handling, metal shaping, and assembling), the number of institutes is in the range of 150 to 200. There is frequent cooperation among different institutes in special interdisciplinary research areas (*Sonderforschungsbereiche*) initiated by the German Research Association.

The Fraunhofer Society has a special focus on production technology, with 12 institutes and a staff of about 1,500 people active in the area. The institutes cover all fields of production technology, from machine tools to production management.

A specific problem for German industry in production technology has been the integration of microelectronic devices, because German suppliers adjusted quite late to the growing demand, and German users had to buy foreign products. However, the situation has improved considerably in recent years.

Overall, production technology is a good example of long-standing, close relations between industry and research institutions. In addition, technology transfer in this mature industry is supported effectively by various institutions and associations, based on industrial self-organization.

CONCLUSION:
AN ASSESSMENT OF TECHNOLOGY TRANSFER IN GERMANY

This analysis of the mechanisms and institutions of technology transfer, the examination of the four focal areas, the comparison of the U.S. and German structures, and the assessment of the German panel members allow for some general conclusions.

Because technology transfer is an exchange of technological, technical, and organizational know-how between partners, effective technology transfer needs the particiation of both research institutions and industry. The German companies cover a broad spectrum of technology areas and have both strengths and weaknesses. The four focal areas selected for this study emphasize the weaknesses; had other areas been selected (e.g., telecommunications, transport, or environmental technology), the picture would have been more positive. However, as a general assessment of the *industrial environment,* the weaknesses often concern emerging, future-oriented fields; and in these fields, new research results are insufficiently used for industrial development. In contrast, the diffusion of technical knowledge in mature areas such as production technology is effectively supported by a close network of institutions and associations based on industrial self-organization. The relationships of small and medium-sized enterprises (SMEs) to research institutes are already quite close, but there is still great potential for including more SMEs in technology transfer.

As to the *research environment,* Germany has a broad and diverse system of research institutions. In particular, many university institutes, An-Institutes, state-run institutes, and Fraunhofer institutes focus on applied, industry-oriented activities. German science is internationally competitive, also in newly emerging fields. Unfortunately, it is precisely in these fields that the industrial base is often small, so that scientific institutions sometimes have difficulty in finding appropriate industrial partners. Nor is the cultural environment favorable for research institutions to produce spin-out companies.

Technology transfer in Germany is primarily institutionalized rather than personalized. Its main channels are contract and cooperative research supported by other means such as conferences and informal meetings. Bridging institutions like the Fraunhofer and An-Institutes play a decisive role in technology transfer. Many professors in engineering departments of universities come from industry, which implies a flow of knowledge from industry to university and, later on, close university-industry relations. These relations are documented, among other means of technology transfer, by a high number of master's and Ph.D. theses done in industrial enterprises or in cooperation with companies. In polytechnical schools, the preparation of a thesis in industry is compulsory. The appointment of professors from industry and the preparation of theses in cooperation with industrial enterprises are effective instruments of technology transfer that are not a matter of course in other countries.

Some transfer channels are presently of low or medium importance, but they will play an increasing role in the activities of nonindustrial R&D institutions. These channels comprise the presentation of research results, opportunities at trade fairs, the organization of seminars for industrial researchers, the establishment of sponsor organizations, regional incubator centers, and research parks.

The *cultural environment* in Germany is characterized by a limited entrepreneurial spirit. This situation is due to a low-risk mentality on the individual and societal levels. A visible consequence is the high number of regulations in all areas of entrepreneurial activities, imposing additional costs and a cumbersome bureaucracy. Furthermore, financial incentives for entrepreneurs are low, primarily because of the lack of R&D investment tax credits, as well as unfavorable asset-based financing and revenue taxation. Lack of private venture capital and public offering opportunities results in fewer incentives for new technology-based firms in emerging fields. However, as the comparison with the United States shows, this channel of technology transfer is very effective. In addition, in Germany, the environment for professional mobility is unfavorable, so technology transfer through the movement of individual researchers is less significant than it is in the United States. In particular, personnel exchanges between research institutions and industry are restricted by an inflexible regulatory framework for public institutions. A further problem within the cultural environment is the public's low acceptance of some new technologies (e. g., genetic engineering).

To summarize, many instruments of technology transfer work well. However, a more risk-taking, dynamic spirit is necessary, in particular in emerging fields of technology, if Germany is to maintain its international technological competitiveness.

ANNEX III

Examples of Technology Transfer in Germany

GTS-GRAL: TECHNOLOGY TRANSFER FROM UNIVERSITY TO A NEW TECHNOLOGY-BASED FIRM

Starting in 1976, technology was developed to design a device-, computer-, and application-independent graphics interface. The interface was to allow for reusable graphics applications and to establish a common, state-of-the-art graphics programming technology. This development was initiated by Prof. Encarnação, head of the graphics institute at the Technical University of Darmstadt. He made this an international effort by involving the national standardization bodies of various major countries and the International Standardization Organization (ISO). These standardization activities resulted in a first graphics standard GKS (ISO 7942) in 1985, followed by a number of other graphics standards on CGM (ISO 8632), PHIGS (ISO 9592), and CGI (ISO 9636).

In parallel, Prof. Encarnação initiated the implementation of a prototype at his institute to verify the applicability of the theoretical work to practical situations. The design team worked in the DIN and ISO groups on one side and gathered practical experience on the other. One of this team's members was Günther Pfaff, a research assistant in Prof. Encarnação's group, who also coauthored a book on computer graphics programming with GKS.

In addition to the goals of specifying a graphics standard and accompanying this work with a prototype implementation, a secondary goal was considered as well: to provide a kind of public domain implementation to nonprofit organizations that could also be licensed to industrial parties interested in this technology. The nonprofits would get a time and know-how advantage if they based their own products on such a prototype system. In exchange, they would have to provide funding to implement the prototype.

It became quickly evident that the successful implementation and distribution of a professional graphics library such as GKS could not be done from within the university. Also, third-party funding was not available at short notice. Therefore, in 1984, Günther Pfaff joined the privately held company GRAL, which was started some time before by a former assistant of Prof. Encarnação. GRAL hired two programmers and under the management of Günther Pfaff, an industrial implementation of GKS (GKSGRAL) was developed in a 1-year period. In order to bring in professional marketing and sales experience, GRAL entered into a joint venture with a young sales company (GTS) to form GTS-GRAL. They released the first product version and quickly acquired a number of customers within Germany. However, it was soon recognized that substantially more marketing and development resources were needed for GTS-GRAL to become an internationally recognized player in the market. Late in 1985, a venture capital company came on board and provided around $2 million over a period of 2 years. GTS-GRAL developed the product further to cover all major computer operating systems and graphics peripherals existing in the market. A distribution network was established to cover the European countries as well as the United States. GTS-GRAL grew to a $4 million company by 1987.

After the appearance of X-Windows and its rapid market acceptance, the emphasis of GTS-GRAL was shifted to X-Windows products as a distributor and to graphics packages centering around the CGM and CGI technology (the GRALX product line). This development continued in line with international standards developed after the GKS era. As of 1995, GTS-GRAL had grown to a multimillion-dollar company in Germany. Most recently, it opened subsidiaries in the United Kingdom, France, Benelux, and Switzerland. In addition, an office was opened in San Jose, California, to adapt the GRALX graphics technology to the U.S. market.

Looking back to the relation of GTS-GRAL with the original research activities at the university and the technology transfer involved, three aspects are important:

1. the intellectual property rights related to the functional specification;
2. the ownership of actual software developed; and
3. the know-how acquired by students and research assistants during their university stay.

The definition of the graphics system in this case resulted in a published ISO standard; in this way, the functional specification became public property. In fact, numerous implementations of GKS were developed and several dozens became known on a broader basis. So, there was no issue of licensing intellectual property rights.

The issue of transferring a software product to the industry for further sublicensing is more interesting. In this case, the prototype developed in parallel with the standardization process was mostly intended to cover and prove the applicability of subaspects of the overall system definition. These modules were far from a salable product or even from forming a basis for completion as a full product. If more resources, in the form of computer equipment and a professional software development environment, had been available, a sublicensing to original equipment manufacturers (OEMs) could have been meaningful.

What remains is the time-to-market advantage of perhaps 1 year for the people involved with the specification process. This represented the most critical factor for GTS-GRAL's success with GKS.

CTS-GRAL, Dr. G.E. Pfaff, Darmstadt, February 1996

CO_2 DYEING PROCESS: INDUSTRIAL COOPERATIVE RESEARCH

The original idea of using supercritical fluids as media for dyeing processes came from the *Deutsches Textilforschungszentrum Nord-West* (DTNW) in Krefeld. The concept is based upon the fact that the textile finishing industry is a large consumer of water, and therefore new technologies requiring little or no water consumption would be highly desirable. The advantage of using supercritical media (the process has been termed SFD [Supercritical Fluid Dyeing]) is that no water is used; therefore, no waste water problem exists.

The *Forschungskuratorium Gesamttextil,* an industrial research association, asked for public support of a related research project and received a grant through the Association of Industrial Research Organizations (*Arbeitsgemeinschaft Industrieller Forschungsvereinigungen,* AiF) (AiF-No. 8666).

All research activities in this field have been carried out under the guidance of DTNW. One doctoral thesis, one university diploma work, one work for a college diploma, and several additional laboratory investigations have been completed. The first breakthrough occurred after completion of the AiF research project.

The research results are published in scientific papers and have been presented at national and international meetings and in exhibitions on environmental technologies (21 published papers, 5 others in press).

The possibility of constructing an SFD apparatus, which was central to the research project, was demonstrated at the technical level. A prototype apparatus was shown at the ITMA '91 exhibition in Hannover. Following this, some medium-scale industrial machines have been built and introduced in practice. Based on further results and experience, a completely new construction was presented at the ITMA '95 in Milan.

Research in supercritical CO_2 represents a high ecological standard; investigations in the industry indicate that SFD can be handled very economically. At present, research is under way to create new products using the new technology. In the longer term, besides their use in new dyeing processes, supercritical fluids may facilitate other textile finishing processes, for example, impregnation or even biochemical or enzyme catalysis.

Dr. Eckhard Schollmeyer, DTNW, Krefeld, August 1995

PRODUCTION AUTOMATION: TRANSFER FROM A FRAUNHOFER INSTITUTE TO INDUSTRY

Tasks and Targets of the Fraunhofer Institute for Production Technology and Automation (IPA)

The main emphasis of the institute's R&D work lies in solving the organizational and technological problems posed in industrial production.

The IPA's R&D projects are aimed at pointing out and exploiting reserves in automation and at rationalizing the plants in order to maintain and enhance competitiveness and jobs through improved, more cost-effective and environment-friendly manufacturing sequences.

In the course of realizing these targets, methods, components, and appliances—even complete machines and plants—are developed, tested, and employed for demonstration purposes. The projects are mainly commissioned by industrial firms. Projects promoted by state research programs are also carried out.

One specialty of the institute deals with customer-specific solutions in the area of material-handling automation, (e.g., palletizing, commissioning, conveying, magazining, handling, and transporting). The services offered range from planning to the realization of partial systems up to complete material-handling systems.

The activities of the institute will be illustrated by the examples of material-handling automation in palletizing and commissioning.

Automatic Chaotic Palletizing (ACP)

The palletizing of packing units from a random product range into a store assortment, ready for dispatch, is carried out exclusively manually at present. Economic and work-ethic reasons argue for automating this process. However, for a random packing unit mix a suitably fast palletizing algorithm was lacking, one that, for example, would give an industrial robot the position for the next packing unit on a pallet already partially loaded with packages. With the institute's own funds (basic funding), an algorithm, capable of going on-line, was developed for palletizing cube-shaped packages in a random assortment. Operability was proved by experimental means.

A major commercial enterprise (Würth GmbH & Co.) was interested in utilizing the developed process. By means of a provisional prototype, built in the company's own distribution center, the technical feasibility and the capacity of the algorithm to meet the specific demands were shown. A prototype facility was realized in the next project phase, together with a manufacturer of material handling systems (Beumer KG). At present, a large-scale facility, consisting of five robot palletizing stations for a new goods distribution center, is being built. At the same time, Beumer is marketing the ACP to their own plans or specifications support as a licensee of IPA. The algorithm is being further developed in response to customer- or task-specific demands.

The future situation looks bright for IPA:

- The institute has gained the Würth Group, a well-known, growth-oriented enterprise, as a key, trailblazing customer.
- Beumer KG, a major company in the material-handling field, is marketing the ACP product worldwide to their clients. This means a wide market access for IPA.

Automatic Commissioning with the Help of Robots

Along similar lines, IPA is at present working on system solutions for the automated commissioning of easily separable and handled packages (e.g. bags, cube-shaped or cylindrical packages). The pathway from the idea to the well-financed work area should resemble ACP's.

Technical feasibility has already been confirmed by simulation studies and a system constructed for the Würth Group. A prototype should clarify the suitability of the process. Customer-specific solutions will be worked out in collaboration with a material-handling-systems manufacturer.

The target is the technological leadership for difficult commissioning jobs, in cooperation with leading machine and plant manufacturers. With the help of this cooperation, we can gain access to important customers of these systems' firms with their often worldwide organization and professional marketing.

FhG-IPA, Dr. M. Hägele, Stuttgart, February 1996

MEDIGENE: ESTABLISHMENT OF A START-UP COMPANY IN BIOTECHNOLOGY

A group of scientists of the Gene Center Munich (University Munich) and a representative of the chemical industry founded the start-up company, MediGene, in June 1994, with the goal of exploiting basic research results and leading technologies developed at the Gene Center. MediGene is a venture-capital-backed company (DM 3 million). Further financing (DM 3 million) is coming form the Federal Research Ministry and the State of Bavaria.

The company is collaborating with the Gene Center Munich and many other universities and clinics in Germany and abroad. In addition, it has also entered strategic alliances with partners from the pharmaceutical industry.

MediGene expects to produce its first products in about 5 years. Because MediGene is involved in biopharmaceutical research, all products developed have to pass through clinical trials before being commercialized. MediGene's first clinical trial was slated to begin in 1996.

After the company was formed, the intellectual property rights of scientists at universities were transferred to MediGene and patents were applied for. The patent holders are the inventors and the company. MediGene holds exclusive rights, and the inventors will, in the case of commercial success, be paid according to the "German inventor law."

Because the firm is in the early development stage, it is not yet possible to assess whether MediGene will be successful commercially.

Dr. Peter Heinrich, MediGene, Munich, August 1995

TECHNOLOGY LICENSING BUREAU (TLB) OF THE HIGHER EDUCATION INSTITUTIONS IN BADEN-WÜRTTEMBERG

Introduction

The TLB (Technology Licensing Bureau) of the Higher Education Institutions in Baden-Württemberg is a facility of the University of Karlsruhe that for several years has been exploiting, on behalf of several universities in the region, technologies that are generally the basis for commercial patents or applications for patents. The following examples describe in brief the course of some TLB technology transfer activities in a variety of technology fields.

Innovation Award for a Topical Subject

The Karlsruhe University, the National Research Center Karlsruhe (FZK), and the German Cancer Research Center (DKFZ) collaboratively developed a new method for cancer diagnosis and therapy, based on DNA encoding variant CD44 surface proteins. The initial invention on which all the subsequent work is based was made in 1990, and this work in its entirety was awarded the Baden-Württemberg research award in 1995.

One of the inventors approached the Technology-Licensing-Bureau (TLB), as he had already reached an agreement with a large German pharmaceutical company enterprise (Boehringer Ingelheim) on the private sale of the invention for DM 50,000. In view of his involvement with the FZK and the presence of a co-inventor, an employee of the university, he wanted to settle this matter as quickly and smoothly as possible.

In the FZK, university professors who work as heads of FZK institutes are obliged to assign their rights to research results to the FZK, according to a clause in their contracts on extra jobs (sidelines), so contact was made with the FZK patent office. Since the patent office of the FZK is also bound by a framework agreement to handle the patent business of the DKFZ, it was agreed to let the FZK take over responsibility for the case. Since this invention did not relate to mainstream R&D activities of the FZK, a patent attorney experienced in the field of gene technology applied for a joint patent for all three institutions.

After the original German patent application had been filed, negotiations were conducted with the above-mentioned enterprise and, as a result, a long-term research cooperation (budgeted at more than DM 500,000) between the FZK and the company began, which included the granting of an exclusive license by the university, the FZK, and the DKFZ. Efforts to introduce the first products to the market include submission of permit applications with, for example, the German Federal Health Office and the U.S. Food and Drug Administration (FDA), which are regularly required for products relating to this type of invention.

Measuring Instrument to Determine the Diameter of Particles Transported in Currents

The exact measurement of the diameter of particles that are transported in fluids is an important precondition for further development in many technology fields, including

- mechanical engineering,
- spray techniques,
- energy production,
- "clean-room" technology, and
- protection of the environment.

Research into laser methods to measure currents and particles has been carried out for years at the Institute of Hydromechanics of the University of Karlsruhe. A process to render the flow field visible in real time was patented only a few months after the patent was applied for and shortly afterward was licensed to an American company that is one of the world's leading manufacturers of this type of equipment.

Another recently patented invention by the hydrodynamics institute makes it possible to determine in real time the diameter of particles in gas or liquid current flows without physical contact.

The Karlsruhe PALAS (PArticle-LASer) GmbH was one of the founding firms in the Karlsruhe Technology Factory (a business incubation center) and has made a name for itself even outside Germany, especially in the field of particle supply and analysis. In the past, PALAS repeatedly has prepared the way for scientific instruments to gain access to the commercial market by producing tailor-made prototypes and by skillful marketing measures.

The university and the company agreed to do joint prototype development on the basis of an option agreement and to establish a formal licensing agreement after the preparatory work has been completed.

Diesel Soot Elimination

A medium-sized company situated in Cologne and part of a U.S.-Japanese international concern is active in the field of dust elimination and milling technology. Among the inventions of the Institute of Mechanical Engineering of the University of Karlsruhe handled by the TLB, there are at present two for which contracts have been signed with this company, including the "electrocyclone for diesel soot elimination."

The problem of soot elimination is one of the most difficult in the field of dust elimination because of unfavorable physical parameters. Despite decades of international research efforts, no suitable process had been developed. A starting point was the electrocyclone, which was taken up again in work on a Ph.D. thesis in the institute. It was possible to obtain satisfactory results for soot elimination through suitable modifications in the construction parameters of electrocyclones.

This invention was promoted by the European Research Center project on measures to clean the air. According to the rules of the sponsoring agency, the University of Karlsruhe applied for a patent. In spite of considerable efforts to find industrial partners, no starting point for awarding a license or the start of a cooperation for further (commercial) development could be found initially.

The doctoral candidate took a job with the company after finishing his thesis. The company had been trying for some time, more or less without success, to eliminate the soot from standing diesel installations, as customers in the shipping industry must conform to new exhaust emission norms when sailing in harbor. After the firm was informed of the invention of the University of Karlsruhe and was convinced that the new employee would build his efforts on the entire know-how gathered at the university during the development process, an option on the further development and future exploitation of the invention was signed in July 1995.

The second contract, with this same company, concerns the invention of a process to eliminate the smallest particles, which opens up a realm of possible applications in gas purifying technology and the synthesis of valuable materials. The high technical potential of the invention was confirmed by the firm's initial steps to upgrade the process to large scale.

Optimization of Component Parts Using Finite-Element Method Software—Company Set-Up

At the Institute for Machine Construction of the university, a doctoral candidate initiated the so-called Computer-Aided Optimization System Sauter (CAOSS), a finite-element method (FEM) optimization program that can be used as an additional module for most of the already-known software packages on the market, such as ANSYS and NASTRAN. It can solve various component part optimization tasks and, in the long run, leads to savings in materials while simultaneously improving the strength or life cycle of the components. CAOSS was given the "European Academic Software Award" in 1994.

For some time, the doctoral candidate has been contemplating exploiting CAOSS commercially in a start-up company. Since the beginning of 1995, he has done this with the support of Baden-Württemberg's program for start-up companies created by university graduates, which is funded by the Ministry for Science and Research.

The TLB has supported the candidate in the past, not only in reaching a software licensing agreement with the university that allows him to commercialize the CAOSS copyright, which fell to the university, but also in comprehensive counseling when applying to the promotion program mentioned above.

At present, TLB assistance is concentrated on the possible application for trademark protection, as well as drawing up possible sublicensing and distribution agreements with firms that should back up the new firm, especially in the fields of national and international distribution.

Efforts Do Not Always Meet with Success—Solar A.C.-D.C. Converter

Within a promotion project of the Energy Research Foundation Baden-Württemberg, a completely new solar a.c.-d.c. converter system was developed at the Electrotechnical Institute of the university. This system promised functional improvements and at the same time price reductions compared with state-of-the-art techniques.

After a rough market analysis and preliminary discussions with several parties interested in licensing the invention, the institute together with the inventor decided not to apply for a patent, as this could have been counterproductive to the efforts to commercialize the invention. Small and medium-sized enterprises in the electrotechnical branch do not necessarily like using patent rights as a security measure, since these entail an obligation to publish, and the smaller enterprises fear that they would not be able to defend themselves against imitators from big industry.

Negotiations were held with a number of medium-sized enterprises, and several expressed interest in an exclusive license agreement and made offers. Then, the consent of the Energy Research Foundation to an exclusive license agreement for a limited period of time had to be negotiated, as this case was not foreseen in the statutes of the Energy Research Foundation.

It took about 5 months to obtain this consent; then, the best offer got the license. The money involved was not the only deciding factor; the size and the know-how of the enterprise in the field of electro-solar technology were also taken into consideration.

Immediately, a company that had not been awarded the license took the university to court, claiming damages of DM 750,000 because of lost profits, alleging that the licensing agreement had already been promised to them and claiming to have already made a considerable investment with a view to going into production. The university was involved in considerable administrative work for approximately 10 months, and the case was finally dismissed on all counts.

In the meantime, the licensee had great difficulties in introducing the product to the market. This was in part because of the negative public views of solar energy. This negativity resulted when the government promoted the technology for several years but then permitted electricity suppliers or big industrial firms to buy up the production plants for solar cells and then transfer them abroad, due to a lack of returns.

The promotion of the technology in Germany was no more than the proverbial "drop in the ocean." Whereas in Japan, 100,000 roofs were equipped with solar systems at public expense, in Germany, only 1,000 roofs were equipped by the national government, although some federal states set up additional smaller programs.

The solar industry would need a quantum leap in production volume in order to be able to offer attractive prices. Only if such an increase were guaranteed would chances improve for selling the solar-current converter of the University of Karlsruhe.

It can be noted that, in the meantime, technology transfer has been achieved, even in a medium-sized enterprise in Baden-Württemberg, albeit at great cost (court case, negotiating new conditions for public support rules). But the desired final result, a product in the market on a large scale, still remains in the distant future, if it can ever be achieved at all. The assistance provided by TLB to the industrial partner is now focused mainly on the attempt to find distribution partners for the licensee in southern Europe or more especially in the United States, where market conditions are more favorable right now.

Summary

The brief case studies presented here illustrate the manifold mechanisms and courses of technology transfer. Although the examples do not provide a basis for generally acceptable conclusions in the strict scientific sense, they provide insights in the following areas:

- legal framework conditions;
- receptiveness to technology transfer of large as opposed to small firms and start-ups;
- time and effort to market; and
- framework conditions conducive to success (personnel transfer).

LEGAL FRAMEWORK CONDITIONS

There is reason to believe that it is neither the consistent reduction in price nor simplifying the access to technologies from the public sector that leads to successful technology transfer, but rather the creation of possibilities to exclude competitors that seems to motivate companies to adopt and actively pursue marketing. It is apparently helpful if the transfer is achieved, not by simply ceding the rights to use, but by allowing the research institutions to remain rightful owners and so retain a permanent right to decide on the future exploitation of the technology.

The cooperation between national research centers and universities in technology transfer seems to work, although complex legal framework conditions in the patent area and in the area of framework conditions issued by sponsoring agencies do not really encourage cooperation in individual cases.

RECEPTIVENESS TO TECHNOLOGY TRANSFER OF LARGE AS OPPOSED TO SMALL FIRMS

Not only large concerns—sometimes operating multinationally—but also small and medium-sized firms are to be found among the participants in technology transfer with public research institutions and universities. There are grounds for the theory that large enterprises generally come better equipped to cooperate on a high-technology level with public research bodies. Small firms are only then successful if they themselves have emerged from the research environment, that is, if they have a product range oriented to modern technologies, understand the aims of the public research institution system, and are capable of introducing their technology-oriented products to the market.

TIME AND EFFORT TO MARKET

The phrase "time to market" is understood universally as the key to the survival in industrial competition. There are grounds for the theory that the results of publicly funded research cannot be utilized to shorten drastically innovation cycles, as is often deemed necessary, especially within political circles. Feeding fundamental research results into the markets via technology transfer measures is, depending on the technology area, a mid- to long-term task that regularly needs substantially more resources than the investment in the originating R & D process.

FRAMEWORK CONDITIONS CONDUCIVE TO SUCCESS

Results of publicly promoted research that are amenable to technology transfer do not find users by themselves. The matching of inventors and suppliers to potential producers is an important task, one in which there is plenty of room for systemization and further development.

From the point of view of someone who has to market the results of technological research, the question arises whether it is satisfactory to commercialize a technology simply because the inventor joined a company that by chance needed this invention. Would this company have been identifiable under different circumstances? Could it have been motivated to transfer technology without this personal factor? Or, to put the question another way: What role do these and other framework conditions have in the success of technology transfer as a whole?

The theories and questions outlined here should be checked continuously in the future, using broader experiences and data bases.

TLB, Thomas Gering, Ph.D., Karlsruhe, August 1995

APPENDIXES

Notes

PART I:
OVERVIEW AND COMPARISON

1. Technological innovation has been defined as "the processes by which firms master and get into practice product [or process] designs that are new to them, whether or not they are new to the universe, or even to the nation" (Nelson and Rosenberg, 1993). These processes integrate multiple functions, including organized R&D, design, production engineering, manufacturing, marketing, and other value-adding activities in a complex web containing multiple feedback loops (Kline, 1990; Kline and Rosenberg, 1986).

2. In the case of publications and workshops, it is very difficult to determine whether and to what extent information transferred is used for specific purposes.

3. For further discussion of the many factors that have shaped the development of the German and American innovation systems, see Ergas (1987), Keck (1993), and Mowery and Rosenberg (1993).

4. In 1994, Germany had a population of roughly 81.4 million and a workforce of 39.6 million, compared with a U.S. population of approximately 260.7 million and a U.S. workforce of 132.5 million. In 1995, the German gross national product was $2,420.5 billion, and that of the United States was $6,981.7 billion (Organization for Economic Cooperation and Development, 1996b).

5. In this context, synergy is the mutual stimulation of researchers, working in different related areas.

6. Three clusters of service industries (R&D and testing services; communications services; and computer programming, data processing, and other computer-related engineering services) accounted for the vast majority of nonmanufacturing R&D in 1992. Much of the increase in nonmanufacturing R&D in the United States is accounted for by changes in the National Science Foundation's survey of industrial R&D in 1991, which resulted in an upward estimate of such R&D, and a reclassification of R&D activities from several R&D-intensive manufacturing industries to the service sector. Nevertheless, the structural change in the U.S. industrial R&D base during the past 20 years has been significant. (National Science Foundation, 1992, 1995c)

7. Because of its sheer volume, U.S. defense R&D and procurement pushed and pulled technology development in these fields during their early stages much more extensively than in any other

industrialized country. The importance of defense R&D and procurement for the technological development of most of these industries, with the exception of aerospace, has declined dramatically in recent decades. For further discussion, see Alic et al. (1992).

8. This problem is explained in more detail in Part III, *Universities, Statistics on General Research Structures*. However, according to BMBF data, about 24 percent of the university funds are related to health and about 20 percent to engineering. This leads to estimates of about 13 percent for the German health sector (instead of 3.3 percent according to the official statistics) as a whole, and about 15 to 20 percent for industrial development (instead of 12.7 percent). There are no data that reveal how well engineering science relates to other objectives, such as energy or environmental technology. These adjustments to the official data have been introduced in the table; consequently, the category "general university funds" has a share of 22 to 27 percent (instead of 38.8 percent).

9. The financial contribution of state and local governments to U.S. academic research is understated by the data in Table 1.5. If the share of general-purpose funds provided by state and local governments and used by universities for separately budgeted research or to cover unreimbursed overhead costs associated with research were added to the states' targeted support of academic research, the percent share attributed to state and local governments would increase, perhaps by as much as 5 to 10 percentage points.

10. For more details, see Part III, *Universities, Statistics on General Research Structures*.

11. U.S. academic researchers compete for a larger share of their total direct research funding in a centralized "national" peer review system than do their German counterparts. Moreover, receipt of the vast majority of U.S. academic research overhead funds is wholly dependent on the aggregate success of individual principal investigators competing for funds at the federal level, whereas overhead funds are in effect guaranteed independently of competitive performance in the German system.

12. For more details, see Part III, *Universities, Technology Transfer in the Four Focal Areas*.

13. For an evaluation of the Engineering Research Centers program, see National Academy of Engineering (1989); for an evaluation of the Science and Technology Center programs, see Committee on Science, Engineering, and Public Policy (1996); and for assessments of the Industry-University Cooperative Research Centers program, see Gray et al. (1986, 1988) and Hetzner et al. (1989). Industry-University Cooperative Research Centers receive start-up funding from NSF that tapers down over 5 years. For years 6 and beyond, NSF funding continues at only a token level ($25,000 to $50,000 per year).

14. The 15-percent figure includes an unidentified number of An-Institutes in the social sciences and humanities.

15. For more details, see Part III, *Universities, An-Institutes and Other External Institutes*.

16. For an overview of institutional forms of industry collaboration at universities, see Table 3.6 in Part III.

17. The Bayh-Dole Act of 1980 gave nonprofit organizations such as universities and small businesses the right to patent inventions they developed with federal support; granted government-owned and operated laboratories the authority to grant exclusive licenses to inventions which they patented; and prevented public disclosure of information about inventions to allow for patent applications to be filed. Although Bayh-Dole did not originally apply to any of the DOE laboratory management and operations contractors, the law was subsequently amended to include them.

18. With external funds from the BMBF and other sources accounting for a growing share of total academic research funding, the exploitation privilege of German academic researchers is increasingly circumscribed by the intellectual property claims of external funders. See also the section *Selected Technology Transfer Issues in a Comparative Context*, below.

19. Reviewing existing research on the topic, Stankiewicz (1994) notes that what high-tech start-ups usually spin off from universities are "not technologies-as-products but rather R&D and problem-solving capabilities." For further discussion, see *Technology Transfer from Higher Education to Industry* in Part II.

20. The German figure includes laboratories at Helmholtz Centers, departmental research institutes, and Blue List institutes.

21. The Blue List institutes are independent research institutes with supraregional importance; heterogeneous structure, legal status, and technical importance; and are supported almost entirely by public funding, half from the federal government and half from state governments. For further discussion of these institutes and the departmental research institutes, see the section *Technology Transfer from Public Intermediate Institutions* in Part III.

22. U.S. panel member Albert Narath notes that DOE's success with CRADAs has fostered a growing volume of industry-funded "work-for-other" business (i.e., industry-sponsored contract research) in some federal laboratories.

23. The Stevenson-Wydler Technology Innovation Act mandated that federal laboratories actively seek cooperative research with state and local governments, academia, nonprofit organizations, and private industry and disseminate information about their activities and research. It established the Center for the Utilization of Federal Technology (CUFT) at the National Technical Information Service and required the establishment of an Office of Research and Technology Applications (ORTA) at each federal laboratory, setting aside 0.5 percent of each laboratory's budget to fund laboratory technology transfer activities. The act also established the National Medal of Technology.

The Federal Technology Transfer Act of 1986 (P.L. 99-502) amended Stevenson-Wydler to accelerate technology transfer by requiring that personnel evaluations of federal laboratory scientists and engineers include information about their support of technology transfer activities and that government-owned, government-operated (GOGO) laboratories pay inventors a minimum 15-percent share of any royalties generated by the licensing of their inventions. It gave directors of GOGO laboratories authority to enter into CRADAs, to license inventions that might result from such CRADAs, to exchange laboratory personnel, services, and equipment with research partners, and to waive rights to lab inventions and intellectual property under CRADAs. The act allows for federal employees to participate in commercial development with private firms if there is no conflict of interest, and created a charter for and funded the Federal Laboratory Consortium.

The National Competitiveness Technology Transfer Act of 1989 (P.L. 101-189) further amended the Stevenson-Wydler Act to allow for the protection, in CRADA arrangements, of information, inventions, and innovations, against disclosure under the Freedom of Information Act for a period of 5 years. It also established a technology transfer mission for the nuclear weapons laboratories and clarified that government-owned contractor-operated laboratories could execute CRADAs and enter into other technology transfer activities.

U.S. panel member Albert Narath estimates that Sandia National Laboratories receives over $1 million in royalties on patents each year. Based on trends over the past 5 years, Narath estimates that royalty revenues will represent a significant fraction of some federal laboratories' budgets 10 years from now.

24. For further details, see the discussion of technology transfer by Helmholtz Centers in the section *Technology Transfer from Public Intermediate Institutions* in Part III.

25. Max Planck institutes and individual departments within them are generally established around the work of an outstanding scientist. This personality-centered form of organization (i.e., the Harnack principle), first used by the institutes of the Max Planck Society's predecessor organization, the Kaiser Wilhelm Society, can explain the finite lifetime of MPIs or departments within them. If a departing head scientist is not replaced by an equivalent successor, the research focus of the institute or department may be changed (depending on the new leader) or even dissolved. For further discussion of the Max Planck Society, see the section *Technology Transfer from Public Intermediate R&D Institutions* in Part III.

26. For further discussion of issues relevant to the future of U.S. federal laboratories, see the section *U.S. Federal Laboratories and Technology Transfer to Industry* in Part II.

27. For example, German Helmholtz Centers had close ties to the nuclear energy industry but have relatively few industrial contacts in their new areas of activity.

28. Fraunhofer institute directors are generally part-time employees of the Fraunhofer Society and part-time civil servants of their institute's host state. In addition, there is a cooperation contract between the university and the Fraunhofer institute to avoid conflicts of interest.

29. The seven largest engineering-oriented institutes are Battelle Memorial, IIT Research, Midwest Research, Research Triangle, Southern Research, Southwest Research, and SRI International. Their research ranges from basic research to development and is largely comparable to that of the Fraunhofer institutes. These U.S. engineering institutes have built up highly competitive competencies in a variety of specific technical areas and serve national and international markets. Most of them, however, with the exception of Southwest Research Institute and SRI International, perform 70 percent or more of their research for government clients—a larger share than is true for the Fraunhofer institutes. Unlike the Fraunhofer Society, the independent R&D institutes receive no public base-institutional funding and so must rely exclusively on contract research. Therefore, for example, the Southwest Research Institute can use only about 1.5 percent of its revenues for self-initiated research. Due to the lack of public base funding, the engineering-oriented institutes have to generate a significant portion of their budgets by selling testing and technical services to industrial clients.

30. For the U.S. situation, see Carr and Hill (1995). No data are available regarding the total volume of cooperative industrial R&D in the Germany. The fact that more than 10 percent of the industrial R&D budget in Germany is spent externally indicates that the total volume of industrial cooperative research is significantly greater than the 1 percent of industrial research associations within the AiF.

31. The National Cooperative R&D Act of 1984 (NCRA; P.L. 98-462) provides exemption from treble damages in antitrust lawsuits to companies that register their joint R&D ventures with the U.S. Department of Justice. In so doing, the act offers a clear signal from the federal government in support of industrial cooperative research. Vonortas (1996) notes that fragmentary early evidence regarding strategic alliances in R&D, including consortia, shows that cooperative R&D activity began increasing prior to passage of the NCRA. In 1993, the NCRA was amended by the National Cooperative Research and Production Act (NCRPA; P.L. 103-42), which extended exemption from treble damages to registered joint production ventures involving firms also engaged in joint R&D activity. For further discussion, see the section *Technology Transfer by Privately Held, Nonacademic Organizations*, in Part II, and Hagedoorn (1995).

32. During the early 1990s, the Department of Energy embraced a relatively expansive view of its laboratories' role in supporting the R&D/technology needs of civilian industry. However, subsequent review and action by the federal government, as exemplified by the so-called Galvin Report (Secretary of Energy Advisory Board, 1995) has tended to moderate the proliferation of federal laboratory-industry partnerships. For further discussion, see Part II, *U.S. Federal Laboratories and Technology Transfer to Industry*.

33. For further discussion, see Part II, *Technology Transfer by Privately Held, Nonacademic Organizations*.

34. For further discussion of the many factors that have contributed to the special role of high-tech start-up companies in the United States, see Mowery and Rosenberg (1993) and National Academy of Engineering (1995c).

35. However, the coming opening of a European stock exchange, EASDAQ, modeled on the American NASDAQ, might improve the situation.

36. As discussed in this section (*Technology Transfer from Higher Education Institutions*), German universities have little incentive to devote resources to patent licensing and marketing activities, since under German law the right to exploit inventions resulting from university-based research supported by base-institutional funds resides exclusively with the individual researchers involved.

37. Technologically mature industries are defined by the binational panel as industries in which technological advance is predominantly incremental and moderately paced. By contrast, technologically dynamic, or revolutionary, industries are characterized by very rapid and frequently radical or "breakthrough" technological change.

38. For an overview of the industry-government-university consortium, the American Textile Partnership (AMTEX), and its many research and outreach activities, see the AMTEX home page at <http://amtex.sandia.gov/>.

39. See, for example, recent publications and current research initiatives of the Massachusetts Institute of Technology (MIT) International Motor Vehicle Program on the program's home page, http://web.mit.edu/org/c/ctpid/www/imvp/index.html. For a recent assessment of the government-industry Partnership for New Generation Vehicles, see National Research Council (1997).

40. For further discussion of these and other industry-led initiatives aimed at the manufacturing technology needs of small and medium-sized firms in more technologically mature U.S. industries, see Part II, *Technology Transfer by Privately Held, Nonacademic Organizations*.

41. See, for example, the discussion of NIST's Manufacturing Extension Partnerships, or of state technology extension deployment programs such as the Thomas Edison Institute in Ohio or the Ben Franklin Partnership in Pennsylvania, in Part II, *Technology Transfer in the United States*, as well as Coburn (1995).

42. However, exceptions to this general rule are possible.

43. For more details, see the section, *Technology Transfer from Public Intermediate R&D Institutions* in Part III.

44. According to the German/European ruling, every publication of an invention prior to the filing of a patent application—including publications of the inventor—is considered a prejudicial disclosure opposed to its novelty.

45. For further discussion of the importance of day-to-day personal contact for technology transfer as demonstrated empirically by the revived importance of regional agglomerations of industrial skills and comparative advantage, see Brooks (1996), David et al. (1992), Pavitt (1991), and Reger (1997).

46. See, for example, Secretary of Energy Advisory Board (1995); Executive Office of the President, Office of Science and Technology Policy (1995); and National Academy of Sciences et al. (1995); and, for Germany, Weule et al. (1994).

47. For data regarding the movement of U.S. scientists and engineers between government, industry, and academia, see National Science Foundation (1994b) and Part II, *Technology Transfer from Higher Education to Industry*. Unfortunately, no comparable data exist with which to assess quantitatively the relative job mobility of scientists and engineers in Germany and the United States.

48. The U.S. delegation's definition of "infrastructural" and "pathbreaking" R&D draws on the taxonomy of technologies developed by Alic et al. (1992). Infrastructural R&D is directed at the discovery and development of infrastructural technologies—technologies generally low in technical risk and difficult to appropriate privately, but which enhance the performance of a broad spectrum of firms and industries. Pathbreaking R&D is directed at the discovery and development of pathbreaking technologies—technologies high in technical risk that create new industries or transform existing industries.

49. The 1997 tax law introduces an amendment consistent with the delegation's recommendation. Nevertheless, the decision of the Federal Financial Court makes clear that the tax status of research performed by public and semipublic institutions in Germany is still disputed.

50. Since 1989, the U.S. Department of the Air Force had been preparing detailed technology road maps for defense contractors for all areas of science and technology related to the department's mission as part of its Technology Area Plans.

51. See, also, the recently released technology road map for the U.S. chemical industry, *Technology Vision 2020*, authored by the American Chemical Society, Chemical Manufacturers Association, American Institute of Chemical Engineers, Council for Chemical Research, and the Synthetic Organic Chemical Manufacturers Association (American Chemical Society et al., 1996). This road map effort was launched, in part, by a request from the White House Office of Science and Technology Policy.

52. The Government-University-Industry Research Roundtable, in collaboration with the private-

sector Industrial Research Institute and the Council on Competitiveness, is currently hosting a series of regional workshops on industry-university research collaboration.

53. See the German-American collaborative study, *Conflict and Cooperation in National Competition for High-Technology Industry* (Hamburg Institute for Economic Research, Kiel Institute for World Economics, and National Research Council, 1996) for an extensive treatment of these and related issues. The report is a constructive first step toward articulating principles for international cooperation in science and technology.

PART II:
DEFINING THE U.S. TECHNOLOGY TRANSFER ENTERPRISE

1. Technology transfer is a person-to-person activity, or a body-contact sport. Inventions and new technologies spring from and reside in the human mind. Written descriptions, samples, or even working prototypes rarely convey all that is to be known about a new technology. The developer's knowledge and intuition about further potential must be transferred via personal contact between individuals.

While the transfer of intellectual property is often thought of as the essence of technology transfer, such a view is misleading. Signing of license agreements, payments of royalties, and transfers of intellectual property are among the few elements of technology transfer that lend themselves to quantification, and thus they form the majority of available metrics of technology transfer. But unpatented know-how, ideas, and suggestions often constitute information of considerable value, however difficult to measure and evaluate. Among companies, mergers and acquisitions often have important technology components, but the value of technology is rarely visible in the public data on such events. Furthermore, other less formal mechanisms such as conferences, meetings, and even personal relationships among technologists make an important but largely unmeasured contribution.

In addition, a semantic problem has arisen in recent years. The very term "technology transfer" has fallen out of favor among many who view the term as outmoded, too narrow in scope, and too closely linked with the "linear" model of innovation. Others prefer technology collaboration, technology deployment, technology utilization, or other terms. The term is sufficiently imprecise that a general definition of technology transfer brief enough to be useful is impossible to develop. Operational definitions of technology transfer are easier to devise in a specific context and are best constructed in terms of specific mechanisms of transfer. The authors of the U.S. report define the term technology transfer broadly, incorporating the following mechanisms:

- Formation of new technology-based companies from R&D organizations (spin-offs and others)
- Licensing of patents, software and technical know-how, prototypes, biological materials
- Performing contract R&D for clients and transferring the results
- Sharing information in interactive events (conferences, workshops, briefings, and visits)
- Performing cooperative R&D
- Forming R&D or technology transfer consortia
- Providing technical assistance
- Employing unique R&D facilities and capabilities
- Activities that catalyze or facilitate any of the above

2. Federally funded research and development centers (FFRDCs) are defined as contractor-operated and mostly contractor-owned research facilities established at the request of federal agencies with congressional authorization. FFRDCs draw over 70 percent of their funding from the federal government.

3. The National Science Foundation (NSF) classifies research and development into three categories: basic research, applied research, and development. Basic research seeks to advance scientific or technical knowledge or understanding of a particular phenomenon or subject without specific applications in mind. Applied research recognizes a specific need and seeks new knowledge or understanding in order to meet that need. Development is "the systematic use of the knowledge or

understanding gained from research directed toward the production of useful materials, devices, systems, or methods, including design and development of prototypes and processes" (National Science Board, 1993).

4. In addition to the $20.3 billion of industrial R&D funded directly by federal agencies through contracts, several federal agencies (most prominently the Department of Defense) reimbursed U.S. companies for roughly $3 billion of the $4.4 billion spent by private-sector contractors on independent research and development (IR&D) and the preparation of bids and proposals (B&P) during fiscal 1995, an activity that frequently involves technical work. IR&D and B&P expenses are treated by federal procurement regulations as indirect or overhead costs (i.e., expenses that increase a firm's total costs, but cannot be attributed to specific contracts). All intellectual property resulting from IR&D belongs to the performing firm. The $4.4 billion of industry IR&D and B&P expenditures are included in the $101.7 billion of R&D funding attributed to industry in 1995 in federal R&D data (Defense Contract Audit Agency, 1997; National Science Board, 1996).

Prior to the early 1990s, all reimbursable IR&D projects were to have "potential military relevance." The National Defense Authorization Act for Fiscal Years 1992 and 1993 (P.L. 102-190) provided for the gradual removal of limitations on the amount DOD reimburses contractors for IR&D expenditures and partially eliminated the need for advance agreements and technical review of IR&D programs. Reimbursement is now allowable for a broader range of IR&D projects of interest to DOD including those designed to develop dual-use technologies, enhance industrial competitiveness or to develop technologies for environmental concerns. For further information regarding IR&D and B&P, see Alic et al. (1992) and National Science Board (1993).

5. The current administration's explicit intention is to reduce the share of public spending for defense R&D to equal that devoted to federal nondefense missions.

6. Until recently, most federally funded R&D performed by U.S. companies was concentrated in a few "dual-use" industries such as electronics and aerospace. As late as 1988, 61 percent of all federal R&D support of industry went to aerospace and 14 percent went to the electronics and communications sector. This federal contribution represented 76 percent and 38 percent, respectively, of total R&D spending in these sectors (National Academy of Engineering, 1993; National Science Foundation, 1996b).

7. It is worth noting the large contrast in the distribution of effort between publicly funded defense and nondefense research and development. Ninety percent of public funding for defense-related R&D is for development, testing, and evaluation, with applied research, basic research, and R&D plant accounting for the remaining 10 percent. In contrast, public nondefense R&D spending is divided more evenly among the 3 major categories, with 30 percent for development, 30 percent for applied research, and 30 percent basic research, with the remaining 10 percent for R&D plant (National Science Board, 1996).

8. Alic et al. (1992) note that most of the university-based engineering research sponsored by DOD, DOE, the Atomic Energy Commission, and the National Aeronautics and Space Administration "was 'engineering science'—i.e., investigations of natural phenomena underlying engineering practice—rather that engineering design, manufacturing operations, or the construction and testing of prototype equipment."

9. If Department of Energy (DOE) laboratories that focus primarily on nuclear weapons research are added to those of DOD, the national security mission laboratories account for roughly 55 percent of total federal laboratory expenditures and 60 to 70 percent of the total number of laboratory researchers. At present, slightly less than half of all DOE laboratory resources are dedicated to weapons research (Committee on Science, Engineering, and Public Policy, 1992).

10. This sharp division has frequently been stricter in theory and rhetoric than in practice since World War II. See Brooks (1986), Cohen and Noll (1991), Kash (1989), Mowery and Rosenberg (1989), National Academy of Engineering (1993), and Nelson (1989).

11. For further discussion of the Bayh-Dole Act and its implications for university-industry technology transfer, see pp. 98–99, 102–108.

12. For further discussion of CRADAs and other federal laboratory technology transfer efforts, see pp. 135–151.
13. For further discussion of NCRA and the growth of industrial R&D consortia, see pp. 156–162.
14. For further discussion of NSF university-industry research centers, see pp. 111–118.
15. See discussion of SEMATECH, pp. 157–159. For information on the TRP, see also Annex II, p. 208.
16. The phase-out of Department of Defense funding of SEMATECH was negotiated voluntarily, not mandated.
17. By 1994, 40 states had technology extension programs (Coburn, 1995) in addition to other types of programs that also assist manufacturers. Approximately half of the state programs were operated by educational institutions, with the balance managed by nonprofit organizations or state agencies. These programs offer different types of services, including supply of technical information, seminars and workshops, demonstrations, referral of consultants and other experts, and in-plant consultation. However, intensive field assistance (generally agreed to be the most effective technique) was provided by only a few programs (Shapira et al., 1995).
18. Although not captured in national industrial R&D statistics, many small and medium-sized companies perform product and process design that would likely be reported as R&D in the more organized setting of a large firm.
19. The National Science Board (1996) notes that a large share of the R&D spending in computer software and communication services was spent by companies formerly classified as manufacturing industries. This is not surprising given the growing importance of software and other information technologies relative to "hardware" in most industries.
20. Between 1991 and 1995, industry's share of total basic research performed in the United States declined from 29.5 percent to 24.2 percent (National Science Board, 1996).
21. In 1992, researchers at the Georgia Institute of Technology surveyed members of the Industrial Research Institute (generally large, research-intensive firms), and asked respondents to indicate their most significant sources of external technology. The results indicated that other companies (U.S. and foreign) were the most significant external technology source, with universities second, private databases third, and federal laboratories fourth.
22. See pp. 90–124 for further discussion of these trends.
23. See pp. 135–139 for further discussion of these trends.
24. A U.S. affiliate of a foreign-owned firm is a company located in the United States in which a foreign person or business has a "controlling" stake (i.e., 10 percent or more of the company's voting equity).
25. See Annex II, *Case Studies in Technology Transfer: Software,* for discussion of factors that have facilitated the rapid proliferation of software start-up companies in recent decades.
26. For further discussion of the role high-tech start-ups have played in the development of the software and biotechnology industries, see Annex II.
27. Technological uncertainty—and therefore technological risk—shapes the opportunity for start-ups and smaller, technologically oriented companies. This observation is crucial to understanding why start-ups and entrepreneurs dominate technology-intensive sectors of the economy. Typical early barriers (barriers to entry or "start-up") derive less from the need to command massive resources than from the ability to bear risk, be creative technologically, and make forward-looking decisions. This also explains why larger competitors are usually not first to exploit these opportunities (National Academy of Engineering, 1995c).
28. For further discussion of the many factors that have contributed to the special role of high-tech start-up companies in the United States, see Mowery and Rosenberg (1993) and National Academy of Engineering (1995c).
29. This perception is reinforced by a recent report by the Office of Science and Technology Policy (1995) that identifies 27 technologies in 7 major areas seen as crucial to "develop and further

long-term national security or economic prosperity in the United States." The report finds that the United States leads or has parity with Japan and Europe in each of the 27 technologies.

30. For a comparison of U.S. and German patent specialization using European Patent Office statistics see Part III, Figures 3.3 and 3.4.

31. Approximately 1,000 companies were surveyed. The survey response rate was 57 percent, and approximately one-third of firms responding indicated that they had introduced a new product or process during the 1990–1992 period or were planning to introduce a new product between 1993 and 1995 (National Science Board, 1996).

32. This does not include R&D performed by university-administered FFRDCs, which performed over $5 billion of R&D in 1995. For further discussion of the role of FFRDCs in U.S. technology transfer, see pp. 125–126.

33. National shares of world scientific and technical literature were determined by a review of some 3,500 major U.S. and international technical journals. U.S. academic and nonacademic researchers collectively accounted for nearly 34 percent of scientific and technical articles published in all U.S. and international journals, and more than 33 percent in all major fields with the exception of chemistry (23 percent) and physics (27 percent). U.S. academic researchers accounted for more than 70 percent of all articles published in U.S. science and engineering journals. By way of comparison, German researchers accounted for nearly 7 percent of articles in the world's science and engineering journals, with the largest shares in two fields, chemistry (9 percent) and physics (8 percent) (National Science Board, 1996).

34. Federal funds include grants and contracts for academic R&D (including direct and reimbursed indirect costs) by agencies of the federal government. State/local funds include funds for academic R&D from state, county, municipal, or other local governments and their agencies, including funds for R&D at agricultural and other experiment stations. Industry funds include all grants and contracts for academic R&D from profit-making organizations, whether engaged in production, distribution, research, service, or other activities. Academic institutional funds include institutional funds for separately budgeted research and development, cost-sharing, and under-recovery of indirect costs; they are derived from (1) general-purpose state or local government appropriations; (2) general-purpose grants from industry, foundations, and other outside sources; (3) tuition and fees, and (4) endowment income. Other nonprofit sources include grants for academic R&D from foundations and voluntary health agencies, as well as restricted individual gifts (National Science Board, 1996).

35. In 1862, Congress passed the Morrill Act. This act provided resources for the establishment of state universities (land grant colleges) to pursue research and education in the "agricultural and mechanical arts." Subsequent acts of Congress, including the Hatch Act of 1887 and the Adams Act of 1906, expanded federal support for agricultural research and a national system of agricultural extension centers. See Mowery and Rosenberg (1993).

36. Despite this general shift in orientation, many institutions retained a commitment to industrial extension service and regional economic development even as they took on support of federal agency missions. Among these are many of the original land grant colleges (see note 35, above), as well as other institutions such as Georgia Institute of Technology, Rensselaer Polytechnic Institute, Pennsylvania State University, and University of Minnesota Mines Experimental Station.

37. Matkin (1990) points out that this divergence in research cultures began in the United States prior to the turn of the century. For further discussion of academic and industrial research cultures and their interaction, see Dasgupta and David (1994).

38. For an informative discussion of the origins and consequences of Bayh-Dole, see Wisconsin BioIssues (1994).

39. For further information regarding these centers, see National Science Foundation website URLs http://www.eng.nsf.gov/eec/i-ucrc.htm; http://www.cise.nsf.gov/asc/STC.htm; http://www.eng.nsf.gov/eec/erc.htm; and http://www.nsf.gov/mps/dmr/mrsec.htm. See also note 53, below.

40. See *New Federal Industrial R&D Initiatives,* Part II, and Annex II, p. 208, for further information regarding these two initiatives.

41. See, for example, the report of the Government-University-Industry Research Roundtable (1992).

42. The following discussion of university technology transfer mechanisms draws upon the general taxonomy used by Matkin (1990).

43. Surveys by Morgan et al. (1994a,b) indicate that 87 percent of U.S. university engineering faculty have been consultants to industry or government (National Science Board, 1996).

44. The involvement of graduates and research staff in the spin-off of new technology-based firms has been studied at individual institutions, however. See, for example, the studies by BankBoston (1997) and Roberts (1991).

45. Roughly 16,660 high-tech companies were established in the United States between 1980 and 1994—546 in the field of biotechnology, 5,196 in software, 1,907 in computer hardware, 1,293 in electronic components, 1,933 in telecommunications, 1,917 in automation, 487 in advanced materials, 507 in photonics and optics, and 4,874 in other high-tech fields (National Science Board, 1996).

46. The Association of University Technology Managers survey covers data on sponsored research, licensing, start-ups, gross royalties, invention disclosures, patents applied for and issued, legal fees, and staffing. The survey population for fiscal year 1995 consisted of 279 institutions, including 196 U.S. universities, 53 hospitals and research institutes, 25 Canadian institutions, and 5 third-party patent management firms. 62 percent of these institutions responded to the survey, including 127 U.S. universities (roughly 65 percent of U.S. universities contacted). The response rate for the top 100 U.S. research universities (ranked by volume of federal research monies received) was 87 percent.

47. H. Wiesendanger, Office of Technology Licensing, Stanford University, personal communication to Simon Glynn, research associate, National Academy of Engineering, August 10, 1993.

48. The patents on recombinant DNA techniques by Boyer and Cohen is an example: Income from the Cohen-Boyer patents for 1991 amounted to $14.6 million, or 58 percent of total income from all patents held by Stanford (H. Wiesendanger, Office of Technology Licensing, Stanford University, personal communication to Simon Glynn, National Academy of Engineering, August 10, 1993).

49. For discussion of the growing involvement of universities in venture finance, see Feller (1994) and Matkin (1990).

50. For further discussion of the role of venture capital firms in technology transfer, see pp. 172–173.

51. This section draws heavily on data reported in Cohen et al. (1994).

52. See, for example, National Academy of Engineering (1989) on Engineering Research Centers; National Research Council (1996a) on Science and Technology centers; and Gray et al. (1986; 1988) and Hetzner et al. (1989) on Industry/University Cooperative Research Centers.

53. The Material Processing Center Industry Colloquium at MIT was established subsequent to the launch of the Institute's Materials Processing Center in order to organize the center's relations with industry (Matkin, 1990).

54. For further discussion of the role of business incubators in technology transfer, see pp. 167–169.

55. See, for example, Cohen et al. (1994, 1995); Dasgupta and David (1994); Feller (1994); Government-University-Industry Research Roundtable (1991); Henderson et al. (1995); Mansfield (1996); and Rees (1991).

56. By way of comparison, U.S. universities and colleges employed roughly 150,000 Ph.D. scientists and engineers and another 16,000 individuals with professional, masters, or bachelors degrees in R&D activities in 1993. In addition, it is estimated that some 90,000 full-time graduate students were involved in university-based research that year. See Part II, *Technology Transfer from Higher Education to Industry.*

57. Multiprogram laboratories are large labs with diverse core competencies and resources that permit scientific and engineering work across a wide spectrum of technologies.

58. FFRDCs are defined in regulation. Most of the well-known FFRDCs do research for the Defense Department. Some confusion exists about the distinction between GOCO federal laboratories and FFRDCs. FFRDCs are rigorously defined in criteria published by the Office of Federal Procurement Policy and an official list of FFRDCs, based on those criteria, is maintained by the NSF. Most are single-office facilities employing a small number of researchers; a small percentage are large organizations that employ thousands of scientists and engineers. FFRDCs do not have a specific prescribed management structure, but they must engage in research based on a specific or general request from the federal government, must receive more than 70 percent of their financial support from the government, and must have been brought into existence at the request of the government with congressional authorization.

59. CRADAs have tended to supplant other types of R&D agreements in the Department of Energy (DOE), the Department of Defense, and some other agencies because they offer intellectual property protection lacking in earlier agreements. DOE made a decision in the early 1990s that all new cooperative R&D agreements would be CRADAs. Hence, CRADAs are both supplanting other forms of cooperative agreements and generating new cooperative R&D activity as agencies and laboratories promote CRADAs more actively. Estimating the total number of CRADAs at any point in time is made difficult by differences among agencies with respect to how CRADAs are defined and counted. For example, NASA Space Act agreements are frequently counted as CRADAs, but not always. The Department of Agriculture has a number of cooperative R&D agreement types established in legislation that are similar to CRADAs, but not always counted as such. NIH uses material transfer agreements, which they counted as CRADAs at one time (Personal communication, Robert K. Carr to Proctor Reid, July 4, 1997).

60. According to the U.S. Department of Energy (1994) these percentages are based on the number of CRADAs, not CRADA dollars or duration of CRADA agreements. Therefore, these data are of little informational value.

61. In 1993, the Department of Energy took steps to streamline the CRADA negotiation process in response to criticisms that its procedures were too bureaucratic and time consuming. For further discussion, see U.S. Department of Energy (1993).

62. Mowery and Ziedonis (1997) estimate that 160 new technology-based firms were spun-off from Lawrence Livermore National Laboratory between 1985 and 1995. They defined a spin-off as a firm "founded by anyone with a prior or current employment relationship with the Laboratory." Using a somewhat less restrictive definition of spin-offs (including enterprises "founded by laboratory consultants or nonemployees that seek to commercialize innovations drawing on laboratory technologies"), Markusen and Oden (1996) estimated that less than 100 firms were spun-off from DOE's Los Alamos and Sandia National Laboratories and the U.S. Airforce Phillips Laboratory, all located in New Mexico, between 1980 and 1994.

63. Mowery and Ziedonis (1997) found that 40 percent of the 160 Lawrence Livermore National Laboratory (LLNL) spin-offs identified listed their primary activity as consulting. Seventy-five percent of all spin-offs were owned by current LLNL employees (as of 1995). Of the 37 firms responding to the Mowery and Ziedonis survey, two-thirds stated that LLNL technologies were of "little or no" importance to their establishment and operation. Only 8 of 37 respondents stated that their companies were commercializing LLNL technologies. At the same time, Mowery and Ziedonis observe that "virtually all of these firms' founders noted the importance of the Laboratory as a source of generic expertise and skilled employees."

64. For further discussion of NASA's recent entry into the technology incubator business, see pp. 149, 168–189.

65. Personal correspondence from panel member Albert Narath, Lockheed Martin Corp., to Proctor Reid, July 16, 1997.

66. After remarks by Albert Narath, Lockheed Martin Corp., at the May 7–8, 1996, meeting of the U.S. delegation to the binational panel in Washington, D.C.

67. For further discussion of the economic performance and reciprocity requirements embodied in recent U.S. technology transfer legislation, see National Academy of Engineering (1996b).

68. Knowledge is the coin of technology transfer. Knowledge may be embedded in formal intellectual property documents, such as licenses; may reside in scientists and engineers from the public and private sectors who interact; may be created by cooperative R&D programs; may be embedded in transferred materials, processes, and prototypes; and may move in many other ways. Because the medium of technology transfer is some form of knowledge, to measure the economic value of technology transfer is to measure the economic value of knowledge. This is an old conundrum. Economists and others have struggled with the problems of defining and measuring the economic value of knowledge for many years, without particularly satisfying results. Economic analyses which require dollar valuations of knowledge are often forced to employ surrogates, sometimes crude surrogates, to produce that value. The use of such surrogates reduces the outcome to a (sometimes misleading) approximation. (See U.S. Congress, Office of Technology Assessment, 1986, regarding the measurement of the results of research.)

69. These data are the basis of Figure 2.15 and Table 2.16. However, some of the OMB data are generally considered problematic, since agencies have interpreted the reporting requirements in different ways, particularly the requirements for budgetary information on cooperative R&D.

70. While the 104th Congress was generally favorably inclined toward federal laboratory technology transfer activities (including CRADAs) that were funded out of the regular program activities of federal agencies, the Department of Energy's Technology Transfer Initiative funding for CRADAs, which was a separate line item in DOE's budget, was attacked by Congressional leaders hostile toward specially funded programs that might enter into the "industrial policy" arena.

71. See DOE's home page on the World Wide Web <http://www.dtin.doe.gov/htmls/common/objective.html>.

72. On the other hand, the Galvin Report strongly encourages industrial partnerships as a derivative mission (i.e., partnerships that contribute to DOE's historic mission responsibilities). This "dual-benefit" requirement is not a serious constraint in most cases.

73. This is to be accomplished through enhanced monitoring of contractor-developed technologies, as well as commercialization requirements.

74. Recent efforts along these lines include the so-called Galvin Report (Secretary of Energy Advisory Board, 1995) and the report of the Committee on Criteria for Federal Support of Research and Development (1995).

75. See, for example, the chairman's summary report on the National Academy of Engineering workshop on *Defense Software Research, Development and Demonstration: Capitalizing on Private Sector Capabilities* (National Academy of Engineering, 1996a).

76. The "fourth category" of institutions excludes

   (a) Private firms transferring technology internally and among themselves (except in a few specifically defined cases);

   (b) Federal agencies and laboratories (including FFRDCs) transferring technology to the private sector or elsewhere, and state and local technology organizations transferring technology to and working with the private sector; and

   (c) Universities transferring technology to the private sector (including from university-based technology centers and university-owned technology transfer organizations).

Some types of institutions whose primary activities lie outside the panel's operational definition of technology transfer were also excluded. These include organizations that primarily deliver or produce education and training; after-sale technical services; testing and quality control; published materials and other one-way (i.e., noninteractive) communications; and training in support of hardware production. Institutions engaging primarily in international technology transfer activities are also excluded, because while international technology transfer is important to U.S. industry, it is too diverse and distinct from the principal focus of this survey. Furthermore, the traditional dissemination of research results through publication in professional journals and discussion in open conferences is not generally considered to be part of the technology transfer universe and is not included.

In addition, the emphasis in this section is placed on technology transfer to the private sector as opposed to the development of products and systems for transfer to the government via procurement or other acquisition mechanisms. Finally, funding sources were not used as a criterion for inclusion or exclusion, as government funding for R&D activities is so pervasive that almost all independent R&D and technology transfer institutions are direct or indirect beneficiaries. Ownership and control were more important in defining "independent."

77. Most affiliated R&D institutes are "affiliated" with research universities, research hospitals, and other medical research institutes. Affiliated R&D institutes are very similar to the independent group except for their formal ties to a parent institution and lack of independent legal status (i.e., independent institutes are independent corporate entities with their own governing boards—affiliated institutes are not). Hence, even though most of the department heads at Massachusetts General Hospital and Brigham and Women's Hospital are professors at Harvard Medical School, these hospitals are classified as "independent" teaching hospitals.

78. The National Science Board (1996) estimates that nonprofit institutions, which account for one-half to three-quarters of all R&D performed by fourth-sector organizations, conducted about $5.1 billion worth of R&D in 1995. A lack of consistent estimates of R&D performed by consortia and the fact that R&D performed by affiliated institutes cannot be separated from that of their parent organizations make it difficult to estimate total R&D investment by these organizations. For these reasons, quantitative comparison of the fourth sector's R&D performance with that of the three other principal segments of the U.S. R&D and technology transfer enterprise are also deficient.

79. This does not include $800 million in government-funded R&D performed by federally funded research and development centers administered by nonprofit institutions.

80. The Universities Research Association, Inc., is a consortium of research universities and private nonprofit corporations. However, it serves primarily as a contractor to the federal government for the operation of major scientific facilities, including the Fermi National Accelerator Laboratory (FermiLab), a GOCO and a leader in superconductivity research.

81. Despite the fact that SRI researchers work for multiple clients simultaneously, the institute has never had complaints regarding conflicts of interest or breaches of confidentiality (remarks by H. N. Abramson at meeting of the Binational Panel on Technology Transfer Systems in the United States and Germany, November 7, 1995, Freising, Germany).

82. Industrial consortia first appeared in the United Kingdom early in this century. The concept was transplanted to post-war Japan and to the United States in the 1980s, although there were earlier examples of joint research in the context of specific industries. In the early 1900s, various industry groups such as the American Iron & Steel Institute, the Portland Cement Association, and the American Petroleum Institute established research programs focused on their industry's problems. In the 1970s, prompted by the energy crisis, the Electric Power Research Institute and the Gas Research Institute were formed. Finally, in the 1980s, Japan's rapid development in high-tech industries (particularly electronics and semiconductors) as well as other competitive concerns led to the creation of the Semiconductor Research Corporation, the Microelectronics and Computer Technology Corporation, the National Center for Manufacturing Sciences, SEMATECH, and many other consortia. (Carr and Hill, 1995)

83. Each SEMATECH member calculates its own "return on investment" (ROI) by estimating returns in the form of improvements in manufacturing processes, savings on in-house R&D, etc. and dividing them by the costs they incur through participation in the consortium, i.e., dues paid and other administrative costs associated with participation in SEMATECH programs.

84. These two programs provide funding for industry-related R&D, and since 1994 both have been threatened repeatedly with elimination by the Republican-controlled Congress.

85. Vonortas (1996) notes that other forms of interfirm alliances predominate in these fields, including technology swaps, licensing, mergers and acquisitions, and marketing agreements.

86. A survey of consortia taken for the NSF in 1974 estimated that year's budget for collaborative R&D was $125 million. Using the NSF figures for total U.S. spending on R&D in 1974 ($32,863 million), the $125 million represented only a 0.4-percent share. In that year, consortia conducting energy-related R&D accounted for nearly half of collaborative R&D (Wolek, 1977).

A more recent estimate of collaborative R&D investment was provided by Albert Link in 1989. On the basis of survey data, he found that among manufacturing industries (which do the lion's share of U.S. industrial R&D) the mean expenditure on collaborative research was 7.3 percent of industry-financed R&D. Using NSF's 1989 figures ($72.1 billion for industrially financed R&D) this estimate produced a figure of $5.3 billion spent in collaborative R&D activities in that year.

Gibson and Rogers (1994) estimated that in 1994 1 percent of U.S. research spending went for collaborative R&D. Using the NSF figure for total R&D in 1994 ($172,550 million), this means collaborative R&D consumed an estimated $1,726 million that year.

In the private sector, levels of R&D expenditure, particularly at a project level, are often treated as proprietary information. This is less true with nonprofit and government members of consortia, but data are nonetheless difficult to come by. Definitional problems also complicate the calculation of a national total for collaborative R&D. For example, by many definitions, Bellcore, the research arm of the regional Bell operating companies, is the country's largest research consortium. Its annual budget is close to $1.5 billion. Gibson and Rogers do not include Bellcore in their list of large U.S. consortia. If they did, the estimated percentage of U.S. investment in collaborative R&D would double.

87. For a useful discussion of organizational factors that have been shown to be important in transferring knowledge between members of R&D consortia, see Smilor and Gibson (1991).

88. Although publishers are generally excluded from this survey since their activities are largely noninteractive, it is worth noting that there are a number of publications that track developing technologies in specific fields. These highly focused publications (both traditional and electronic) use a number of public sources (such as papers in technical journals, patents, technical meetings, and press releases) to locate new technologies. Such publications can serve as a very valuable source of information, especially in the private sector.

89. As one might imagine, data on the technology-related activities of law firms are not readily available, in part due to privacy concerns surrounding the attorney-client privilege. However, a search through the Martindale-Hubble database on the Lexis/Nexis service revealed that 484 U.S. law firms include the word technology to describe their practice. A more specific search for two or more technology-transfer-related key words resulted in only 144 "hits." Some firms may not yet include technology terms in their Martindale-Hubble listing. (Indeed, some states do not allow such specificity.) Thus, it is difficult to be certain that one has identified all the technology-oriented law firms. Whatever the actual number of technology-oriented law firms, it is still a small subset of the over 800,000 firms listed by Martindale-Hubble.

90. For further discussion of university-affiliated incubators, see p. 121.

91. The role of federal-laboratory-affiliated incubators is also addressed on p. 149.

92. For selected findings from NBIA's 10th Anniversary Survey of Business Incubators, including data on incubator types, clientele by industry type, types of services offered, as well as estimates of the average number of firms served, firms "graduated," and the average number or FTE jobs created per incubator, see the NBIA website, <http://www.nbia.org/facts.htm>.

93. One study of the risk and reward of venture financing determined that of the 1,004 investments made by 40 venture partnerships between 1985 and 1992, 17 percent were total write-offs, 29 percent yielded returns that were below or at cost, 38 percent yielded returns at 1 to 5 times cost, 8 percent at 5 to 10 times cost, and another 8 percent yielded returns that were over 10 times cost (Horsley, 1997). For further discussion of recent trends in U.S. venture and equity capital markets, see National Research Council and Committee on Science, Engineering, and Public Policy (1997).

94. See pp. 76–80, 88–90, and Annex II, pp. 201–204, for further discussion.

95. For an overview of the industry-government-university consortium, the American Textile Partnership (AMTEX), and its many research and outreach activities, see the AMTEX home page at <http://amtex.sandia.gov/>.

96. See, for example, recent publications and current research initiatives of the Massachusetts Institute of Technology (MIT) International Motor Vehicle Program on the program's home page <http://web.mit.edu/org/c/ctpid/www/imvp/index.html>. For a recent assessment of the government-industry Partnership for New Generation Vehicles, see National Research Council (1997).

97. For further discussion of these and other industry-led initiatives aimed at the manufacturing technology needs of small and medium-sized firms in more technologically mature U.S. industries, see Part II, *Technology Transfer by Privately Held, Nonacademic Organizations*.

98. The chemical industry road map, *Technology Vision 2020*, was authored by the American Chemical Society, Chemical Manufacturers Association, American Institute of Chemical Engineers, Council for Chemical Research, and the Synthetic Organic Chemical Manufacturers Association (American Chemical Society et al., 1996).

99. See, for example, the discussion of NIST's Manufacturing Extension Partnerships, or of state technology extension deployment programs such as the Thomas Edison Institute in Ohio or the Ben Franklin Partnership in Pennsylvania, pp. 77–79, and Annex II, pp. 204–207, as well as Coburn (1995).

Annex II:
Case Studies in Technology Transfer

1. In semiconductors and flat panel displays, for example, U.S. companies face severe competition from Japanese companies that focused their efforts on commercialization of technology that originated in the United States.

2. The six interagency initiatives were biotechnology research, funded at $4.3 billion; advanced materials and processing, at $2.1 billion; global environmental change research, at $1.5 billion; advanced manufacturing technology, at $1.4 billion; high-performance computing and communications, at $1.0 billion; and science, mathematics, engineering, and technology education, at $2.3 billion (National Science Board, 1993).

3. The effect of these incentives are qualified. First, the royalty system does elicit technology disclosures, but it does not elicit the breakthrough observation. Second, views are divided on the income-generating aspects of technology transfer. In a GAO survey of the top 35 universities, average income for licenses was $1.6 million; 9 universities reported income in excess of $1.0 million and only six reported income in excess of $2.0 million. The GAO concluded that "there is a reasonably high probability that many universities that 'invest' in expanded technology licensing operations to produce income [will fail]" (U.S. General Accounting Office, 1992).

This is consistent with the view expressed by an observer at Stanford that "a technology licensing office requires a minimum critical mass of at least $40-50 million to be justified on economic grounds" (Neils Reimers, personal communication, 1993).

4. See <http://www.covesoft.com/biotech>.

5. "Herb Boyer who was then an assistant professor at UCSF . . . presented his work with . . . a restriction enzyme, and I found that interesting. That night, we took a long walk and ended up near a kosher delicatessen near Waikiki Beach. During that particular discussion, eating overstuffed corned beef sandwiches, I proposed a collaboration with Herb that led to the discovery of recombinant DNA" (Stanley Cohen, remarks, Committee on Technological Innovation in Medicine: The University Industry Interface and Medical Innovation, Stanford University, February 21, 1993).

6. This section is derived from pp. 55–95 in Borrus (1988).

7. It is interesting to observe that, although Japan does not currently challenge U.S. dominance in software or hardware, Japan has nonetheless established a dominant position in the area of embedded

software, especially so-called fuzzy systems. Japanese applications of fuzzy logic currently extend to more than 100 product areas, from video cameras to elevators and subway trains. In 1990, revenues from Japanese consumer goods incorporating fuzzy logic microcontrollers exceeded $1.5 billion (U.S. Department of Commerce, 1994).

8. Estimates of annual revenues were calculated by multiplying 6-month data by a factor of 2.2; these data are not comparable to IDC data cited earlier.

PART III:
TECHNOLOGY TRANSFER IN GERMANY

1. To understand R&D structures, it is interesting to look at relative indicators, especially the national expenditure on R&D in relation to the gross domestic product. In Germany, this so-called GERD factor started at the beginning of the 1980s at a level of 2.45 percent and reached nearly 2.9 percent in 1989. Between 1990 and 1994, this factor declined to 2.3 percent, which can be explained by the unification of West and East Germany and the resulting economic problems.

2. The BMBF was established at the end of 1994 by merging the former Ministry of Education and Science and the Ministry for Education, Science, Research and Technology. This merger documents the growing interest of the federal government in a closer linkage of science and technology.

3. For the universities, only research activities are covered; education activities are excluded.

4. For country comparisons, the analysis of patent applications at the European Patent Office leads to meaningful results, because European applications represent a selection of inventions characterized by their high quality; domestic distortions, which can be observed at national patent offices, play a negligible role (Grupp et al., 1996; Organization for Economic Cooperation and Development, 1994c; Schmoch and Kirsch, 1994).

5. In order to achieve a more differentiated picture, all European patents were classified according to a scheme of 30 technology fields. This classification has been elaborated by the Fraunhofer Institute for Systems and Innovation (FhG-ISI) in cooperation with the French Observatoire des Sciences et des Techniques and the French Patent Office. Because of the different patent volumes in different fields of technology, analysis based on absolute numbers of patents can be misleading. Therefore, a specialization indicator, called revealed patent advantage (RPA), was calculated. The RPA indicates a country's share of patents in a particular field compared with the average of the rest of the world. Positive RPA values indicate above-average activities, and negative ones indicate below-average activities.

6. In Figure 3.3, the criterion for assigning a patent to Germany is the address of the inventor (i.e., the location of the laboratory, not the address of the applicant). Therefore, the U.S.-based activities of German companies are not included. For further details, see Part III, *Technology Transfer in Biotechnology*.

7. The statement of high costs seems not to be true for universities, since they generally do not calculate overhead costs (see Part III, *Statistics on General Research Structures*). Therefore, this impediment is primarily an indication of the limited financial resources of SMEs.

8. For further details, see Part III, *Federation of Industrial Research Associations*.

9. The exact number of science associations is very difficult to determine. Schimank (1988b) identified 374, based on Vademecum (1985). A manual for 1995 (Hoppenstedt Verlag, 1995) records a list of about 400 technical or scientific associations, also including small industry-oriented associations. The major methodological problem with clearly determining an exact count relates to the heterogeneity of organization and targets of the different associations.

10. The impact on research and technology development in Greece, for example, is further demonstrated in a study by Kuhlmann (1992).

11. The following description is based largely on the very comprehensive description of Keck (1993).

12. This figure does not include donations through industry-related foundations (e.g., the Volkswagen Foundation).

13. The data of the German Science Council are based on a survey of a sample of R&D-performing firms. The BMBF data are based on a full survey of German universities. Therefore, the BMBF figures seem to be more realistic. Nevertheless, the data of the German Science Council are documented as they provide a consistent data series of the situation before 1990. In contrast, the BMBF data reflect the development in recent years.

14. These percentages are confirmed by a detailed analysis of the school of mechanical engineering at the University of Karlsruhe.

15. The available data unfortunately put biological sciences and geography into one category.

16. They are located in Cologne, Heidelberg, Munich, Stuttgart, Hamburg (two), Berlin, and Düsseldorf.

17. These figures include an unidentified number of An-Institutes in the social sciences and humanities. Since the An-Institutes are legally independent bodies, their expenditures are not included in the universities' budgets.

18. In Karlsruhe, the latter probably come from the national research center Forschungszentrum Karlsruhe [FZK].

19. The FhG-ISI and the National Academy of Engineering (NAE) reached a preliminary agreement on the focal areas in spring 1995. The binational panel agreed to this suggestion in June 1995, but chose the broader area of information technology instead of software. At that time, the survey was already nearly finished.

20. The persons questioned were asked to limit their answers to research activities in the focal areas.

21. That is, in terms of personnel, not money.

22. A clear delimitation between the different types of research is often not possible, and the respondents may have different perspectives. Nevertheless, the different compositions of the areas are obvious.

23. The nearly 1,000 UIRCs in the United States account for almost 70 percent of industry's support for academic R&D (Cohen et al., 1995).

24. This assumption is confirmed by a manual assignment of professor-related patents of 1985 and 1993, published in Becher et al. (1996). According to this analysis, about 80 percent of patent applications with professors as applicants or inventors actually trace back to universites.

25. In 1994, 26 of the 59 members (44 percent) of the Senate (without guests) were Max Planck scientists; 5 members (8 percent) came from other scientific institutions (Max-Planck-Gesellschaft, 1994a).

26. In 1995, as in previous years, outstanding MPG scientists were honored with Nobel prizes in medicine and chemistry.

27. This figure is based on the publication list of each institute, which might contain fewer than the actual number of recipients of doctoral degrees. In any case, the actual number of doctoral students working at MPG institutes is much higher.

28. To become a full-time professor, it is necessary to write a habilitation thesis, which is a kind of second doctoral thesis. The time needed to research and write this required paper varies. At universities, about 5 years is estimated to be appropriate.

29. In particular, the German Science Council has made various suggestions for improved methods of technology foresight (Wissenschaftsrat, 1994).

30. For more details, see Bundesministerium für Forschung und Technologie (1993a), Fraunhofer-Gesellschaft (1985, 1993), Frisch et al. (1982) , Hohn (1989), Imbusch and Buller (1990), Krupp and Walter (1990), and Syrbe (1989).

31. According to the new Frascati definitions, this type is called basic research (Organization for Economic Cooperation and Development, 1994a).

32. In recent years, about 250 spin-offs of Fraunhofer institutes, employing a total of about 1,000 workers, have been established.

33. Processes exclusively based on empirical knowledge such as traditional beer brewing are not included.

References

Abendroth, D. 1993. Ziele und organisatorische Ausgestaltung des Modellversuchs TOU/NBL. Pp. 23–28 in Statusseminar zum Modellversuch Technologieorientierte Unternehmensgründungen in den neuen Bundesländern. G. Bräunling and F. Pleschak, eds. Karlsruhe: FhG-ISI.

Abernathy, F. H., J. T. Dunlop, J. H. Hammond, and D. Weil. 1995. The information-integrated channel: A study of the U.S. apparel industry in transition. Pp. 175–246 in Brookings Papers on Economic Activity: Microeconomics 1995, M. N. Baily, P. C. Reiss, and C. Winston, eds. Washington, D.C.: Brookings Institution.

Aerospace Industry Association. 1994. Aerospace Facts and Figures, 1994–95. Washington, D.C.: Aerospace Industry Association.

AiF-Verwaltung. 1995. Unpublished data of the AiF management. Köln: AiF.

Alic, J. A., L. M. Branscomb, H. Brooks, A. B. Carter, and G. L. Epstein. 1992. Beyond Spinoff: Military and Commercial Technologies in a Changing World. Boston, Mass.: Harvard Business School Press.

Allesch J., D. Preiß-Allesch, and U. Spengel. 1988. Hochschule und Wirtschaft. Bestandsaufnahmen und Modelle der Zusammenarbeit. Köln: Verlag TÜV Rheinland.

American Association for the Advancement of Science (AAAS). 1995. AAAS Report XX: Research & Development FY 1996. Table I-10 (reporting NSF data).

American Chemical Society, American Institute of Chemical Engineers, Chemical Manufacturers Association, Council for Chemical Research, Synthetic Organic Chemical Manufacturers Association. 1996. Technology Vision 2020: The U.S. Chemical Industry. Washington, D.C.: American Chemical Society.

Amgen, Inc. Various years. Annual Report. Thousand Oaks, Calif.: Amgen, Inc.

Arbeitsgemeinschaft der Großforschungseinrichtungen (AGF). 1994. Programmbudget 1994. Bonn: AGF.

Arbeitsgemeinschaft der Großforschungseinrichtungen. 1995. Programmbudget 1995. Bonn: AGF.

Arbeitsgemeinschaft industrieller Forschungsvereinigungen (AiF). 1991. Handbuch 1991. Köln: AiF.
Arbeitsgemeinschaft industrieller Forschungsvereinigungen. 1992. Report. Transfer und Umsetzung von Ergebnissen der industriellen Gemeinschaftsforschung. Köln: AiF.
Arbeitsgemeinschaft industrieller Forschungsvereinigungen. 1994. Forschungsreport. Öffentlich finanzierte Vorhaben der industriellen Gemeinschaftsforschung. Köln: AiF.
Arbeitsgemeinschaft industrieller Forschungsvereinigungen. No Year. Die Unternehmen in den Mitgliedsvereinigungen der AiF. Daten und Strukturen. Köln: AiF.
Armstrong, J. 1997. Institutional cultures and individual careers. The Bridge 27, 1: 4-10.
Association of University Related Research Parks. 1997. What Is a Research Park? [Online]. Available: http://www.siue.edu/AURRP/whatis.html [September 16, 1997].
Association of University Related Research Parks. 1995. 1994/1995 Worldwide Research and Science Park Directory. Washington, D.C.: AURRP.
Association of University Technology Managers (AUTM). 1993. The AUTM Licensing Survey, Fiscal Years 1991 and 1992. Cranbury, N.J.: AUTM.
Association of University Technology Managers. 1994. The AUTM Licensing Survey: Fiscal Years 1993, 1992, and 1991. Norwalk, Conn.: AUTM.
Association of University Technology Managers. 1995. Working with federal labs. Chapter 2 in AUTM Manual, Part VI. Norwalk, Conn.: AUTM.
Association of University Technology Managers. 1996. AUTM Licensing Survey FY 1991–FY 1995. Norwalk, Conn: AUTM.
BankBoston. 1997. MIT: The Impact of Innovation. A BankBoston Economics Department Special Report. March. Boston, Mass.: BankBoston.
Becher, G., G. Gielow, R. Herden, S. Kuhlmann, and U. Kuntze. 1989. FuE-Personalkostenzuschüsse: Strukturentwicklung, Beschäftigungswirkungen und Konsequenzen für die Innovationspolitik. Final report to the BMWi. Karlsruhe/Berlin: FhG-ISI/ DIW.
Becher, G., T. Gering, O. Lang, and U. Schmoch. 1996. Patentwesen an Hochschulen. Eine Studie zum Stellenwert gewerblicher Schuztrechte im Technologietransfer Hochschule-Wirtschaft. Bonn: Bundesministerium für Bildung, Wissenschaft, Forschung und Technologie.
Beise, M., G. Licht, and A. Spielkamp. 1995. Technologietransfer an kleine und mittlere Unternehmen—Analyse und Perspektiven für Baden-Württemberg. Baden-Baden: Nomos Verlag.
Bianchi-Streit, M., N. Blackburne, R. Budde, H. Reitz, B. Sagnell, H. Schmied, and B. Schorr. 1984. Economic Utility Resulting from CERN Contracts (Second Study). Geneva: CERN.
Bierhals, R., and U. Schmoch. 1997. Aktive Patentpolitik an Einrichtungen der Ressortforschung. Report to the BMWi. Karlsruhe: FhG-ISI.
Bild der Wissenschaft. 1994a. bdw-Patentindex Gentechnik, 11/94:6.
Bild der Wissenschaft. 1994b. Von wegen zweckfrei, 11/94:108–109.
Bilstein, R. E. 1989. Orders of Magnitude: A History of the NACA and NASA, 1915–1990. NASA Special Publication 4406 in the NASA History Series. Washington, D.C.: National Aeronautics and Space Administration.
Blumenthal, D. 1992. Academic-industry relationships in the life sciences. Journal of the American Medical Association 268:3344–3349.

Blumenthal, D., M. Gluck, K. S. Louis, and D. Wise. 1986a. Industrial support of university research in biotechnology. Science 231:242–246.

Blumenthal, D., M. Gluck, K. Louis, M. Stoto, and D. Wise. 1986b. University-industry research relationships in biotechnology: Implications for the university. Science 232:1361–1366.

Borrus, M. 1988. Competing for Control: America's Stake in Microelectronics. Cambridge, Mass.: Ballinger Publishing.

Boston Consulting Group. 1993. The Changing Environment for U.S. Pharmaceuticals: The Role of U.S. Pharmaceutical Companies in a Systems Approach to Healthcare. Boston, Mass.: BCG.

Böttger, J. 1993. Forschung für den Mittelstand: Die Geschichte der Arbeitsgemeinschaft industrieller Forschungsvereinigungen "Otto von Guericke" e.V. (AiF) im wirtschaftspolitischen Kontext. Köln: Deutscher Wirtschaftsdienst.

Bozeman, B., K. Coker, and M. Papadakis. 1995. Industry Perspectives on Commercial Interactions with Federal Laboratories. Atlanta: School of Public Policy, Georgia Institute of Technology.

Brandt, R. 1991. Can the U.S. stay ahead in software? Business Week (March 11):98.

Brooks, H. 1986. National science policy and technological innovation. Pp. 119–167 in The Positive Sum Strategy, R. Landau and N. Rosenberg. eds. Washington, D.C.: National Academy Press.

Brooks, H. B. 1996. The problem of attention management in innovation and sustainability. Technological Forecasting and Social Change 53:21–26.

Bundesministerium für Bildung, Wissenschaft, Forschung und Technologie (BMBF). 1995a. Förderfibel 1995. Förderung von Forschung, Entwicklung und Innovation. Bonn: BMBF.

Bundesministerium für Bildung, Wissenschaft, Forschung und Technologie. 1995b. Gründung der Helmholtz-Gemeinschaft Deutscher Forschungszentren (HGF). Press release of June 29, 1995. Bonn: BMBF.

Bundesministerium für Bildung, Wissenschaft, Forschung und Technologie. 1996. Bundesbericht Forschung 1996. Bonn: BMBF.

Bundesministerium für Bildung, Wissenschaft, Forschung und Technologie. 1997. Zur technologischen Leistungsfähigkeit Deutschlands. Aktualisierung und Erweiterung 1996. Bonn: BMBF.

Bundesministerium für Forschung und Technologie (BMFT). 1988. Bundesbericht Forschung 1988. Bonn: BMFT.

Bundesministerium für Forschung und Technologie. 1992. Situation der Großforschungseinrichtungen, Pressedokumentation 18/92. Bonn: BMFT.

Bundesministerium für Forschung und Technologie. 1993a. Bundesbericht Forschung 1993. Bonn: BMFT.

Bundesministerium für Forschung und Technologie. 1993b. Ratgeber Forschung und Technologie, Ausgabe 1993/94. Köln: Deutscher Wirtschaftsdienst.

Bush, V. 1945. Science—The Endless Frontier. A Report to the President on a Program for Postwar Scientific Research (July 1945). Reprinted by the National Science Foundation, Washington, D.C., 1990.

Carr, R. K. 1995. U.S. Federal Laboratories and Technology Transfer. Prepared for the Binational Study on Technology Transfer Systems in the United States and Germany: Lessons and Perspectives. National Academy of Engineering, Washington, D.C.. [Online]. Available: http://millkern.com/rkcarr/fedlbpap.html [September 15, 1997].

Carr, R. K., and C. T. Hill. 1995. R&D and technology transfer in the United States: The least-known piece of the puzzle. Paper prepared for the Binational Study on Technology Transfer Systems in the United States and Germany: Lessons and Perspectives. Washington, D.C.: National Academy of Engineering. [Online]. Available: http://millkern.com/rkcarr/fourth.html [September 15, 1997].

Coburn, C., ed. 1995. Partnerships: A Compendium of State and Federal Cooperative Technology Partnerships. Columbus, Ohio: Battelle Press.

Cohen, W. M., R. Florida, and W. R. Goe. 1994. University-Industry Research Centers in the United States. Pittsburgh, Pa.: Center for Economic Development, Carnegie Mellon University.

Cohen, W. M., R. Florida, and L. Randazzese. 1995. American University-Industry Research Centers in Biotechnology, Computers, Software, Semiconductors and Manufacturing. Report to the National Academy of Engineering. Pittsburgh, Pa.: Carnegie Mellon University.

Cohen, L. R., and R. G. Noll. 1991. The Technology Pork Barrel. Washington, D.C.: The Brookings Institution.

Cohen, W. M., and L. Randazzese. 1996. Eminence and enterprise: The impact of industry support on the conduct of academic research in science and engineering. Paper presented to the NBER Technology Policy Workshop, August 14, 1996. Pittsburgh, Pa.: Department of Social and Decision Sciences, Carnegie Mellon University.

Commission of the European Communities. 1992. Large Scientific Installations in the Community and the Development of Advanced Technologies. Report of a study group. Brussels: Commission of the European Communities.

Committee on Criteria for Federal Support of Research and Development. 1995. Allocating Federal Funds for Science and Technology. National Academy of Sciences, National Academy of Engineering, Institute of Medicine, and National Research Council. Washington, D.C.: National Academy Press.

Committee on Science, Engineering, and Public Policy. 1992. The Government Role in Civilian Technology: Building a New Alliance. Washington, D.C.: National Academy Press.

Committee on Science, Engineering, and Public Policy. 1996. An Assessment of the National Science Foundation's Science and Technology Centers Program. Washington, D.C.: National Academy Press.

Competitiveness Policy Council. 1993. Reports of the Subcouncils. March. Washington, D.C.

Council on Competitiveness. 1996. Electronics. Pp. 89–103 in Endless Frontier, Limited Resources: U.S. R&D Policy for Competitiveness. Washington, D.C.: Council on Competitiveness.

Council of Consortia CEOs. 1997. Home page. [Online]. Available: http://www.oai.org/CofC/index.html [June 15, 1997].

Creedon, J. F. 1992. NASA Technology Transfer. Report of the Technology Transfer Team, J. F. Creedon, chair. Washington, D.C.: National Aeronautics and Space Administration.

Dasgupta, P., and P. David. 1994. Toward a new economics of science. Research Policy 23:487–521.

David, P. A., D. C. Mowery, and W. E. Steinmuller. 1992. Analyzing the economic payoffs from basic research. Economics of Innovation and New Technology 2:73–90.

Defense Contract Audit Agency. 1997. Independent Research and Development and Bid and Proposal Costs Incurred by Major Defense Contractors in the Years 1995 and 1996. April 1997. Washington, D.C.: Defense Contract Audit Agency.

Defense Science Board. 1994. Defense Science Board Task Force on Lab Management (DSB-LM), Interim Report, 14 April 1994. [Online]. Available: http://www.dtic.dla.mil/labman/projects/interim.html [1997, June 15].

Deutsche Forschungsgemeinschaft. 1993. Jahresbericht 1993, Band 1: Aufgaben und Ergebnisse. Bonn: Deutsche Forschungsgemeinschaft.

Deutsches Patentamt (DPA). 1993. Jahresbericht 1993. München: DPA.

Deutsches Patentamt. 1995. Jahresbericht 1995. München: DPA.

Dibner, M. D. 1988. Biotechnology Guide U.S.A. Basingstoke/New York: Macmillan Publishers.

Dickens, C. H. 1996. The academic engineering research enterprise: Status and trends. Pp. 69–132, in Forces Shaping the U.S. Academic Engineering Research Enterprise, National Academy of Engineering. Washington, D.C.: National Academy Press.

Doctors, S. I. 1971. The NASA Technology Transfer Program: An Evaluation of the Dissemination System. Westport, Conn.: Praeger Publishers.

Dolata, U. 1995. Nachholende modernisierung und internationales innovationsmanagement. Strategien der deutschen chemie- und pharmakonzerne in der neuen biotechnologie. Pp. 456–480 in Biotechnologie - Gentechnik. Eine Chance für neue Industrien. T. Von Schell and H. Mohr, eds. Berlin/Heidelberg/New York: Springer Verlag.

Dose, N. 1993. Alte und neue handlungsformen staatlicher steuerung im bereich von forschung und technologie. Pp. 399–452 in Instrumente und Formen staatlichen Handelns. K. König and N. Dose, eds. München: Heymann.

Eberhardt, D. 1989. Campusnahe entwicklungszentren (Beispiel: Science-park in der wissenschaftsstadt ulm). Pp. 207–222 in Aus der Praxis des Technologietransfers Wissenschaft / Wirtschaft. Bericht der Arbeitsgruppe Technologietransfer der Kanzler und leitenden Verwaltungsbeamten der wissenschaftlichen Hochschulen der Bundesrepublik Deutschland, G. Selmayer, ed. Karlsruhe: Universität Karlsruhe.

Economist (The). 1994. Technology in Utah: Software valley. (April 3):69.

Eichborn, v. J. F. 1985. Die wirtschaftliche nutzung der biotechnologie. In Biotechnologie. Herrschaft oder Beherrschbarkeit einer Sclüsseltechnologie. Proceedings of a conference on November 23/24, 1994. München: Hans-Böckler-Stiftung.

Ergas, H. 1987. Does technology policy matter? Pp. 191–245 in Technology and Global Industry: Companies and Nations in the World Economy. H. Brooks and B. Guile, eds. Washington, D.C.: National Academy Press.

Ernst and Young. 1993. Biotech 94. Long-Term Value, Short-Term Hurdles. San Francisco: Ernst and Young.

Etzkowitz, H. 1988. Making of an entrepreneurial university: The traffic among MIT, industry and the military, 1860–1960. In Science, Technology and the Military, Vol. 12, E. Mendelsohn, M. R. Smith and P. Weingart, eds. Norwell, Mass.: Kluwer Academic Publishers.

EUREKA. 1993. Evaluation of EUREKA industrial and economic effects. Paris: EUREKA.

EUREKA. 1995. Technologische Zusammenarbeit in Europa. Dokumentation 1995. Köln: EUREKA.

Executive Office of the President. Office of Science and Technology Policy. 1995. Interagency Federal Laboratory Review. Final Report. Washington, D.C.: U.S. Government Printing Office.

Federal Coordinating Council for Science, Engineering, and Technology. 1993. FCCSET Initiatives in the FY 1994 Budget. Washington, D.C.: Executive Office of the President, Office of Science and Technology Policy.

Feller, I. 1990. Universities as engines of R&D-based economic growth: They think they can. Research Policy 19:335–348.

Feller, I. 1994. The university as an instrument of state and regional economic development: The rhetoric and reality of the U.S. experience. Paper presented at the CERP/AAAS Conference on "University Goals, Institutional Mechanisms, and the Industrial Transferability of Research," Stanford, Calif.: Stanford University, March 18–20.

Ferguson, C., and C. Morris. 1993. Computer Wars: The Fall of IBM and the Future of Global Technology. New York: Basic Books.

Ferne, G., and P. Quintas. 1991. Software Engineering: The Policy Challenge. Paris: Organization for Economic Cooperation and Development.

FhG-Zentralverwaltung. 1995. Unpublished data of the central administration of the Fraunhofer Society. München: FhG.

Florida, R., and D. F. Smith. 1993. Venture capital and industrial competitiveness. A research report to the U.S. Economic Development Administration. Pittsburgh, Pa: Carnegie Mellon University.

Fonds der Chemischen Industrie. 1995. Leistungsbilanz. Förderaufwendungen 1950 bis 1995. Frankfurt: Fonds der Chemischen Industrie.

Forschungszentrum Karlsruhe. 1996. Forschungszentrum Karlsruhe. Technik und Umwelt. Auf einen Blick. Karlsruhe: Forschungszentrum Karlsruhe.

Fraunhofer-Gesellschaft (FhG). 1985. Modell Fraunhofer Gesellschaft, Entwicklung, Analyse, Fortschreibung. München: FhG.

Fraunhofer-Gesellschaft. 1988. Fraunhofer-Verbund Mikroelektronik. München: FhG.

Fraunhofer-Gesellschaft. 1993. Jahresbericht 1993. München: FhG.

Fraunhofer-Gesellschaft. 1994. Jahresbericht 1994. München: FhG.

Freeman, C. 1968. Chemical process plant: Innovation and the world market. National Institute Economic Review 45(August):29–57.

Friedland, J. 1993. Japan's soft spot. Far Eastern Economic Review (August 5):54.

Frisch, F., A. Imbusch, and W. Zitzelsberger. 1982. Die vertragsforschung in der FhG: Ein unternehmerisches konzept und seine realisierung. FhG-Berichte 4/82:4–10.

Gaden, E. L., Jr. 1991. Biotechnology. Pp.103–109 in National Interests in an Age of Global Technology, T. H. Lee and P. P. Reid, eds. Washington, D.C.: National Academy Press.

Gale Research. 1996. Research Centers Directory, 1996–1997. Detroit, Mich.: Detroit, Mich.: Gale Research.

Gehrke, B., and H. Grupp. 1994. Innovationspotential und Hochtechnologie. Technologische Position Deutschlands im internationalen Wettbewerb. 2. Auflage. Heidelberg: Physica-Verlag.

Geimer, H., and R. Geimer. 1981. Research Organisation and Science Promotion in the Federal Republic of Germany. München/New York/London: Saur-Verlag.

Gibson, D. V., and E. M. Rogers. 1994. R&D Collaboration on Trial. Cambridge, Mass.: Harvard Business School Press.

Government-University-Industry Research Roundtable. 1991. Industrial Perspectives on Innovation and Interaction with Universities. Washington, D.C.: National Academy Press.

Government-University-Industry Research Roundtable, Industrial Research Institute, and Council on Competitiveness. 1996. Industry-University Research Collaborations. Washington, D.C.: National Academy Press.

Gray, D. O., W. A. Hetzner, J.D. Eveland, and T. Gidley. 1986. NSF's industry/university cooperative research centers program and the innovation process: Evaluation-based lessons. Pp. 175–193 in Technological Innovation: Strategies for a New Partnership, D. O. Gray, T. Solomon, and W. A. Hetzner, eds. Amsterdam: Elsevier Science Publishers B. V. (North Holland).

Gray, D., T. Gidley, and N. Koester. 1988. Evaluation of the NSF Industry/University Cooperative Research Centers: Longitudinal Analyses of Outcome and Process. Washington, D.C.: National Science Foundation.

Grindley, P. 1996. The future of the software industry in the United Kingdom: The limitations of independent production. Pp. 197–239 in The International Computer Software Industry: A Comparative Study of Industrial Evolution and Structure, D. Mowery, ed. New York: Oxford University Press.

Grupp, H., ed. 1993. Technologie am Beginn des 21. Jahrhunderts. Heidelberg: Physica-Verlag.

Grupp, H., G. Münt, and U. Schmoch. 1995. Wissensintensive Wirtschaft und ressourcenschonen Technik, Teile D and E: Abgrenzung der Technik- und Wissenschaftsgebiete und Potentialanalyse. Report to the BMBF. Karlsruhe: FhG-ISI.

Grupp, H., G. Münt, and U. Schmoch. 1996. Assessing different types of patent data for describing high-technology export performance. Pp. 271–287 in Innovation, Patents and Technological Strategies. Organization for Economic Cooperation and Development (OECD), ed. Paris: OECD.

Hagedoorn, J. 1995. Strategic technology partnering during the 1980s: Trends, networks, and corporate patterns in non-core technologies. Research Policy 24:207–231.

Hamburg Institute for Economic Research, Kiel Institute for World Economics, and National Research Council. 1996. Conflict and Cooperation in National Competition for High-Technology Industry. Washington, D.C.: National Academy Press.

Harhoff, D., G. Licht, and M. Smid. 1995. Innovationsverhalten der deutschen Wirtschaft. Die Innovationsaktivitäten kleiner und mittlerer Unternehmen. Report to the BMBF. Mannheim: ZEW.

Hartl, M., and W.-P. Hentschel. 1989. Abwicklung von drittmittelaufträgen über private institute von hochschullehrern und wissenschaftlichen mitarbeitern. Pp. 11–54 in Aus der Praxis des Technologietransfers Wissenschaft / Wirtschaft. Bericht der Arbeitsgruppe Technologietransfer der Kanzler und leitenden Verwaltungsbeamten der wissenschaftlichen Hochschulen der Bundesrepublik Deutschland, G. Selmayer, ed. Karlsruhe: Universität Karlsruhe.

Häusler, J. 1989. Industrieforschung in der Forschungslandschaft der Bundesrepublik: Ein Datenbericht. Discussion Paper. Köln: Max-Planck-Institut für Gesellschaftsforschung.

Henderson, R., A. B. Jaffe, M. Trajtenberg. 1994. Numbers up, quality down? Trends in university patenting 1965–1992. Presentation at the CEPR/AAAS Conference "University Goals, Institutional Mechanisms, and Industrial Transferability of Research," Stanford Calif.: Stanford University, March 18–20.

Henderson, R., A. Jaffe, and M. Trajtenberg. 1995. Universities as a Source of Commercial Technology: A Detailed Analysis of University Patenting, 1965–88. NBER Working Paper No. 5068. Cambridge, Mass.: National Bureau of Economic Research.

Herden, R. 1992. Technologieorientierte Außenbeziehungen im betrieblichen Innovationsmanagement. Heidelberg: Physica-Verlag.

Herrmann, W.A. 1995. Industrie und hochschule. Gemeinsam sind wir stärker. Chemie heute edition 95/96:58–61.

Hetzner, W. A., T. R. Gidley, and D. O. Gray. 1989. Cooperative research and rising expectations: Lessons from NSF's Industry/University Cooperative Research Centers. Technology in Society 11:335–345.

Hill, C. T. 1995. Review of R&D Collaboration on Trial: The Microelectronics and Computer Technology Corporation, by D. V. Gibson and E. M. Rogers. Issues in Science and Technology 11(3):87–90.

Hilts, P. 1993a. Lab limits plan to give company discoveries. Science Times: New York Times, July 13. Sec. C, p.12.

Hilts, P. 1993b. Research group's tie to drug maker is questioned. New York Times. June 18. Sec. A, p.24.

Hochschulrektorenkonferenz (HRK). 1996. Zur Finazierung der Hochschulen. Dokumente zur Hochschulreform 110/1996. Bonn: Hochschulrektorenkonferenz.

Hohmeyer, O., B. Hüsing, S. Maßfeller, and T. Reiß. 1994. Internationale Regulierung der Gentechnik. Praktische Erfahrungen in Japan, den USA und Europa. Heidelberg: Physica-Verlag.

Hohn, H. W. 1989. Forschungspolitik als Ordnungspolitik. Das Modell Fraunhofer-Gesellschaft und seine Genese im Forschungssystem der Bundesrepublik Deutschland. Köln: Max -Planck-Institut für Gesellschaftsforschung.

Hooper, L. 1993. The creative edge: Nurturing high-tech talent requires a delicate balancing act; but the payoff can be huge. Wall Street Journal (May 24):R6.

Hoppenstedt Verlag. 1995. Verbände, Behörden, Organisationen der Wirtschaft. Darmstadt: Verlag Hoppenstedt.

Horsley, P. 1997. Trends in private equity. Presentation to the Committee on Science Engineering and Public Policy/National Research Council Workshop "The Role of Private Finance in Capitalizing on Research," Washington, D.C., National Academy of Sciences, April 21.

Humphrey, A. E. 1993. Engineering challenges in rDNA technology. Annuals of the New York Academy of Sciences 721:1–11.

Humphrey, A. E. 1995. R&D and technology transfer in biotechnology. Presentation to the National Academy of Engineering/Fraunhofer Society binational panel on Technology Transfer Systems in the United States and Germany, Freising, Germany, November 8.

Imbusch, A., and U. Buller. 1990. Die funktionen der Fraunhofer-Gesellschaft im innovationsprozeß. Pp. 373–388 in Handbuch des Wissenschaftstransfers, H. J. Schuster, ed. Berlin/Heidelberg/New York: Springer-Verlag.

Internal Revenue Service. 1995. 1992, S Corporation—All Returns: Balance Sheet by Major Industrial Group. [Online]. Available: http://www.irs.gov/plain/tax_stats/soi/corp_gen.html [September 15, 1997].

JESSI. 1995. Excerpts from the April Review 1995. München: JESSI.

Kantzenbach, E., and M. Pfister. 1995. Nationale Konzeptionen der Technologiepolitik in einer globalisierten Weltwirtschaft. Der Fall Deutschlands und der Europäischen Union. HWWA-Report No. 154. Hamburg: HWWA.
Kash, D. E. 1989. Perpetual Innovation: The New World of Competition. New York: Basic Books.
Keck, O. 1993. The national system of technical innovation in Germany. Pp. 115–157 in National Innovation Systems. A Comparative Analysis, R. R. Nelson, ed. New York/Oxford: Oxford University Press.
Kircher, M. 1993. Zur situation allgemeiner und angewandter gentechnik in Deutschland. BioEngineering 2:16f.
Kline, S. J. 1990. Models of Innovation and Their Policy Consequences. Report INN-4B. Stanford, Calif.: Stanford University.
Kline, S. J., and N. Rosenberg. 1986. An Overview of Innovation. Pp. 275–305 in Positive Sum Strategy, R. Landau and N. Rosenberg, eds. Washington, D.C.: National Academy Press.
Klodt, H. 1995. Technologiepolitik in Europa. Konflikte zwischen nationaler und gemeinschaftlicher kompetenz. Wissenschaftsmanagement 1/3:122–125.
Knorr, C., U. Schmoch, P. Keen, and D. Agrafiotis. 1996. Legal and Institutional Constraints and Opportunities for the Dissemination and Exploitation of R&D Activities, Scientific Institutions in Plant Biotechnology. Report to the Commission of the European Union. Karlsruhe: FhG-ISI.
Kommission der Europäischen Gemeinschaften. 1990. Forschungs- und Technologieförderung der EG. Brussels: Kommission der Europäischen Gemeinschaften.
KoWi (Koordinierungsstelle EG der Wissenschaftsorganisationen). 1992. Forschungs- und Technologieförderung der Europäischen Gemeinschaft: Ziele, Instrumente, Prinzipien. Brussels/Bonn: KoWi.
Krüger, H. 1995. Institut an der universität. Wissenschaftsmanagement 1:42–43.
Krull, W., and F. Meyer-Krahmer. 1996. Science, technology, and innovation in Germany—Changes and challenges in the 1990s. Pp. 3–29 in Science and Technology in Germany, W. Krull and F. Meyer-Krahmer, eds. London: Cartermill Publishing.
Krupp, H., and G. Walter. 1990. Die Fraunhofer-Gesellschaft in den 90er Jahren - gesellschaftliche Verantwortung, Internationalität und Mitbestimmung bei Forschung und Entwicklung. Karlsruhe: FhG-ISI.
Kuhlmann, S. 1991. Report: Federal Republic of Germany. Pp. 75–142 in The University-Industry and Research-Industry Interfaces in Europe, S. Kuhlmann, ed. Brussels/Luxembourg: Commission of the European Communities.
Kuhlmann, S. 1992. Thematic Evaluation of Community Support Frameworks for Research and Technology Development in Greece. Karlsruhe: FhG-ISI.
Kuhlmann, S., and U. Kuntze. 1991. R&D Cooperation by Small and Medium-sized Companies. Pp. 709–712 in Proceedings of PICMET 1991. Portland International Conference on Management of Engineering and Technology, Portland, Oregon.
Kulicke, M. 1990. Regionale Beteiligungsgesellschaften in der Bundesrepublik Deutschland 1990 - ihre Geschäftsfelder, Beteiligungsstrategien und jüngeren Strategieänderungen. Karlsruhe: FhG-ISI.
Kulicke, M. 1993. Chancen und Risiken junger Technologieunternehmen. Heidelberg: Physica-Verlag.

Kulicke, M. 1994. Haben wir einen Mangel an technologieorientierten Unternehmen in Deutschland im internationalen Vergleich? Karlsruhe: FhG-ISI.

Kulicke, M., and U. Wupperfeld. 1996. Beteiligungskapital für junge Technologieunternehmen. Ergebnisse des Modellversuchs "Beteiligungskapital für junge Technologieunternehmen." Report to the BMBF. Karlsruhe: FhG-ISI.

Lageman, B., W. Friedrich, M. Körbel, A. Oberheitmann, and F. Welter. 1995. Der volkswirtschaftliche Nutzen der industriellen Gemeinschaftsforschung für die mittelständische Industrie. Essen: RWI.

Lampe, D., and S. Rosegrant. 1992. Route 128: Lessons from Boston's High-Tech Community. New York: Basic Books.

Licensing Executives Society. 1993. Consultants & Brokers in Technology Transfer. Hartford, Conn.

Lundgren, P. 1979. Technisch-wissenschaftliche vereine zwischen wissenschaft, staat und industrie, 1860–1914: Umrisse eines Forschungsfeldes. Technikgeschichte 46:181–191.

Lütz, S. 1993. Die Steuerung industrieller Forschungskooperationen. Funktionsweise und Erfolgsbedingungen des staatlichen Förderinstrumentes Verbundforschung. Frankfurt/New York: Campus-Verlag.

Management-Informationen. 1995. Großforschungseinrichtungen und industrie. Management-Informationen/Politik, April 10, 1995, Nr. 644/655:2–3.

Mansfield, E. 1988. Industrial R&D in Japan and the United States: A comparative study. American Economic Review 78(2):223–228.

Mansfield, E. 1995. Academic research underlying industrial innovations: Sources, characteristics, and financing. The Review of Economics and Statistics (February).

Markusen, A, and M. Oden. 1996. National laboratories as business incubators and region builders. The Journal of Technology Transfer 21(1-2):93–108.

Matkin, G. W. 1990. Technology Transfer and the University. American Council on Education and Macmillan Publishing Co. New York: Macmillan Publishing Co.

Max-Planck-Gesellschaft. 1994a. Jahrbuch 1994. Veröffentlichungen. Göttingen/ München: Verlag Vandenhoek & Rupprecht.

Max-Planck-Gesellschaft. 1994b. Jahrbuch der Max-Planck-Gesellschaft 1994. Göttingen/München: Verlag Vandenhoek & Rupprecht.

Max-Planck-Gesellschaft. 1994c. Jahresbericht 1994. München: MPG.

Max-Planck-Gesellschaft. 1995. Wissen für das 21. Jahrhundert. München: MPG

Max-Planck-Gesellschaft. Various years. Jahrbuch der Max-Planck-Gesellschaft. Göttingen/München: Verlag Vandenhoek & Rupprecht.

Mayntz, R. 1991. Scientific research and policy invention: The structural development of public-financed research in the Federal Republic of Germany. Pp. 45–60 in The University Within the Research System: An International Comparison, A. Orsi Battaglini and F. Roversi Monaco, eds. Baden-Baden: Nomos-Verlag.

McNaughton, W. P., and B. Dooley. 1995. Demonstration projects for diffusion of multifaceted technology: Characterization and strategy. International Journal of Technology Management 10(2/3).

Merrill Lynch. 1996. Investment insights: Biotechnology—On the verge. [Online] Available: http://merrill-lynch.ml.com/ [March 27, 1996].

Meusel, E.-J. 1990. Einrichtungen der großforschung und wissenstransfer. Pp. 359–371 in Handbuch des Wissenschaftstransfers, H. J. Schuster, ed. Berlin/Heidelberg/New York: Springer-Verlag.

Meyer-Krahmer, F. 1990. Science and Technology in the Federal Republic of Germany. Harlow: Longman.

Meyer-Krahmer, F. 1996. Fraunhofer 2000. Strategies of applied research. Pp. 145–162 in Science and Technology in Germany, W. Krull and F. Meyer-Krahmer, eds. London: Cartermill Publishing.

Ministry of International Trade and Industry. 1989. Small Business in Japan 1989: White Paper on Small and Medium Enterprises. Tokyo, Japan.

Morgan, R. M., D. E. Strickland, N. Kannankutty, and J. Grillon. 1994a. How engineering faculty view academic research: Part II. ASEE Prism (November).

Morgan, R. M., D. E. Strickland, N. Kannankutty, and J. Grillon. 1994b. Summary of responses to engineering research in U.S. universities: Survey of engineering faculty. Unpublished paper of preliminary survey results. April.

Mossinghoff, G. 1992. Parental Guidance for Orphan Drugs. Letter to the Editor, Wall Street Journal, April 17. Sec. A, p.11.

Mowery, D. 1996. Introduction. Pp. 3–14 in The International Computer Software Industry: A Comparative Study of Industrial Evolution and Structure, D. Mowery, ed. New York: Oxford University Press.

Mowery, D., and R. Langlois. 1996. The federal government role in the development of the U.S. software industry. Pp. 53–85 in The International Computer Software Industry: A Comparative Study of Industrial Evolution and Structure, D. Mowery, ed. New York: Oxford University Press.

Mowery, D. C., and Nathan Rosenberg. 1989. Technology and the Pursuit of Economic Growth. New York: Cambridge University Press.

Mowery, D. C., and N. Rosenberg. 1993. The U.S. national innovation system. Pp. 29–75 in National Innovation Systems: A Comparative Analysis, R. R. Nelson, ed. New York: Oxford University Press.

Mowery, D. C., and A. A. Ziedonis. 1997. The commercialization of national lab technology through the formation of "spinoff" firms: Evidence from Lawrence Livermore National Laboratory. Unpublished paper. Haas School of Business, University of California at Berkeley.

National Academy of Engineering. 1989. Assessment of the National Science Foundation's Engineering Research Centers Program. Washington, D.C.: National Academy Press.

National Academy of Engineering. 1993. Mastering a New Role: Shaping Technology Policy for National Economic Development. Washington, D.C.: National Academy Press.

National Academy of Engineering. 1995a. Forces Shaping the U.S. Academic Engineering Research Enterprise. Washington, D.C.: National Academy Press.

National Academy of Engineering. 1995b. Revolution in the U.S. Information Infrastructure. Washington, D.C.: National Academy Press.

National Academy of Engineering. 1995c. Risk and Innovation: The Role of Small High-Tech Companies in the U.S. Economy. Washington, D.C.: National Academy Press.

National Academy of Engineering. 1996a. Defense Software Research, Development, and Demonstration: Capitalizing on Continued Growth in Private-Sector Investment. Summary of a Workshop, July 17, 1995. Washington, D.C.: National Academy of Engineering.

National Academy of Engineering. 1996b. Foreign Participation in U.S. R&D: Asset or Liability? Washington, D.C.: National Academy Press.

National Academy of Sciences, National Academy of Engineering, Insitute of Medicine, National Research Council. 1995. Allocating Federal Funds for Science and Technology. Washington, D.C.: National Academy of Press.

National Aeronautics and Space Administration. 1994. NASA Commercial Technology: Agenda for Change. Washington, D.C.:

National Business Incubator Association. 1992. The State of the Business Incubation Industry, 1991. Athens, Ohio: NBIA.

National Business Incubator Association. 1997. [Online] Available: http://www.nbia.org/whatis.htm [September 15, 1997].

National Institute of Standards and Technology. 1997. NIST at a Glance. [Online]. Available: http://www.nist.gov/public_affairs/guide/glpage.htm [September 15, 1997].

National Research Council. 1983. Risk Assessment in the Federal Government: Managing the Process. Washington, D.C.: National Academy Press.

National Research Council. 1987. Agricultural Biotechnology: Strategies for National Competitiveness. Washington, D.C.: National Academy Press.

National Research Council. 1990. Keeping the U.S. Computer Industry Competitive: Defining the Agenda: A Colloquium Report. Washington, D.C.: National Academy Press.

National Research Council. 1992a. Computing the Future: A Broader Agenda for Computer Science and Engineering. Washington, D.C.: National Academy Press.

National Research Council. 1992b. Putting Biotechnology to Work: Bioprocess Engineering. Washington, D.C.: National Academy Press.

National Research Council. 1992c. The Government Role in Civilian Technology: Building a New Alliance. Washington, D.C.: National Academy Press.

National Research Council. 1992d. U.S.-Japan Technology Linkages in Biotechnology: Challenges for the 1990s. Office of Japan Affairs. Washington, D.C.: National Academy Press.

National Research Council. 1993a. Learning to Change: Opportunities to Improve the Performance of Smaller Manufacturers. Washington, D.C.: National Academy Press.

National Research Council. 1993b. Survey of Doctorate Recipients. Unpublished data. Office of Scientific and Engineering Personnel. Washington, D.C.: National Research Council.

National Research Council. 1994. Science and Judgment in Risk Assessment. Washington, D.C.: National Academy Press.

National Research Council. 1995a. Maximizing U.S. Interests in Science and Technology Relations with Japan: Report of the Defense Task Force. Washington, D.C.: National Academy Press.

National Research Council. 1995b. Unit Manufacturing Processes. Manufacturing Studies Board. Washington, D.C.: National Academy Press.

National Research Council. 1996. An Assessment of the National Science Foundation's Science and Technology Centers Program. Washington, D.C.: National Academy Press.

National Research Council. 1997. Review of the Research Program of the Partnership for a New Generation of Vehicles: Third Report. Washington, D.C.: National Academy Press.

National Research Council and Committee on Science, Engineering, and Public Policy. 1997. Financing Technology-Based Start-Ups. Washington, D.C.: National Academy Press. Forthcoming.

National Science Board. 1989. Science & Engineering Indicators—1989. NSB 89-1. Washington, D.C.: U.S. Government Printing Office.

National Science Board. 1992. The Competitive Strength of U.S. Industrial Science and Technology: Strategic Issues. NSB-92-138. Committee on Industrial Support for R&D. Arlington, Va.: National Science Foundation.

National Science Board. 1993. Science & Engineering Indicators—1993. Washington, D.C.: U.S. Government Printing Office.

National Science Board. 1996. Science & Engineering Indicators—1996. NSB 96-21. Washington, D.C.: U.S. Government Printing Office.

National Science Foundation. 1990. National Patterns of R&D Resources: 1990. NSF 90-316. Washington, D.C.: U.S. Government Printing Office.

National Science Foundation. 1991. Federal Funds for Research and Development: Fiscal Years 1989, 1990, 1991. Washington, D.C.: National Science Foundation.

National Science Foundation. 1992a. National Patterns of R&D Resources: 1992. NSF 92-330. Washington, D.C.: National Science Foundation.

National Science Foundation. 1992b. Research and Development in Industry: 1990. Arlington, Virginia.: National Science Foundation.

National Science Foundation. 1994. Federal R&D Funding by Budget Function: Fiscal Years 1993-95. NSF 94-319. Washington D.C.: National Science Foundation.

National Science Foundation. 1995a. Federal Scientists and Engineers: 1989-93. NSF 95-336. Arlington, Va.: National Science Foundation.

National Science Foundation. 1995b. National Patterns of R&D Resources: 1994. Arlington, VA: National Science Foundation.

National Science Foundation. 1995c. Research and Development in Industry: 1992. NSF 95-324. Arlington, Virginia.: National Science Foundation.

National Science Foundation. 1996a. Academic Science and Engineering: Research and Development Expenditures, FY 1994. [Online]. Available: http://www.nsf.gov./sbe/srs/dexp/94dst/start.htm [September 15, 1997].

National Science Foundation. 1996b. National Patterns of R&D Resources: 1996. Arlington, Va.: National Science Foundation.

National Science Foundation. 1996c. Research and Development in Industry: 1994, (early release tables). Arlington, Va.: National Science Foundation.

National Science Foundation. 1997. National Patterns of R&D Resources: 1997. Arlington, Va.: National Science Foundation.

National Technical Information Service/National Technology Transfer Center. 1994. Directory of Federal Laboratory and Technology Resources, 5th Edition. [Online] Available: http://iridium.nttc.edu/brs.html [October 2, 1995].

Nelson, R. R. 1989. What is private and what is public about technology? Science, Technology, and Human Values 14(Summer):229-241.

Nelson, R. R., and R. Levin. 1986. The Influence of Science University Research and Technical Societies on Industrial R&D and Technical Advance. Policy Discussion Paper Series Number 3. Research Program on Technology Change. New Haven, Conn.: Yale University.

Nelson, R. R., and N. Rosenberg. 1993. Technical innovation and national systems. Pp. 3–21 in National Innovation Systems: A Comparative Analysis, R. R. Nelson, ed. New York: Oxford University Press.

NERAC, Inc. 1997. home page. [Online] Available: http//www.nerac.com.80/about/about.html [June 15, 1997].

Office of Science and Technology Policy. 1994. High Performance Computing and Communications: Towards a National Infrastructure. Washington, D.C.: U.S. Government Printing Office.

Office of Science and Technology Policy. 1995. Critical Technologies Report. Washington, D.C.: U.S. Government Printing Office.

Olson, S. 1986. Biotechnology: An Industry Comes of Age. Washington, D.C.: National Academy Press.

Organization for Economic Cooperation and Development. 1985. Software: An Emerging Industry. Paris: Organization for Economic Cooperation and Development.

Organization for Economic Cooperation and Development (OECD). 1994a. Frascati Manual. Proposed Standard Practice for Surveys of Research and Experimental Development. Paris: OECD.

Organization for Economic Cooperation and Development. 1994b. Using Patent Data as Science and Technology Indicators. Patent Manual 1994. Paris: OECD.

Organization for Economic Cooperation and Development. 1995. Main Science and Technology Indicators. 1995/2. Paris: OECD.

Organization for Economic Cooperation and Development. 1996a. Main Science and Technology Indicators, 1996/2. Paris: OECD.

Organization for Economic Cooperation and Development. 1996b. OECD in Figures: Statistics on the Member countries. 1996 Edition. Published with The OECD Observer, No. 200, June/July.

Organization for Economic Cooperation and Development. 1996c. Research and Development in Industry 1973–93. Paris: OECD.

Pavitt, K. 1991. What makes basic research economically useful? Research Policy 20:109–119.

Perry, W. J. 1995. Memorandum on DOD Domestic Technology Transfer/Dual Use Technology. U.S. Department of Defense Website http://www.dtic.dla.mil/techtransit/techtransfer/perry.html.

Pfirrmann, O., U. Wupperfeld, and J. Lerner. 1997. Venture Capital and New Technology-Based Firms. A U.S.-German Comparison. Heidelberg: Physica-Verlag.

Püttner, G., and U. Mittag. 1989. Rechtliche Hemmnisse der Kooperation zwischen Hochschulen und Wirtschaft. Baden-Baden: Nomos Verlagsgesellschaft.

Rabbitt, M. C. 1997. The United States Geological Survey: 1879-1989, USGS Circular 1050 [Online]. Available: http://www.usgs.gov/reports/circulars/c1050/05-20-97_c1050.html [September 15, 1997].

Randazzese, L. P. 1996. Exploring university-industry technology transfer of CAD technology. IEEE Transactions on Engineering Management 43(4):393–401.

Rappa, M., and K. Debackere. 1992. Technological communities and the diffusion of knowledge. R&D Management 22/3:209–220.

Rea, D., H. Brooks, R. M. Burger, and R. LaScala. 1996. The semiconductor industry—A model for industry/university/government cooperation. Research and Technology Management (forthcoming).

Read, J., and K. Lee, Jr. 1994. "Datawatch." Health care innovation: Progress report and focus on biotechnology. Health Affairs (Summer):215–225.

Rees, J. 1990. Industry Experience with Technology Research Centers. Washington, D.C.: U.S. Economic Development Administration.

Reger, G. 1997. Koordination und strategisches Management internationaler Innovationsprozesse. Heidelberg: Physica-Verlag.

Reger, G., and S. Kuhlmann. 1995. European Technology Policy in Germany. The Impact of European Community Policies upon Science and Technology in Germany. Heidelberg: Physica-Verlag.

Reingold, J. 1995. Under watchful eyes: What's behind the sudden improvement in the FDA's notoriously slow drug approval process? Financial World (August 1):40–41.

Reinhard, M., and H. Schmalholz. 1996. Technologietransfer in Deutschland. Stand und Reformbedarf. Berlin/München: Duncker & Humblot.

Reiss, T., and B. Hüsing. 1992. Potentialanalyse der Auftragsforschung in der Biotechnologie. Report to the BMBF and the board of directors of the Fraunhofer Society Karlsruhe. Karlsruhe: FhG-ISI.

Roberts, E. 1991. Entrepreneurs in High Technology: Lessons from MIT and Beyond. New York: Oxford University Press.

Roessner, J. D. 1993. Patterns of Industry Interaction with Federal Laboratories. Atlanta: School of Public Policy, Georgia Institute of Technology.

Rosenberg, N., and R. Nelson. 1994. American universities and technical advance in industry. Research Policy 23:323–348.

Rosenbloom, R., and W. Spencer. 1996. The transformation of industrial research. Issues in Science and Technology (Spring):68–74.

Saccocio, D. 1996. Advanced displays and visual systems. Included in Small Companies in Six Industries: Background Papers for the NAE Risk and Innovation Study. Washington, D.C.: National Academy Press.

Scherer, F. M. 1996. The size distribution of profits from innovation. Presentation at the International Conference on the Economics and Econometrics of Innovation, Strasbourg, June 3.

Schiele, O. H. 1993. The importance of cooperative research for small and medium-sized enterprises (SMEs) as the foundation for a successful free enterprise system. Speech at the 20th Annual Apparel Research Conference, Atlanta, Georgia. December 1–2, 1993.

Schimank, U. 1988a. Institutionelle Differenzierung und Verselbständigung der deutschen Großforschungseinrichtungen. Köln: Max-Planck-Institut für Gesellschaftsforschung.

Schimank, U. 1988b. Wissenschaftliche Vereinigungen im deutschen Forschungssystem: Ergebnisse einer empirischen Erhebung. Max-Planck-Institut für Gesellschaftsforschung. Köln. Knowledge in Society 1/2:69–85.

Schimank, U. 1990. Technology policy and technology transfer from state-financed research institutions to the economy: Some German experiences. Science and Public Policy 17/4:219–228.

Schmoch, U., and N. Kirsch. 1994. Analysis of International Patent Flows. Report to the OECD. Karlsruhe: FhG-ISI.

Schmoch, U., E. Strauss, and T. Reiss. 1992. Patent law and patent analysis in biotechnology. Biotechnology Forum Europe 9/6 (June):379–384.

Schmoch, U., K. Koschatzky, M. Kulicke, T. Laube, and D. von Wichert-Nick. 1996a. Freie Erfindungen erfolgreich verwerten. Köln: Verlag TÜV Rheinland.

Schmoch, U., S. Hinze, G. Jäckel, N. Kirsch, F. Meyer-Krahmer, and G. Münt. 1996b. The role of the scientific community in the generation of technology. Pp. 1–138 in Organisation of Science and Technology at the Watershed. The Academic and Industrial Perspective, G. Reger and U. Schmoch, eds. Heidelberg: Physica-Verlag.

Schuster, H. J., ed. 1990. Handbuch des Wissenschaft Transfers. Berlin: Springer Verlag.

Secretary of Energy Advisory Board. 1995. Alternative Futures for the Department of Energy National Laboratories. (The Galvin Report.) Washington, D.C.: U.S. Department of Energy.

Selmayr, G. 1986. Hemmnisse und Barrieren bei der Kooperation zwischen Wissenschaftlern und Praxis. Pp. 73–89 in Hochschullehrer und Praxis, J. Allesch, R. Amann, and D. Preiß-Allesch, eds. Berlin: Weidler Buchverlag.

Selmayr, G., ed. 1987. Organisationsformen des Technologietransfers Wissenschaft / Wirtschaft, Band II, Bericht der Arbeitsgruppe Technologietransfer der Kanzler und leitenden Verwaltungsbeamten der wissenschaftlichen Hochschulen der Bundesrepublik Deutschland. Karlsruhe: Universität Karlsruhe.

Selmayer, G. 1989. Drittmittel aus der wirtschaft. umfrage der arbeitsgruppe technologietransfer der kanzler und leitenden verwaltungsbeamten der wissenschaftlichen hochschulen der Bundesrepublik Deutschland. Pp. 225–243 in Aus der Praxis des Technologietransfers Wissenschaft / Wirtschaft. Bericht der Arbeitsgruppe Technologietransfer der Kanzler und leitenden Verwaltungsbeamten der wissenschaftlichen Hochschulen der Bundesrepublik Deutschland, G. Selmayer, ed. Karlsruhe: Universität Karlsruhe.

Shapira, P. 1990. Modernizing Manufacturing: New Policies to Build Industrial Extension Services. Washington, D.C.: Economic Policy Institute.

Shapira, P. 1997. Manufacturing extension services: Performance, challenges, and policy issues. Unpublished working paper. [Online]. Available: http://www.cherry.gatech.edu/mod/pubs/mespcp.htm [June 15, 1997].

Shapira, P., D. Roessner, and R. Barke. 1995. New public infrastructures for small firm industrial modernization in the USA, Entrepreneurship and Regional Development 7:63–84.

Smilor, R. W., and D. B. Gibson. 1991. Technology transfer in multi-organizational environments: The case of R&D consortia. IEEE Transactions on Engineering Management 38(1):3–13.

Software Productivity Consortium. 1996. Home page. [Online] Available: http://www.software.org [April 29, 1996].

Stankiewicz, R. 1994. Spin-off companies from universities. Science and Public Policy 21(2) (April):99–107.

Steinmueller, W. E. 1996. The U.S. software industry: An analysis and interpretive history. Pp. 15–52 in The International Computer Software Industry: A Comparative Study of Industrial Evolution and Structure, D. Mowery, ed. New York: Oxford University Press.

Straus, J. 1997. The present state of the patent system in the European Union as compared with the situation in the United States of America and Japan. EUR 17014 EN. Luxembourg: European Commission.

SV-Wissenschaftsstatistik. 1994. Forschung und Entwicklung in der Wirtschaft 1991 - mit ersten Daten bis 1993. Essen: SV - Gemeinnützige Gesellschaft für Wissenschaftsstatistik.

SV-Wissenschaftsstatistik. 1996. Forschung und Entwicklung in der Wirtschaft 1993 - mit ersten Daten bis 1995. Essen: SV - Gemeinnützige Gesellschaft für Wissenschaftsstatistik.

Syrbe, M. 1989. Wissenschaft als unternehmen, Fraunhofer-Gesellschaft 1949–1989. FhG-Berichte 3/89:6–14.

Tettinger, P. J. 1992. Das forschungsinstitut an der unversität. Eine effiziente organisationseinheit zur förderung des wissenstransfers im geiste der "public private partnership." Wissenschaftsrecht-Wissenschaftsverwaltung-Wissenschaftsförderung 26/1:1–12.

Tornatzky, L. G., Y. Bats, N. E. McCrea, M. L. Shook, and L. M. Quittman. 1996. The Art and Craft of Technology Business Incubation: Best Practices, Strategies and Tools from 50 Programs. Research Triangle Park, N.C., and Athens, Ohio: Southern Technology Council and the National Business Incubator Association.

Tregarthen, S. 1992. Prescription to stop drug companies' profiteering. Counterpoint, Wall Street Journal, March 5. Sec. A, p.15.

Universität Karlsruhe. 1995. Unpublished data of the central administration. Karlsruhe: Universität Karlsruhe.

U.S. Centers for Disease Control and Prevention. 1996. CDC National AIDS Clearinghouse home page. [Online]. Available: http://www.cdcnac.org/ [April 26, 1996].

U.S. Congress, Office of Technology Assessment (OTA). 1986. Research Funding as an Investment: Can We Measure the Returns? Washington, D.C.: Office of Technology Assessment.

U.S. Congress, Office of Technology Assessment. 1993. Background Paper: Advanced Network Technology. Washington, D.C.

U.S. Department of Commerce. 1989. Manufacturing Technology 1988. Current Industrial Reports SMT(88)-1. Washington, D.C., May.

U.S. Department of Commerce. 1993. U.S. Industrial Outlook 1993. Washington, D.C.: U.S. Government Printing Office.

U.S. Department of Commerce. 1994. U.S. Industrial Outlook. Washington, D.C.: U.S. Government Printing Office.

U.S. Department of Commerce. 1996. Foreign Direct Investment in the United States: Operations of U.S. Affiliates of Foreign Companies, Preliminary 1994 Estimates. Washington, D.C.: U.S. Dept. of Commerce, July 1996.

U.S. Department of Defense, Advanced Research Projects Agency. 1995. The Technology Reinvestment Project: Dual-Use Innovation for a Stronger Defense. [Online] Available: http://www.trp.arpa.mil/trp/annual_95/cover.html [June 15, 1997].

U.S. Department of Energy. 1993. DOE-Approved CRADA Language and Guidance. Washington, D.C.: U.S. Department of Energy.

U.S. Department of Energy. 1994. Our Commitment to Change: A Year of Innovation in Technology Partnerships. Washington, D.C.: U.S. Department of Energy.

U.S. General Accounting Office. 1983. The Federal Role in Fostering University-Industry Cooperation. GAO/PAD-83-22. Washington, D.C.: U.S. General Accounting Office.

U.S. General Accounting Office. 1992. University Research: Controlling Inappropriate Access to Federally Funded Research Results. GAO/RCED-92-104. Washington, D.C.: U.S. General Accounting Office.

U.S. General Accounting Office. 1995. Manufacturing Extension Programs: Manufacturers' Views of Services. GAO/GGD-95-216BR. Washington, D.C.

U.S. Patent and Trademark Office. 1995. Number of Utility Patent Applications Filed in the United States, by Country of Origin, Calendar Year 1965 to Present. Washington, D.C.: U.S. Patent and Trademark Office.

U.S. Small Business Administration. 1988. Capital Formation in the States. Washington, D.C.: U.S. Small Business Administration.

Vademecum. 1985. Vademecum deutscher Lehr- und Forschungsstätten. Stuttgart: J. Raabe Verlag.

Vademecum. 1993. Vademecum deutscher Lehr- und Forschungsstätten. Stuttgart: J. Raabe Verlag.

VDE/VDI, Gesellschaft für Mikroelektronik. 1994. Technologietransfer für Anwender der Mikroelektronik. Frankfurt: VDE/VDI.

Venture Economics. 1994. The National Venture Capital Association 1993 Annual Report. Washington, D.C.: National Venture Capital Association.

VentureOne. 1997. National Venture Capital Association 1996 Annual Report. San Francisco: VentureOne Corporation.

Vonortas, N. S. 1996. Research Joint Ventures in the United States. Unpublished paper. Center for International Science & Technology Policy and Department of Economics. Washington, D.C.: The George Washington University.

Washington Post. April 25, 1996. The Road to a Human Gene Map [Figure]. Completion of Yeast Cells' Genetic Blueprint is Hailed as Scientific "Milestone." Sec. A, p.10.

Wegner, G. 1995. Grundlagenforschung zur Verwissenschaftlichung technischer Prozesse. MPG Spiegel. 1/95:38-45.

Weule, H., et al. 1994. Zusammenarbeit GFE/Industrie. Stuttgart: Daimler-Benz.

Weule-Kommission. 1994. Zusammenarbeit GFE / Industrie. Stuttgart: Daimler-Benz.

Whitehead, J. 1992. Parental guidance for orphan drugs. Letter to the Editor, Wall Street Journal, April 17. Sec. A, p.11.

Whiteley, R. L., A. S. Bean, and M. J. Russo. 1996. Meet your competition: Results from the 1994 IRI/CIMS annual survey. Research Technology Management (January-February).

Wisconsin BioIssues. 1994. Bayh-Dole, tech transfer and the public good: An interview with Howard Bremer. March. Madison: University of Wisconsin.

Wissenschaftsrat. 1986. Stellungnahmen zur Zusammenarbeit zwischen Hochschule und Wirtschaft. Köln: Wissenschaftsrat.

Wissenschaftsrat. 1992. Zur Förderung von Wissenschaft und Forschung durch wissenschaftliche Fachgesellschaften. Köln: Wissenschaftsrat.

Wissenschaftsrat. 1993a. Daten und Kennzahlen zur finanziellen Ausstattung der Hochschulen. - Alte Länder 1980, 1985, 1990. Köln: Wissenschaftsrat.

Wissenschaftsrat. 1993b. Drittmittel der Hochschulen 1970 bis 1990. Köln: Wissenschaftsrat.

Wissenschaftsrat. 1994. Empfehlungen zu einer Prospektion der Forschung, 1945/94. Köln: Wissenschaftsrat.

Wolek, F. W. 1977. The Role of Consortia in the National R&D Effort. Report submitted to the National Science Foundation, July 1977.

Wolff, H., G. Becher, H. Delpho, S. Kuhlmann, U. Kuntze, and J. Stock. 1994. FuE-Kooperation von kleinen und mittleren Unternehmen. Heidelberg: Physica-Verlag.

Wupperfeld, U. 1994. Strategien und Management von Beteiligungsgesellschaften im deutschen Seed-Capital-Markt. Ergebnisse einer empirischen Untersuchung von 33 Beteiligungsgesellschaften und Banken. Karlsruhe: FhG-ISI.

Wüst, J. 1993. Technologietransfer am Rande gemeinnütziger Forschungseinrichtungen. Unpublished lectures. Dresden: TU Dresden.

Zucker, L., M. Darby, and M. Brewer. 1994. Intellectual Capital and the Birth of U.S. Biotechnology Enterprises. Working Paper No. 4653. Cambridge, Mass.: National Bureau of Economic Research.

Biographical Information for the Binational Panel

H. NORMAN ABRAMSON retired as the executive vice president of Southwest Research Institute in 1991 at the completion of 35 years of service in increasingly responsible positions. Abramson has B.S. and M.S. degrees in mechanical engineering and engineering mechanics from Stanford University and a Ph.D. in engineering mechanics from the University of Texas-Austin. He is internationally known in the field of theoretical and applied mechanics and particularly for his expertise in the dynamics of contained liquids in astronautical, nuclear, and marine systems. Abramson is a past vice president and past governor of the American Society of Mechanical Engineers, a past director of the American Institute of Aeronautics and Astronautics, and has served as an officer or director of several other professional societies. He is a member of the National Academy of Engineering (NAE) and was an elected member of its Council during 1984–1990. He has also served on a variety of NAE and National Research Council committees and panels, including the Committee on Science, Engineering, and Public Policy; U.S. National Committee on Theoretical and Applied Mechanics; Committee on Computational Mechanics; Committee on Earthquake Engineering Facilities; Ship Structure Committee; Research and Technology Coordinating Committee for FHWA; and Committee on Technology Policy Options in a Global Economy. He has also served on advisory boards to various governmental agencies and as a consultant to a number of organizations.

JOSÉ ENCARNAÇÃO is a professor of computer science at the Technical University of Darmstadt, head of its Interactive Graphics Research Group, chair of the board of the Darmstadt Computer Graphics Center, and director of the Darmstadt R&D Institute of the Fraunhofer Research Society. He serves as a consultant to government, industry, and several international institutions, and was a founder of Eurographics. Encarnação is author of a large number of publications in internationally reviewed journals, and he is also author or coauthor of four textbooks in German and four in English. He is editor or co-editor of several books and many proceedings dealing with computer graphics and related applications. Encarnação is member of the editorial board of various professional journals such as *IEEE CG&A, Computer Graphics Forum, Visual Computer,* and *Computer-Aided Geometric Design,* and is the editor-in-chief of *computers & graphics,* an international journal on computer graphics published by Pergamon Press. In addition, he is the managing editor of the English technical book series *Computer Graphics-Systems and Applications,* which is published by Springer-Verlag. Encarnação holds a Diplom-Ingenieur (Dipl.-Ing.; diploma) and a Doktor-Ingenieur (Dr.-Ing.; doctorate) in electrical engineering from the Technical University of Berlin. He is a member of the Gesellschaft für Informatik (GI), the Verband deutscher Elektrotechnik (VDE), the Association of Computing Machinery (ACM), and is the German representative at Task Committees of the International Federation of Information Processing (IFIP).

ALEXANDER H. FLAX was Home Secretary of the National Academy of Engineering (NAE) from 1984–1992 and subsequently served as NAE senior fellow. His career includes positions as chief scientist of the Air Force, vice president and technical director of the Cornell Aeronautical Laboratory, assistant secretary of the Air Force for R&D, director of the National Reconnaissance Office, and president of the Institute for Defense Analyses (from 1969 until his retirement in 1983). Flax has served on numerous government advisory boards, committees, and panels for agencies, including the Office of the Secretary of Defense, the Air Force, the Defense Intelligence Agency, the National Aeronautics and Space Administration, and the White House, and has served on advisory bodies on engineering programs at Princeton and Stanford Universities as well. He is an honorary fellow of the American Institute of Aeronautics and Astronautics (AIAA), and is a member of the International Academy of Astronautics. He has received many awards and honors, including the Lawrence Sperry Award of the AIAA and the General Thomas D. White Space Trophy, and was named Elder Statesman of Aviation by the National Aeronautic Association in 1992. Flax received a B.S. in aeronautical engineering from the Guggenheim School of Aeronautics of New York University and a Ph.D. in physics from the University of Buffalo.

ROBERT C. FORNEY is a retired executive vice president, member of the board of directors, and member of the executive committee of E.I. du Pont de Nemours & Co. During his almost 40-year career with du Pont, he held a wide variety of research, manufacturing, engineering, marketing, and general management positions, the first 27 years of which were in man-made fiber activities. He is a member of the National Academy of Engineering and serves on the board of several for-profit and not-for-profit organizations. Forney received a B.S. and Ph.D. in chemical engineering and an M.S. in industrial engineering, all from Purdue University.

HERBERT GASSERT studied at the Technical University of Stuttgart, where he received the degree of Dr.-Ing. in 1964. He entered Brown, Boveri & Cie AG, Mannheim, in 1963, where he became head of the business section "Large Machines" in 1972 and a member of the executive board in 1976. From 1980 to 1987, he was director of the executive board of Brown Boveri, Mannheim, and a member of the board of the corporation Brown Boveri; from 1988 to 1994, he was a member of the advisory board of Asea Brown Boveri AG, Mannheim; since 1991, he has been a member of the advisory board of ABB Kraftwerke AG. Gassert holds many posts in industry and science, including director of the advisory board of Henningsdorfer Stahl GmbH i.L., honorary senator of the Universities of Stuttgart and Mannheim, member of the German Science Council, chairman of the German Association of Technical-Scientific Societies (DVT), and chairman of the advisory board of the Max Planck Institute for Metal Research, Stuttgart. Gassert has published numerous articles on economic, societal, and environmental problems

DAVID A. HODGES is the Daniel M. Tellep Distinguished Professor of Engineering at the University of California at Berkeley. He earned a B.E.E. at Cornell University and an M.S. and Ph.D. from the University of California at Berkeley. From 1966 to 1970, he worked at Bell Telephone Laboratories. He is currently professor of electrical engineering and computer sciences at UC Berkeley, where he has been a member of the faculty since 1970. He served as Dean of the College of Engineering from 1990 to 1996. Since 1970, Hodges has been active in teaching and research on microelectronics technology and design. Since 1984, his research has centered on semiconductor manufacturing systems. He is the founding editor of the *IEEE Transactions on Semiconductor Manufacturing*, a past editor of the *IEEE Journal of Solid-State Circuits*, and a past chairman of the International Solid-State Circuits Conference. Hodges is the recipient of the 1997 IEEE Education Medal. He is a fellow of the IEEE and a member of the National Academy of Engineering. He serves as a director of Mentor Graphics Corporation, Silicon Image, Inc., and the International Computer Science Institute.

BERND HÖFFLINGER studied physics at the University of Göttingen and Munich, Germany, where he received his Ph.D. in 1967. After a tenure on the scientific staff of the Siemens Research Laboratory in Munich from 1964 to 1967, he served as assistant professor in the School of Electrical Engineering at Cornell University in Ithaca, N.Y. Returning to Munich in 1970, he worked as the manager of the MOS Integrated Circuit Division of the Siemens Components Group. In 1972, Höfflinger founded the Department of Electrical Engineering at the University of Dortmund, where he held the Chair for Electron Devices until 1981. His 1979 sabbatical was spent at the University of California, Berkeley. Between 1981 and 1985, he held positions as head of Electrical Engineering at the University of Minnesota and later at Purdue University. At Minnesota, he was also codirector of the Microelectronics and Information Sciences Center. Since 1985, he has been director of the Institute for Microelectronics Stuttgart. The institute develops, manufactures, tests, and qualifies new application-specific microchips for industrial applications and participates in many international microelectronics research programs. He is also in charge of the electronics manufacturing program at the University of Stuttgart. Höfflinger received the Prize of the German Communication Society in 1968, the Outstanding Paper Award of the IEEE Circuits Conference in 1969, the Darlington Prize Paper Award of the IEEE Circuits and Systems Society in 1980, and the Electronics Letters Premium of the British Institution of Electrical Engineering in 1982. He has been a member of the Düsseldorf Academy of Sciences since 1981. He has authored or coauthored two books on microelectronics, as well as over 200 scientific publications.

PETER HANS HOFSCHNEIDER is head of the Department of Virus Research and a director at the Max Planck Institute for Biochemistry in Martinsried, Germany, where he also served as Managing Director from 1980 to 1985 and from 1989 to 1992. He is a professor in the Medical Faculty of the University of Munich. His main scientific interest is in virology and molecular medicine, including gene therapy. In 1954, Hofschneider received a Ph.D. in psychology from the University of Heidelberg, and a year later earned an M.D. degree from the University of Tübingen. In 1957, following a 2-year period of internship at hospitals in Zürich, Freiburg, and Basle, he joined the scientific staff of the Max Planck Institute for Biochemistry in Munich. Between 1958 and 1972, the year he was appointed a director of the institute, Hofschneider worked as a visiting scientist in institutes such as the Laboratoire de Biophysique in Geneva and the Roche Institute of Molecular Biology in Nutley, New Jersey. In 1967, he was appointed head of the newly created Department of Virus Research at the Max Planck Institute for Biochemistry. Hofschneider has co-edited several scientific journals, such as *Intervirology, Nucleic Acids Research, Current Topics in Microbiology and Immunology*, and *Comprehensive Virology*, and since 1985 he has been a member of the Editorial Board of the medical weekly, *Münchner Medizinische Wochenschrift*. He is a past member of the Medical Advisory Board of the German Cancer Aid Association, of the Board of Governors of the Berlin Centre of Arts and Sciences for Social Research (WZB), and of the Board of Directors of the Society for Promoting Biomedical Research. From 1974 to 1978, he was chairman of the Genetics Society of Germany. At present, Hofschneider is a member of the Scientific Council of the Max Planck Society, for which he served as chairman of the Section for Biology and Medicine from 1980 to 1983, and as chairman of the Scientific Council from 1988 to 1991. He is also a member of the Scientific Board of the German Trust for Cancer Research, a member of the Board of Governors of the German Cancer Aid Association, and of the Heinrich Pette Institute Trust. In addition, he is a member of the German Research Association's (DFG) Senate Commission for Cancer Research, of the Academia Scientiarum et Artium Europaea, and of the Deutsche Akademie der Naturforscher Leopoldina. Hofschneider has received several awards from scientific societies in Germany, including an Honorary Award from the Bavarian Academy of Sciences, the Gerhard-Domagk Prize for Cancer Research, the Dr. Friedrich Sasse Honorary Prize for Medicine, and the Jacob Henle Medal.

ARTHUR E. HUMPHREY received B.S. and M.S. degrees in chemical engineering from the University of Idaho, and a Ph.D. in chemical engineering from Columbia University, where he majored in biochemical engineering. He also holds an M.S. degree in food technology from MIT. He taught biochemical engineering at the University of Pennsylvania for 27 years, and, while there, served as chairman of the chemical engineering department for 10 years and as dean of engineering and applied science for 8 years. In 1980, he went to Lehigh University, where he served as provost and academic vice president for 6 years, followed by another 6 years as director of Lehigh's Center for Molecular Bioscience and Biotechnology. Since July 1, 1992, he has been serving as director of the Biotechnology Institute at The Pennsylvania State University. Humphrey has authored 3 books, written over 260 research papers, and has been granted 3 patents. His research is centered on biotechnology, specifically the design and control of bioprocesses. Among his numerous honors and awards are the John Fritz Medal, the AIChE Annual Lecture, the AIChE Professional Progress Award, the AIChE Food, Pharm. & Bioengineering Award, and the SIM Charles Thom Award for Meritorious Research in Industrial Microbiology. Humphrey is past president of AIChE and past director of the United Engineering Trust, is a member of the National Academy of Engineering, and has been a director of three companies, including two biotechnology companies—Fermentation Design and ABEC, Inc. He serves on numerous National Research Council committees, industrial advisory boards, and university visiting committees. He served as chairman of the Industrial Microbiology Subcommittee of the Science and Technology Joint US/USSR Committee under President Nixon, which involved nearly a dozen visits to Russia. In 1984, Humphrey chaired the Research Briefing Panel for the Office of Science and Technology Policy on "Chemical and Process Engineering for Biotechnology," a task that involved briefing many governmental agencies including OSTP, NSF, NBS, NIH, and Congress. He has held lectureships in biotechnology in numerous foreign countries such as China, Czechoslovakia, Guatemala, Hungary, India, and Mexico, and Fulbright Lectureships in Australia and Japan. He has received honorary degrees from the University of Idaho and Lehigh University and is the recipient of the University of Pennsylvania Distinguished Service Medal.

SIGMAR KLOSE studied chemistry in Braunschweig, and did his Ph.D. thesis at the University of Basel/Switzerland. He joined Boehringer Mannheim in January 1972. Since then he has been located in Tutzing, Bavaria. Since joining Boehringer Mannheim, Klose has worked on the development of automated methods for clinical chemistry. In 1988, he became manager of a number of very large international system projects and senior vice president of R&D Labsystems. He regards the closest possible integration of "High-(Bio)Tech" hardware and software as the most exciting challenge of his job.

PETER C. LOCKEMANN received his diploma and doctoral degrees in electrical engineering at the Technical University of Munich in 1958 and 1963, respectively. He joined the California Institute of Technology from 1963 to 1970 as a research fellow in Information Science, after which he spent 2 years with the Gesellschaft für Mathematik und Datenverarbeitung (GMD), Bonn, as a senior scientist. Since 1972, he has been a professor of Informatics at the University of Karlsruhe and, since 1985, he has also been a director of the Computer Science Research Center at Karlsruhe. His research interests are in the areas of engineering databases technology and applications, with a more recent emphasis on constraint enforcement strategies and techniques, object-oriented modeling techniques, active databases, and the integration of telecommunications and database technologies. He is author of four textbooks (one on telecommunications and databases) and more than 100 research papers in journals, conference proceedings, and books, and is an editor of six books. He has received numerous research grants from public institutions and private industry. Lockemann is deeply involved in developing better mechanisms for close interaction between academic or public-supported research institutions and private industry. As a director of an extra academic research institute, he is actively involved in the collaboration between industry and public research and in the transfer of the most recent technologies to industrial and commercial use. He is a member of the IEEE Computer Society, the Association for Computing Machinery, and the German Computer Society. He was a national representative to a Task Committee of the International Federation of Information Processing (IFIP) on Information Systems for over 10 years and is presently president of Very Large Databases (VLDB) Endowment, Inc. Lockemann has served on the supervisory boards and in working groups of national and international scientific organizations, industrial institutions, and governmental committees.

KNUT MERTEN is president and chief executive officer of Siemens Corporate Research, Inc., the American laboratory of Siemens AG's Central Research and Development organization in Munich, Germany. He studied mathematics and physics at the Technical University of Aachen, Germany, and acquired his Ph.D. from Darmstadt University, Germany, in 1975. Prior to joining the data processing group of Siemens, Munich, in 1980, he was professor of applied mathematics at the University of Bremen, Germany. In 1983, he joined the central research division of Siemens, where his responsibilities included IC-CAD development, CMOS submicron and high-performance bipolar technology, and numerical simulation tools for technology and fabrication. The Princeton facility holds responsibilities for Siemens Corporate R&D in key areas of learning systems, imaging and visualization, multimedia, and software engineering.

WILLIAM F. MILLER is Herbert Hoover Professor of Public and Private Management Emeritus, Graduate School of Business, Stanford University. He is also a professor of computer science emeritus, School of Engineering; senior fellow emeritus, Institute for International Studies; and chairman of the Executive Committee, Stanford Computer Industry Project. He is president emeritus of SRI International, having retired from there in 1990. Miller has served on many government commissions, as director of several corporations, and is a member of several honorary and professorial societies. He also serves on international advisory groups and is a past member of the National Science Board. Currently, he is a member of the California Council on Science and Technology, a member of National Academy of Engineering, and a fellow of the American Academy of Arts and Sciences, the Institute of Electrical and Electronics Engineers, and the American Association for the Advancement of Science. He was recognized by Tau Beta Pi as the Eminent Engineer in 1989. Miller is actively engaged in development of the new information infrastructure in Silicon Valley and in California. He speaks and writes on technology development, global changes in business strategy, policies for technology development, local and regional economic development, and the integration of socialist economies into the world economy. He received his B.S., M.S., Ph.D., and Sc.D. honoris causa degrees from Purdue University.

AL NARATH is president, Energy and Environment Sector, Lockheed Martin Corp. From April 1989 until August 1995, Narath was president of Sandia Corp. and director of Sandia National Laboratories. He worked at Sandia in various positions since 1959, except for the period from April 1984 to April 1989, during which time he was vice president, Government Systems, at Bell Laboratories in Whippany, N.J. He is a member of the Center for Strategic and International Studies, the National Research Council Board on Physics and Astronomy, NASA's Advisory Committee on the International Space Station, the Critical Technologies Panel of the Competitiveness Policy Council, the American Physical Society's Physics Planning Committee, the Board of Directors of the Congressional Economic Leadership Institute, the University of New Mexico College of Engineering Advisory Council, and the Coalition to Increase Minority Doctorates. He is on the selection committees for the National Academy of Engineering and the Department of Commerce National Medal of Technology; a member of DOE's Openness Advisory Panel; an ex officio member of DOE's Laboratory

Operations Board; and a participant on various other government panels and study groups. He has served on many government advisory committees and is a member of the National Academy of Engineering and the Phi Beta Kappa honor society, and is a fellow of the American Physical Society (APS) and the American Association for the Advancement of Science. Narath has received several awards, including the U.S. Department of Energy's Secretary's Contractor Manager Award and the APS' George E. Pake Prize. He received a B.S. in chemistry from the University of Cincinnati and a Ph.D. in physical chemistry from the University of California at Berkeley.

WALTER L. ROBB, president of Vantage Management, Inc., a consulting and investment firm, was the General Electric Company's senior vice president for corporate research and development until December 31, 1992. He directed the GE Research and Development Center, one of the world's largest and most diversified industrial laboratories, and served on the company's Corporate Executive Council. Robb is a chemical engineer with a B.S. from The Pennsylvania State University and a Ph.D. from the University of Illinois, and joined GE with the Knolls Atomic Power Laboratory. Prior to returning to the GE R&D Center, he headed GE Medical Systems for 13 years. He directed that organization's growth into the world's leading producer of medical diagnostic imaging equipment, turning it into a billion-dollar-plus per year advanced technology business with more then 10,000 employees worldwide. Presently, he serves on the boards of Marquette Medical Systems, Cree Research, Celgene, Neopath, and Mechanical Technologies, Inc., and on the Advisory Council of the Critical Technologies Institute in Washington, D.C. He is a member and serves on the Council of the National Academy of Engineering. In September 1993, he received the National Medal of Technology from President Clinton for his leadership in the CT and MR imaging industry.

OTTO H. SCHIELE studied at the University of Karlsruhe, where he received the degrees of Dipl.-Ing. and Dr.-Ing. He is professor of mechanical engineering and Dr.-Ing. E.h. of the Technical University at Darmstadt. He has held numerous posts in industry, science, and policy. He is currently vice president of the German Federation of Industrial Research Associations "Otto von Guericke" (AiF); chairman of the technological advisory council of the Federal State of Rheinland-Pfalz; member of the committee of the Fraunhofer Institute for Production Technology and Automation, Stuttgart; member of the board of the German Association of Technical-Scientific Societies (DVT); and member of the senate committee for applied research of the German Research Association (DFG). In addition he is member of several industrial supervisory boards. Schiele was president of the German Machinery and Plant Manufacturer's Association (VDMA) (1983–1986), vice president of the Federation of German Industries (BDI) (1983–1986), member of the Industrial Research & Development Advisory Committee (IRDAC) of the European Commission (1985–1991), member of the senate of the Fraunhofer Society (1986–1993). He has published numerous articles and has been awarded many honors, among them the Honorary Medal of the Association of German Engineers (VDI) and the Great Cross of Merit of the Order of Merit of the Federal Republic of Germany.

GERHART SELMAYR studied law in Munich and Erlangen and received his Ph.D. in 1961. He worked in various institutions of the military administration— 1963–1967 and 1969–1971—with posts in the military district administration of Munich and Ulm, in the NATO embassy in Paris, and the Federal Ministry of Defense in Bonn. In 1968–1969 he was personal adviser to the head of the Federal Chancellery. He was appointed head of the central administration department of the Federal Institute for Vocational Training Research in Berlin (1971–1973) and became chancellor of the College of the Federal Armed Forces, Munich (now the University of the Federal Armed Forces, Munich) (1973–1978). Since 1978, Selmayr has been chancellor of the University of Karlsruhe. He is vice chairman of the board of trustees of the Patent Office for German Research of the Fraunhofer Society, Munich (since 1985); member of the board of trustees of the Hochschul-Informations-System GmbH, Hanover (since 1988); and federal speaker for the German university chancellors (since 1994).

WILLIAM J. SPENCER was named president and CEO of SEMATECH in October 1990 and Chairman of the SEMATECH Board in July 1996. He has held key research positions at Xerox Corp., Bell Laboratories, and Sandia National Laboratories. Before joining SEMATECH in October 1990, he was group vice president and senior technical officer at Xerox Corporation in Stamford, Connecticut. Prior to joining the Xerox Palo Alto Research Center as manager of the Integrated Circuit Laboratory in 1981, Spencer served as director of systems development from 1978 to 1981 at Sandia National Laboratories in Livermore and director of microelectronics at Sandia National Laboratories in Albuquerque from 1973 to 1978. He began his career in 1959 at Bell Telephone Laboratories. Spencer received the Regents Meritorious Service Medal from the University of New Mexico in 1981, the C. B. Sawyer Award for contribution to "The Theory and Development of Piezoelectric Devices" in 1972, and a Citation for Achievement from William Jewell College, where he also received an honorary doctorate degree in 1990. He is a member of the National Academy of Engineering, a fellow of IEEE, and serves on numerous advisory groups and boards. Spencer has an A.B. from William Jewell College in Liberty, Missouri, an M.S. in mathematics from Kansas State University, and a Ph.D. in physics from Kansas State University.

Index

A

Advanced Research Projects Agency, 157, 219, 222
   computer science research funding, 228, 229–230
Advanced Technology Program, 76, 157
Aerospace industry, U.S.
   allocation of public R&D monies, 8
   export trends, 88
   R&D spending trends, 82
Agriculture R&D, U.S.
   extension programs, 204
   government spending, 72
AIDS research, 185, 192, 193
AiF. *See* Federation of Industrial Research Associations
American Society of Heating, Refrigerating and Air-Conditioning Engineers, 170
American Society of Mechanical Engineers, 170
American Supplier Institute, 210–211
Ames Research Center, 129
Amgen, 184–185
An-Institutes, 18–19, 26, 50, 242, 342, 343
   advantages, 288
   budget structures, 289
   challenges for, 289
   function, 288
   scope of research in, 288–289
   technology transfer role, 289–290
Antitrust law, 76, 209, 235–236
Application process, 15
   academic grants, U.S., 94
   European Union programs, 265
   in industry consortia, 27
Applied research
   college and university, U.S., 69, 92, 95
   federal laboratories, U.S., 124
   in Fraunhofer Society, 325
   in German R&D system, 248
   government funding, U.S., 65, 72–73
   industry funding, U.S., 67
   industry trends, U.S., 82–83
   nonprofit organizations, U.S., 70
   in university-industry research centers, 113
Argonne National Laboratory, 128
Asynchronous Transmission Mode, 54

B

Basic research
  allocation of R&D monies, 9
  colleges and universities, U.S., 69, 92, 95
  federal laboratories, U.S., 124
  in Fraunhofer Society, 325
  in German R&D system, 248, 249–250
  government funding, U.S., 65, 72–73
  in Helmholtz Centers, 316
  industry trends, U.S., 67, 82–83
  in Max Planck institutes, 23, 309
  nonprofit organizations, U.S., 70
  software development, 224
  university-industry research centers, U.S., 113
Bayh-Dole Act, 19, 21, 32, 74, 99, 103, 133
  outcomes, 144–145, 191
  provisions, 134, 191
Biotechnology, 6
  case example of German start-up, 353–354
  financial backing, U.S., 181–182
  foreign investment in new U.S. firms, 183, 189
  future prospects, 193
  industrial research association projects, Germany, 338
  intellectual property rights issues, 190–191
  international comparison of R&D activities, 297–298
  licensing revenues for universities, U.S., 187–188
  National Institutes of Health–funded research, 189
  new companies based on, U.S., 180–181, *see also* Amgen, Genentech, Medigene
  nonmedical uses, 179–180
  pharmaceutical industry investments, 182–183
  public funding of R&D, U.S., 184–186, 189
  R&D activities, Germany, 252, 290–292, 343–345
  regulatory issues, U.S., 191–192
  technological scope, 177–178
  technology transfer intermediaries, 190
  therapeutic applications, 178–179
  trade secrecy laws, U.S., 191
  university-industry technology transfer, U.S., 120, 122, 184–187
  university research, U.S., 183
Blue List institutes, 10, 21, 23, 33, 243, 319–320, 344
BMBF. *See* Ministry for Education, Science, Research, and Technology
British Technology Group, USA, 165
Brokers, technology, 164–166, 205
Brookhaven National Laboratory, 128
Budgets, R&D
  administrative structures in universities, Germany, 283–284
  An-Institutes, 289
  biotechnology, 180, 184–186, 189
  Blue List institutes, 319
  colleges and universities, 12–16, 67–69, 91–96, 93–96, 274–282
  computer science, 224–225, 228–229
  contract research institutes, 25, 26
  Department of Agriculture, 132
  Department of Defense, 127
  Department of Energy, 127
  as determinant of technology transfer, 3
  distribution of licensing revenues in universities, 187–188
  Environmental Protection Agency, 131
  EUREKA initiative, 269–270
  European Union, 244, 263–267
  federal laboratories, 20–21, 125–126, 127
  flat-panel display technology, 219
  focal-area distribution, 290–292
  foreign investments, 84
  Fraunhofer Society, 242, 322–324
  funding sources, 4
  German total, 246
  government, 6–9, 63–67, 73–77, 89–90
  health-related, 184
  Helmholtz Centers, 313–314, 316–317
  for industry consortia, 28, 157–158, 335, 336–338

industry-sponsored research in universities, 84, 110–111
international comparison, 4, 5, 6–9, 62–63, 246, 297
manufacturing and production technologies, 195–196, 199, 200
Max Planck institutes, 307–309
National Institutes of Health, 130
orphan drug research, 192–193
portfolio distribution, 6
private nonacademic organizations, 151–152, 153, 155
recommendations for enhancing technology transfer in Germany, 43
semiconductor industry, 214–215
service industries, 81–82
socioeconomic objectives, 9
software development, 224–225
state and local governments, U.S., 67
university-industry research centers, 18, 19, 111, 112–114
U.S. defense, 70–72
U.S. industry, 67
U.S. nonacademic nonprofits, 70
U.S. nondefense, 72–73
U.S. private sector, 79–80, 82–84
U.S. states, 9, 77–79
Business Roundtable, 46

C

Capital markets
as determinant of technology transfer, 3, 36
in Germany, 260–261, 262–263
recommendations for enhancing technology transfer in Germany, 42
See also Venture capital
Carnegie-Mellon University, 112–113
Center for the Utilization of Federal Technology, 135
Chambers of Crafts, 256
Chambers of Industry and Commerce, 30, 256
Chemistry, 250
Civil engineering, 282

Civil Engineering Research Foundation, 170–171
Collegial interchange, 142–143
technology transfer conference organizers, 167
technology transfer in biotechnology, 190
Communications technology. *See* Information and communication technology
Community of Science, 163
Competitive Technologies, Inc., 165
Computer-aided design, 106–107
Computer science
basic research, U.S., 224, 225
defense-related R&D, U.S., 229–231, 232–233
professional associations, U.S., 231
public R&D monies, U.S., 8, 224–225
R&D spending trends, U.S., 82
technology transfer case example, Germany, 349–351
technology transfer mechanisms, 231
See also Microelectronics industry; Software development
Consortia. *See* Joint research ventures, R&D consortia
Consulting
federal laboratories, U.S., 142
manufacturing and production technology transfer, 210
state–sponsored, 205
technology transfer, 16–17, 101–102, 166
by university professors, Germany, 286
Contract research institutes. *See* Private nonacademic R&D organizations
Cooperative Research and Development Agreements (CRADAs), 76, 135, 142, 143–144
advantages, 21, 139
distribution by technology, 138–139
federal laboratory implementation, 139
future prospects, 147, 148
manufacturing and production technology R&D, 201
microelectronics industry, 219

National Institutes of Health, 189
  operations, 21–22
  origins, 137–138
  preference and reciprocity agreements, 143–144
  trends, 84
  utilization trends, 138
Cost of ownership concept, 220–221
Council of Consortia CEOs, 160
Council on Competitiveness, 46
CRADAs. *See* Cooperative Research and Development Agreements
Cree Research, Inc., 104–105

D

Defense spending, 6–8, 63, 70–72
  aerospace R&D, U.S., 82
  computer science R&D, U.S., 229–231, 232–233
  dual-use technologies, 71
  federal laboratory R&D, U.S., 125–126, 127
  in German universities, 278
  in growth of semiconductor industry, U.S., 216–217
  manufacturing and production technology R&D, U.S., 196, 197–198
  See also Department of Defense
Department of Agriculture, 15, 20
  industrial problem-solving initiatives, 76
  research activities, 132–133
  technology transfer activities, 133
Department of Commerce, 77
Department of Defense, 15, 20, 97, 217
  aerospace R&D, 82
  computer science R&D, 229
  future of federal laboratories, 147–148
  industrial development initiatives, 76, 77
  information analysis centers, 54–55
  laboratories, 127
  manufacturing and production technology R&D, 196, 197–198
  microelectronics industry and, 217, 219
  R&D spending, 70–72

Department of Energy, 15, 20, 21, 72
  civilian laboratories, 128
  CRADAs, 138, 139, 142, 148
  federal laboratories, 126, 127–128, 148–149
  manufacturing and production technology R&D, 196, 198, 201
  patent licensing, 137
Department of Health and Human Services, 20, 72
  manufacturing and production technology R&D, 196
  patent licensing, 137
Department of Labor, 77
Department of Transportation, 76
Departmental research institutes, 21, 23, 320
DFG. *See* German Research Association
Diversity, 5–6, 92
Dryden Flight Research Center, 129–130

E

Economic development, U.S., 77–79
Electric Power Research Institute, 237–240
Electronics industry
  export trends, U.S., 88
  technology transfer, 123
  See also Microelectronics industry
Entrepreneurial behavior, 29–30
  as obstacle to technology transfer in Germany, 347–348
  recommendations for enhancing technology transfer in Germany, 43–44
  in software development, 234, 235
Environmental Protection Agency, 77
  R&D budget, 131
  research laboratories, 131–132
Environmental sciences
  government spending, U.S., 72
  international R&D collaboration, 51
Equity stock companies, 260
EUREKA initiative, 244, 268–270, 343
  Joint European Submicron Silicon Initiative, 244, 270–272, 343

European Commission, 14, 34
European Patent Organization, 34
European Union, 5, 38
  EUREKA initiative, 268-270, 343
  recommendations for U.S. collaborations, 48-50
  research funding, 244
  research programs, 263-267, 269, 343
Exchange programs, federal laboratory, 141-142
Extension programs
  agriculture model, 204
  manufacturing and production technology transfer, 204-213

F

Federal laboratories, Germany. *See* Blue List institutes; Helmholtz Centers; Max Planck institutes
Federal laboratories, U.S.
  civilian, 128-133
  collegial interchange activities, 142-143
  conflict-of-interest issues, 144
  consulting activities, 142
  contractor-operated, 125, 126, 128, 129, 142
  Cooperative Research and Development Agreements, 137-139
  defense-related, 126-128
  exchange programs, 141-142
  funding, 20-21
  in future of technology transfer, 147-149
  future prospects, 24-25, 149-151
  GOGOs, 125, 127, 129, 135
  information dissemination activities, 140-141
  legislative mandates for technology transfer, 133-135, 149-150
  limitations to technology transfer, 143-144, 150-151
  management, 125
  manufacturing and production technology transfer, 201-203
  national security issues, 143
  patent licensing, 125, 136-137
  private-sector input to, 48
  R&D expenditures, 125-126, 127
  R&D spending, 65, 67
  reimbursable work in, 142
  start-up/spin-off companies, 139
  structure and operations, 20, 124-125
  technical assistance activities, 141
  technology business incubators and, 168-169
  technology transfer activities, 20, 21-22
  technology transfer challenges, 37
  technology transfer effectiveness, 144-147
  technology transfer mechanisms, 135-143
  work with smaller enterprises, 143
Federal Research in Progress, 140
Federal Technology Transfer Act of 1986, 74-76, 135, 136, 143, 144
Federally Funded Research and Development Centers, 65, 67-69
  Department of Defense, 127
  structure and function, 125
Federation of Industrial Research Associations, 27, 242, 243
  budget and finance, 335, 336-338
  function, 333
  in German R&D system, 248, 249
  origins, 332-333
  research orientation, 338-339
  structure and operations, 333-335
  technology transfer activities, 339-341
  variation by industrial sector, 337-338
FhG. *See* Fraunhofer institutes
Finance. *See* Budgets, Equity stock companies, R&D, venture capital
Flat panel display technology, 214
  future prospects, 223
  international alliances, 223
  R&D, 215-216, 219
  sources of innovation, 222-223
Food and beverage industry, 338-339
Food and Drug Administration, 192, 193
France, 62
Fraunhofer institutes, 10, 25-26, 39-40, 242, 343

advantages, 330–332
budget and finance, 242, 322–324
future prospects, 243, 332
industry relations, 325–326, 328, 329–330
innovation centers, 330
patent licensing activities, 330
public research projects, 328
research orientation, 324–326, 346
structure and function, 242–243, 248, 249, 320–322
technology transfer activities, 326–330
university relations, 328–329

G

Garching Innovation GmbH, 311–312
Genentech, 186–187
Geological Survey, U.S., 133
German-American Academic Council Foundation, 42, 49, 52
German R&D system
  academic funding, 12–16, 43
  academic structure and function, 10–16
  challenges to, 40–41
  contract research institutes, 25–26, 39–40, 44
  departmental research institutes, 21, 23, 320
  European Union programs, 265–267, 279, 343
  external institutions, 287–290
  federal technology transfer initiatives, 257–258
  focal areas, 244, 250–252, 290–292
  government laboratories, 20–21, 23–25
  historical development of technology transfer, 272–274
  human capital characteristics, 39, 42–44, 283, 286–287
  industrial research associations, 27–28, 39, 243–244, 332–341
  intellectual property regime, 33, 44, 300–302
  ministry activities, 248
  new technology-based firms in, 258–260, 261–263
  obstacles to technology transfer, 41, 42
  opportunities for collaboration with U.S., 35
  principal components, 242
  professional/technical associations, 257
  recommendations for enhancing technology transfer, 41–44
  recommendations for fostering international collaboration, 42, 48–52
  small/medium-sized companies in, 30–31, 39, 244–245, 252–256
  spending, 4, 5, 62
  start-up companies in, 29–30, 39, 42
  state-funded initiatives, 257
  structural characteristics, 3, 4, 5–9, 38, 39–40, 242, 246–250
  technology transfer, determinants of success, 358–360
  technology transfer case examples, 349–358
  technology transfer effectiveness, 346–348
  technology transfer intermediaries, 248
  technology transfer mechanisms, 242–245
  university funding, 274–282
  university-industry technology transfer, 242–243, 245, 296–300
  venture capital market, 260–263
  vs. U.S. R&D system, 9–10, 37–40
German Research Association, 14, 248, 276–277
Goddard Space Flight Center, 130
Government-University-Industry Research Roundtable, 46–47
Government role
  development of technology road maps, 45
  in fostering industry-university collaboration, U.S., 99
  in German R&D system, 246, 248
  in growth of microelectronics industry, U.S., 216–217, 219
  international comparison, 40
  manufacturing and production technology transfer, U.S., 204–209

INDEX 415

private-sector input to R&D activities, U.S., 48
R&D employment, U.S., 67
recommendations for enhancing U.S. R&D system, 45
support for long-term R&D projects, 51–52
support for R&D consortia, U.S., 156–157
See also Public monies; *specific government organization*
GTS–GRAL, 349–351

H

Hatch Act, 133
Health-related R&D, 70
  applications of biotechnology, 178–179
  independent R&D organizations, U.S., 152–153
  international collaboration, 51
  public spending, U.S., 8, 72, 184
  technology for diagnosis decision-making, 54
  university patent licensing, U.S., 104–105
Helmholtz Centers, 33, 37, 242, 341
  challenges, 10, 22
  differences among, 316
  function, 10, 243, 312, 313
  funding, 20–21, 22, 243, 312, 313–314, 316–317
  future prospects, 317–319
  industry interaction, 315–316, 317–319
  origins and development, 273, 312–313
  patent licensing activities, 22–23, 316–317
  political environment, 313
  research orientation, 313, 314–315, 316, 343, 344
  technology transfer activities, 312, 315–317
  university collaborations, 317
High Performance Computing and Communications program, 229
Howard Hughes Medical Institute, 70
Human capital
  academic R&D employment, U.S., 70, 92
  Fraunhofer Society–university interaction, 329
  government R&D employment, U.S., 67
  industry R&D employment, U.S., 67
  international comparison of R&D systems, 38–39
  microelectronics research, Germany, 342
  recommendations for enhancing technology transfer, Germany, 42–43
  research universities, Germany, 283
  in semiconductor industry, U.S., 216
  in software development technology transfer, 225
  sources of, 11
  technology transfer via, 3, 36, 225
  in university-industry research centers, U.S., 111
  university-industry technology transfer, 99–102, 286–287
  U.S. industry R&D employment, 79

I

Idaho National Engineering Laboratory, 128
Incubators. *See* Technology business incubators
Industrial development
  government spending, 8, 72, 73–77
  historical university-industry relations, U.S., 96–99
Industrial liaison programs, 118–119
Information analysis centers, 54–55
Information and communication technology
  export trends, U.S., 88
  Fraunhofer Society research, 326
  German R&D activity, 250, 252, 341–342
  for globally active businesses, 53–54
  information analysis centers, 54–55
  for international R&D collaboration, 50, 54–55
  public R&D monies, 8
  R&D spending trends, U.S., 82

technology transfer intermediaries, 163–164
Institutions/organizations, R&D
  diversity, 5–6
  in Germany, 246–250
  nonprofits, U.S., 70
  private nonacademic, 151–162
  similarities of U.S. and German, 9–10, 37
  sources of technology for industry, 90–91
  spending, 4
  structural comparison, U.S. and Germany, 4, 5–9, 38, 39–40
  types of, involved in technology transfer, 2, 62
Instrument manufacturing industry, 82
Integrated Service Digital Network, 53–54
Intellectual property rights
  biotechnology issues, 190–191
  under CRADAs, 21–22
  as determinant of technology transfer, 3, 32–34
  in Germany, 300–302
  in government laboratories, 21–23
  international differences, 33–34
  issues for software development, 236
  recommendations for enhancing technology transfer in Germany, 44
  role of published research, 99–100
  university practices, 19–20
  U.S. law, 74–76
  See also Patent licensing
International collaboration, 34–35
  industrial trends, 84
  information infrastructure, 50, 54–55
  obstacles to, 48–49
  recommendations for, 42, 48–52
  suggested projects, 51, 55–60
International Society of Productivity Enhancement, 167
Internationalization trends
  foreign investment in U.S. biotechnology firms, 183, 189, 193
  information technology for globally active businesses, 53–54
  microelectronics technology transfer, 223
  private-sector R&D, 84
  Internet, 173, 228, 230

J

Japan, 5, 62, 195, 201, 214, 224
JESSI. *See* Joint European Submicron Silicon Initiative
Jet Propulsion Laboratory, 129, 130
Johnson Space Center, 130
Joint European Submicron Silicon Initiative, 244, 270–272
Joint research ventures, 83, 157, 158

K

Kennedy Space Center, 130
Knowledge Express Data Systems, 163

L

Labor markets, 3
  as determinant of technology transfer, 36
Langley Research Center, 129–130
Lawrence Berkeley Laboratory, 128
Lawrence Livermore National Laboratory, 127
Lewis Research Center, 129–130
Life cycle analysis, 3
  for equipment acquisition, 220–221
Life sciences, 14, 73
Los Alamos National Laboratory, 127

M

Machine tool industry, 201, 345–346
Magnetic storage technology, 112–113
Manufacturing and production technologies, 195
  acquisition patterns in smaller firms, 201
  effectiveness of technology transfer programs, U.S., 212–213
  federal R&D, U.S., 196–199

INDEX 417

federal technology transfer programs, U.S., 206–209, 211
Fraunhofer Society research, 324–325
industrial research association projects, Germany, 338
industry networks, U.S., 205
industry profile, U.S., 194
industry R&D, U.S., 195–196
international comparison of R&D, 195, 297–298
obstacles to modernization, U.S., 211–212
R&D activities, Germany, 290–292, 345–346
state–sponsored extension programs, U.S., 204–206, 212–213
supplier development programs, U.S., 210–211
technological scope, 193–194
technology transfer case examples, Germany, 351–353
technology transfer from federal laboratories, U.S., 201–203
technology transfer from universities, U.S., 203–204
technology transfer within private sector, U.S., 209–211
university–industry research centers, U.S., 199, 200
Manufacturing Extension Partnership, 76–77, 90, 203
effectiveness, 212–213
origins and development, 207
structure and operations, 207–208
Market factors
competition in research, 44
cost of ownership concept, 220–221
in Fraunhofer Society research, 328
international comparison, 5, 38
modernization of manufacturing/production sector, 211
new technology–based firms, Germany, 245, 259–260, 262–263
in operations of start-up companies, 85–87
pressures on international businesses, 53

software development, 233
in technology transfer, 3, 29–30, 36
time to market, 359
venture capital firms in technology transfer, 172–173
Marshall Space Flight Center, 130
Massachusetts Institute of Technology, 120
Max Planck institutes, 242, 341
budget and finance, 307–309
distinguishing features, 10
funding, 21
industry grants to, 309
patent licensing, 311–312
research areas, 304, 343, 344
structure and function, 243, 248, 249, 302–304, 305–307
technology transfer activities, 309–312
Mechanical engineering in Germany, 244
Fraunhofer Society research, 324–325
patent licenses, 250
university research funding, 279, 281
MediGene, 353–354
Microelectronics and Computer Technology Corporation, 217, 218–219
Microelectronics industry
consortia, U.S., 217–219
economic significance for U.S., 213–214
Fraunhofer Society research, 325, 326, 329, 343
future prospects, 224
government-industry relationships, U.S., 219
international technology transfer, 223
market characteristics, 214
public R&D monies, 8
R&D activities, 214–216
technological scope, 214
university–industry relationships, U.S., 222
Ministry for Education, Science, Research, and Technology, 14, 317–318
structure and operations, 248
university research funding, 277

Ministry of Defense, 248, 278
Monsanto Corp., 188
Morrill Act, 133
MPG. *See* Max Planck institutes

N

National Advisory Committee on Aeronautics, 133–134
National Aeronautics and Space Administration, 15, 20, 72, 77, 217
   future of technology transfer, 149
   laboratories, 129–130
   legislative history, 133–134
   manufacturing and production technology R&D, 196, 201
   technology business incubators, 169
   technology transfer activities, 134, 140–141
National Association of Manufacturers, 46
National Competitiveness Technology Transfer Act, 135, 143–144
National Cooperative Research Act, 76, 83, 209
National Electronics Manufacturing Initiative, 196–197
National Institute of Standards and Technology, 20, 74
   industrial development programs, 76–77, 90
   laboratories, 129
   manufacturing and production technology R&D, 196, 198–199
   mission, 128–129
National Institutes of Health, 15, 20, 21, 72, 147
   biotechnology research funding, 189
   biotechnology research guidelines, 191–192
   research laboratories, 130–131
   structure and function, 130
National Renewable Energy Laboratory, 128
National Science and Technology Council, 196
National Science Foundation, 15, 47, 72, 76, 184

computer science research funding, 228
manufacturing and production technology R&D, 196, 199
university-industry research center, 18, 99, 115, 199
National security issues, 143
National Technology Transfer Center, 140, 201–203
NERAC, *See* New England Research Applications Center
New England Research Applications Center, 163
Nuclear weapons research, 127

O

Oak Ridge National Laboratory, 128, 198
Organization for Economic Cooperation and Development, 49
Organization for Rationalization of German Industry, 30–31, 257
Orphan Drug Act, 192–193

P

Pacific Northwest Laboratory, 128
Patent and Trademark Amendments. *See* Bayh-Dole Act
Patent licensing
   in Europe, 34
   federal laboratories, U.S., 125
   federal laboratory research, U.S., 136–137
   Fraunhofer Society activities, 330
   in German universities, 300–302, 354–358
   grace period, 33
   Helmholtz Center activities, 22–23, 316–317
   by industry, 88, 250
   international comparison, 19–20, 33, 301–302
   in Max Planck institutes, 311–312
   technology brokers, 164–166
   university-industry technology transfer, U.S., 102–108, 112–113, 190–191
   in U.S. universities, 92, 187–188

See also Intellectual property rights
Performance assessment, technology transfer in federal laboratories, 144–147
Pharmaceutical industry, 82
  biotechnology applications, 178–179
  biotechnology investments, 182–183, 193
  new drug approval process, U.S., 192
  orphan drug research, U.S., 192–193
  R&D activity, Germany, 344–345
  university-industry technology transfer, U.S., 122–123
Physics, 279–280, 281
Policy-making
  biotechnology issues, 190–193
  effects on technology transfer, 3
  obstacles to international collaboration, 48–49
  recommendations, 41–52
  software development issues, 235–236
  university-industry technology transfer issues, 123–124
Political environment
  future of U.S. federal laboratories, 147
  in Helmholtz Centers, 313
Private nonacademic R&D organizations
  affiliated institutes, U.S., 26–27, 155–156
  contributions, 152
  engineering/design/architectural firms, 171–172
  in German R&D system, 39–40, 250, 326. *See also* Fraunhofer institutes
  independent institutes, U.S., 152–155
  manufacturing and production technologies, U.S., 200–201
  principal firms, U.S., 152, 154, 155–156
  professional organizations, U.S., 170–171
  recommendations for enhancing technology transfer in Germany, 43, 44
  research parks, 169–170
  spending, U.S., 151–152, 153, 155

  technology business incubators, U.S., 167–170
  technology transfer effectiveness, U.S., 174–176
  technology transfer intermediaries, U.S., 162–174
  technology transfer mechanisms, U.S., 153–154
  types of, U.S., 151, 152
  See also R&D consortia
Private-sector R&D
  basic and applied research trends, U.S., 82–83
  consortia, 27–29, 39, 119
  contract research institutes, 25–27, 39–40
  cooperative arrangements, 83–84
  cost of ownership calculations, 220–221
  employment, U.S., 67, 79
  Fraunhofer Society interaction, 325–326, 328, 329–330
  German focus, 250–252
  German industrial research associations, 243–244
  government-funded, U.S., 65, 73–77
  government laboratory collaborations, 24, 135
  grants to Max Planck institutes, 309
  Helmholtz Center interaction, 315–316, 317–319
  industrial-nonindustrial linkages, U.S., 89–90
  infrastructural innovations, 88–89
  input to government, 48
  internationalization trends, 84
  manufacturing and production technology, 201, 209–211
  microelectronics industry, 213–224
  new biotechnology companies, 180–183
  nonmanufacturing industries, 81–82
  outsourcing trends, 83–84
  pathbreaking innovations, 89
  recommendations for enhancing, U.S., 45–46, 47, 48
  sectoral distribution, U.S., 80–82
  significance of start-up companies, 84–87

software development firms, 234–235
sources of external technology for, 90–91
spending, 4
spending, U.S., 67, 79–80
structure of U.S. technology transfer system, 62
support for smaller companies, 30–31
technical assistance programs, 119–121
technology road maps for, 45–46
university funding, Germany, 278–279
U.S system strengths and weaknesses, 88–90
See also University–industry relations
Production technology. *See* Manufacturing and production technology
Professional associations, 170–171
computer science, 231
in Germany, 257, 332–333
manufacturing and production technologies, 205, 210
recommendations for enhancing U.S. R&D system, 46
support for smaller companies, 30–31
Public monies
academic research funding, 14–15
allocation of R&D funds, 6–9
biotechnology research funding, U.S., 184–186, 189
defense-related R&D, U.S., 70–72
government funding of industry R&D, U.S., 65, 89–90
health-related R&D spending, U.S., 184
industrial development R&D, U.S., 73–77
for industrial research association projects, Germany, 335
for long-term R&D projects, 51–52
in manufacturing and production technology R&D, U.S., 196–199
nondefense-related R&D, U.S., 72–73
public wage system, Germany, 43–44
R&D objectives, U.S., 70
R&D spending, 40, 63–67, 246
restrictions on academic research, Germany, 282–283
in software development R&D, U.S., 224–225
state industrial technology programs, U.S., 77–79
university research funding, Germany, 274–278

R

R&D consortia, 27–29, 39
government encouragement, U.S., 156–157
industrial trends, U.S., 83
industry, Germany, 243–244, 332–341
international collaboration, 51
legal environment, U.S., 156
manufacturing and production technologies, U.S., 209
microelectronics industry, U.S., 217–219
recommendations for U.S., 45, 46
spending, U.S., 157–158
structure and operations, U.S., 156
technology transfer from, U.S., 159–160, 162
university-industry, U.S., 119
Referral organizations, 162–164
Regulation and legislation
antitrust law, 76, 209, 235–236
barriers to new technology–based firms in Germany, 259–260, 261, 262–263
biotechnology issues, 182, 191–192, 345
challenges to technology transfer system, 41
to encourage industrial development, 74–77
to encourage technology transfer, 32–34
German university research, 274, 282–283
obstacles to international collaboration, 48–49
obstacles to professional mobility in Germany, 42–43

obstacles to technology transfer in Germany, 358–359
orphan drug research, 192–193
protections for R&D consortia, 156
recommendations for enhancing German R&D, 42
recommendations for enhancing U.S. R&D, 45
technology transfer from U.S. federal laboratories, 133–135, 149–150
trade secrecy laws, 191
See also Intellectual property rights; Patent licensing; Taxation in Germany
Research areas/topics
academic distribution, 13–14, 92, 95
allocation of public monies, 6–9
in CRADAs, 138–139
distribution of funding, Germany, 279–280
distribution of government spending, U.S., 72–73
European Union investments, 244
field-specific features of technology transfer, 36
focus of Max Planck institutes, 304
at Helmholtz Centers, 313, 314–315
industry trends, U.S., 80–82
international comparison, 6, 250–252, 252, 296–300
patent licensing activity, 250, 252
spending, Germany, 244, 290–292
spending, U.S., 70
university-industry research centers, U.S., 114
Research Corporation Technologies, 164–165
Research parks, 169–170

S

SAGE. *See* Semi–Automatic Ground Environment air defense system
Sandia National Laboratories, 127, 198, 219
SEMATECH, 76, 152, 157, 158, 209, 217, 218, 220–221, 222

Semi-Automatic Ground Environment air defense system, 231, 232
Semiconductor Research Corporation, 107, 218, 222
Semiconductor technology, 32
German R&D, 252
market characteristics, 214
sources of early innovation, 216–217
technology road maps, 45–46, 222
university R&D, 106–107
U.S. R&D, 214–215
Service Industries, 81–82
Single European Act, 263
Small Business Development Centers, 203
Small Business Innovation Research, 77
Small/medium-sized companies
acquisition of new technologies, 201
challenges to technology transfer system, 41
computer technology for, 54
cooperative research, Germany, 252–258
equity stock companies, 260
federal industrial development initiatives, U.S., 76–77
federal laboratory interaction, U.S., 143
flat-panel display innovation in, 222–223
industrial research associations, Germany, 332
international collaboration, 51
international comparison of R&D activities, 39
obstacles to research collaborations, 254–255
production and manufacturing industry, 194, 211, 345–346
technical assistance programs for, U.S., 119–121
technology transfer needs, 30–32, 90
transfer mechanisms, Germany, 244–245
Smith-Lever Act, 133, 204
Social and cultural factors
challenges to technology transfer system, 41

entrepreneurial risk-taking mentality, 347–348
recommendations for enhancing German R&D, 42
in technology transfer, 36, 41
Social sciences/humanities, 14, 279
Socioeconomic objectives, 9
Software
  computer science R&D and, 233–234
  determinants of R&D activities, 227–229
  entrepreneurs, 234
  future prospects, 236–237
  German R&D, 290–292
  industrial research association projects, Germany, 338
  intellectual property rights, 236
  for internal use, 226–227
  international comparison of R&D, 297–298
  market characteristics, 225–226
  policy issues, 235–236
  R&D structure and spending, 224–225
  sources of innovation, 225
  start-up companies, 234–235
  university-industry technology transfer, 122
  See also Computer science; Microelectronics industry
Software Productivity Consortium, 232–233
Space exploration, 6–8, 72
Start-up companies, 20
  biotechnology, 180–183
  equity ownership by academic institutions, 108–110
  in Germany, 245, 258–263, 353–354
  international comparison, 39
  legal environment, 261
  recommendations for enhancing technology transfer in Germany, 42
  role in technology transfer, 29–30, 84–87, 258
  software, 234–235
  state programs for, 206–209
  technology business incubators, 121
  U.S. federal laboratories and, 139

U.S. trends, 84–85
State and local R&D funds, 67
  for colleges and universities, U.S., 94, 97
  distribution, U.S., 73
  industrial technical assistance programs, U.S., 204–206
  industrial technology programs, U.S., 77–79
  state/university industry research centers, U.S., 199
  for technology transfer, Germany, 257
Steinbeis Foundation, 31, 257
Stennis Space Center, 130
Stevenson-Wydler Technology Innovation Act, 22, 74, 76, 133
  outcomes, 144–145
  provisions, 134–135
Superconducting Supercollider, 184
Supplier development programs, 210–211

T

Taxation in Germany
  recommendations for enhancing technology transfer, 42, 43
  research grants, 285
  venture capital, 259–260, 262
TechLaw Group, 166–167
Technical assistance programs, 119–121
  state–sponsored industrial extension, 204–206, 212–213
Technology business incubators, 121
  federal laboratories and, 168–169
  function, 168
  structure and operations, 167–168
Technology life cycle, 3
Technology Reinvestment Project, 196, 197–198, 208
Technology road maps, 32, 45–46, 176, 222
Technology transfer
  from An-Institutes, 289–290
  in biotechnology industry, 177–193
  from Blue List institutes, 320
  brokers, 164–166, 205
  case examples, Germany, 349–358

channels in German universities, 284–287
from colleges and universities, 99–108, 292–294, 298
conference organizers, 167
consultants, 166
contributions of individuals to, 36
definition, 2–3
determinants of success, at national level, 3, 35–36
determinants of success, Germany, 358–360
differences between U.S.-German systems, 37–40
Fraunhofer Society activities, 326–330
goals, 2
government involvement, 40
Helmholtz Center activities, 312, 315–317
industrial research associations, Germany, 243–244, 339–341
institutional challenges, 37
institutional participants, 2. *see also specific institutional type*
interfirm/intrafirm, 2, 90
intermediary organizations, 162–174
law firms, 166–167
manufacturing and production technologies, 193–213
in mature industries, 30–32
from Max Planck institutes, 309–312
mechanisms, 2–3, 242–245
from private nonacademic organizations, U.S., 151–162, 174–176
from R&D consortia, U.S., 159–160
role of start-up companies, 29–30
in semiconductor industry, 216–217
setting-specific features, 36, 122–123
similarities between U.S. and German systems, 37
size of firm as factor in, 359
in small/medium-sized enterprises, Germany, 244–245, 252–258
system effectiveness, Germany, 346–348
transnational, 34–35

U.S. federal laboratory, future prospects, 147–149
U.S. federal laboratory, historical development, 133–135
U.S. federal laboratory effectiveness, 144–147
U.S. federal laboratory limitations, 143–144
U.S. federal laboratory mechanisms, 135–143
Technology Transfer Act of 1986, 32–33
Technology Transfer Conferences, 167
Textile technology, 351–352
Trade secrecy laws, 191

U

United Kingdom, 62
United States R&D system
  academic employees, 70
  academic research funding, 12–16
  academic structure and function, 10–16, 91–96
  challenges to, 40–41
  college-university activities, 67–70
  contract research institutes, 26–27, 39–40
  defense-related, 70–72
  federally-funded industrial development initiatives, 73–77, 89–90
  government activities, 63–67
  government employees, 67
  government incentives for technology transfer, 32–33
  government laboratories, 20, 21–22, 24–25, 124–133
  government laboratories, future prospects for, 147–149
  government laboratories legislation, 133–135
  government laboratory technology transfer, limitations of, 143–144
  government laboratory technology transfer effectiveness, 144–147
  government laboratory technology transfer mechanisms, 135–143
  human capital characteristics, 38–39

industry employees, 67
industry spending, 67
nonacademic nonprofit organizations, 70
nondefense-related, 72–73
objectives, 70
opportunities for German technology transfer collaboration, 35
private-sector resources, 79–80
R&D consortia, 27, 28–29, 39, 45, 46
recommendations for enhancing technology transfer, 44–48
recommendations for fostering international collaboration, 42, 48–52
responsibility, 38
significance of start-up companies, 29, 39, 84–87
small/medium-sized companies in, 30, 31–32, 39, 45, 90
software development, 227–229, 236–237
spending, 4, 5, 62–63, 67
state programs, 67, 77–79
strengths and weaknesses of industrial enterprise, 88–90
strengths of, 45
structure, 62
university-industry historical relations, 96–99
university-industry technology transfer, 99–124, 296–300
vs. German R&D system, 3, 4, 5–10, 37, 38, 39–40
Universities and colleges
administrative structures, Germany, 282–284
basic/applied research, U.S., 67–69
biotechnology research, U.S., 183
computer science enrollments, U.S., 227
computer science R&D spending, U.S., 225
computer science research funding, U.S., 228–229
computer science technology transfer, U.S., 231
contributions to technology transfer, 11
dissemination of good R&D practices, 46–47
distribution of licensing revenues, U.S., 187–188
distribution of research expenditures by research area, 13–14
diversity, U.S., 92
external institutions, Germany, 287–290
focal research areas, Germany, 14, 279–280
Fraunhofer Society relations, 328–329
funding of research in, 9, 12–16, 65, 69, 274–282
funding sources, U.S., 93–95
funding trends, U.S., 95–96, 98
historical development of technology transfer, Germany, 272–274
innovation incentives for staff, 103, 187
international differences in structure, 12
international R&D collaboration, 50
marketing activities, 19
patent activity, Germany, 300–302
patent licensing, 19–20, 44, 102–106
patent royalties, U.S., 105–108
polytechnical schools, Germany, 273, 286
public vs. private, U.S., 69
R&D challenges for, 37
R&D employment, U.S., 70, 92
R&D spending, 4, 67
recommendations for enhancing technology transfer, 43, 44, 46–47
structure and resources, U.S., 91–92
technology licensing, case examples of, Germany, 354–358
technology transfer organizations in, 103
See also University-industry relations
University-industry relations, 47–48, 84
barriers to, Germany, 294–295
biotechnology research, 186–188
concerns about, 123–124
contract research, 284, 286
contract vs. grant research, 110–111
funding of academic research, 13, 15–16
German R&D structure, 242

German technology transfer
  mechanisms, 242–243, 245
historical development, 96–99, 274
industrial liaison programs, 118–119
industry researchers as faculty, 17
industry-sponsored research, 110–111
intellectual property rights, 190–191
international comparison, 296–300
manufacturing and production industry,
  199, 200, 203–204, 205
microelectronics industry, 222
perception of German university
  institutes, 292–296, 298–299
publication interference, 298–299
research consortia, 119
start-up companies, 20, 108–110
technical assistance programs, 119–121
technology business incubators, 121
technology transfer arrangements, 16–20
technology transfer effectiveness, 121–124
technology transfer mechanisms, 99–101,
  101–108, 284–287
transfer of personnel in Germany, 286–287
University-industry research centers, 18,
  19, 43, 76, 99
  concerns with, 116–118
  definition, 111
  effectiveness, 115–116
  funding, 111, 112–114
  goals and missions, 114–115
  international collaboration, 50
  manufacturing and production
    technology R&D, 199, 200, 203–204,
    206
  U.S. structure and operations, 111–113
  vs. German university-industry
    technology transfer, 296–300

V

Venture capital, 29–30, 86, 104–105,
  172–173
  biotechnology investments, 181–182
  in Germany, 260–263

software development investments, 235

W

World Intellectual Property Organization,
  49
World Trade Organization, 49